罗炳森 黄超 钟侥 著

SQL 优化核心思想

人民邮电出版社

北京

图书在版编目（CIP）数据

SQL优化核心思想 / 罗炳森，黄超，钟侥著. -- 北京：人民邮电出版社，2018.4（2023.5重印）
ISBN 978-7-115-47849-8

Ⅰ．①S… Ⅱ．①罗… ②黄… ③钟… Ⅲ．①SQL语言 Ⅳ．①TP311.138

中国版本图书馆CIP数据核字(2018)第019196号

内 容 提 要

结构化查询语言（Structured Query Language，SQL）是一种功能强大的数据库语言。它基于关系代数运算，功能丰富、语言简洁、使用方便灵活，已成为关系数据库的标准语言。

本书旨在引导读者掌握 SQL 优化技能，以更好地提升数据库性能。本书共分 10 章，从 SQL 基础知识、统计信息、执行计划、访问路径、表连接方式、成本计算、查询变换、调优技巧、经典案例、全自动 SQL 审核等角度介绍了有关 SQL 优化的方方面面。

本书基于 Oracle 进行编写，内容讲解由浅入深，适合各个层次的读者学习。本书面向一线工程师、运维工程师、数据库管理员以及系统设计与开发人员，无论是初学者还是有一定基础的读者，都将从中获益。

◆ 著　　罗炳森　黄　超　钟　侥
责任编辑　胡俊英
责任印制　焦志炜

◆ 人民邮电出版社出版发行　北京市丰台区成寿寺路 11 号
邮编　100164　电子邮件　315@ptpress.com.cn
网址　http://www.ptpress.com.cn
北京七彩京通数码快印有限公司印刷

◆ 开本：800×1000　1/16
印张：20　　　　　　　　　2018 年 4 月第 1 版
字数：445 千字　　　　　　2023 年 5 月北京第 20 次印刷

定价：79.00 元

读者服务热线：(010)81055410　印装质量热线：(010)81055316
反盗版热线：(010)81055315
广告经营许可证：京东市监广登字 20170147 号

前　　言

近年来，随着系统的数据量逐年增加，并发量也成倍增长，SQL 性能越来越成为 IT 系统设计和开发时首要考虑的问题之一。SQL 性能问题已经逐步发展成为数据库性能的首要问题，80% 的数据库性能问题都是因 SQL 而导致。面对日益增多的 SQL 性能问题，如何下手以及如何提前审核已经成为越来越多的 IT 从业者必须要考虑的问题。

现在将 8 年专职 SQL 优化的经验和心得与大家一起分享，以揭开 SQL 优化的神秘面纱，让一线工程师在实际开发中不再寝食难安、谈虎色变，最终能够对 SQL 优化技能驾轻就熟。

编写本书也是对多年学习积累的一个总结，鞭策自己再接再厉。如果能够给各位读者在 SQL 优化上提供一点帮助，也不枉个中辛苦。

2014 年，作者罗炳森与有教无类（网名）联合编写了《Oracle 查询优化改写技巧与案例》一书，该书主要侧重于 SQL 优化改写技巧。到目前为止，该书仍然是市面上唯一一本专门讲解 SQL 改写技巧的图书。

因为《Oracle 查询优化改写技巧与案例》只专注于 SQL 改写技巧，并没有涉及 SQL 优化的具体思想、方法和步骤，本书可以看作是对《Oracle 查询优化改写技巧与案例》一书的进一步补充。

本书共 10 章，各章的主要内容如下。

第 1 章详细介绍了 SQL 优化的基础知识以及初学者切实需要掌握的基本内容，本章可以帮助初学者快速入门。

第 2 章详细讲解统计信息定义、统计信息的重要性、统计信息相关参数设置方案以及统计信息收集策略。

第 3 章详细讲解执行计划、各种执行计划的使用场景以及执行计划的阅读方法，通过定制执行计划，读者可以快速找出 SQL 性能瓶颈。

第 4 章详细讲解常见的访问路径，这是阅读执行计划中比较重要的环节，需要掌握各种常见的访问路径。

第 5 章详细讲解表的各种连接方式、各种表连接方式的等价改写以及相互转换，这也是本书的核心章节。

第 6 章介绍单表访问以及索引扫描的成本计算方法，并由此引出 SQL 优化的核心思想。

第 7 章讲解常见的查询变换，分别是子查询非嵌套、视图合并和谓词推入。如果要对复杂的 SQL（包含各种子查询的 SQL）进行优化，读者就必须掌握查询变换技巧。

第 8 章讲解各种优化技巧，其中涵盖分页语句优化思想、分析函数减少表扫描次数、超大表与超大表关联优化方法、dblink 优化思路，以及大表的 rowid 切片优化技巧。掌握这些调优技巧往往能够事半功倍。

第 9 章分享在 SQL 优化实战中遇到的经典案例，读者可以在欣赏 SQL 优化案例的同时学习罗老师多年专职 SQL 优化的经验，同时学到很多具有实战意义的优化思想以及优化方法与技巧。

第 10 章讲解全自动 SQL 审核，将有性能问题的 SQL 扼杀在"摇篮"里，确保系统上线之后，不会因为 SQL 写法导致性能问题，同时还能抓出不符合 SQL 编码规范但是已经上线的 SQL。

本书对系统面临性能压力挑战的一线工程师、运维工程师、数据库管理员（DBA）、系统设计与开发人员，具有极大的参考价值。

为了满足不同层次的读者需求，本书在写作的内容上尽量由浅入深，前 5 章比较浅显易懂，适合 SQL 优化初学者阅读。通读完前 5 章之后，初学者能够对 SQL 优化有一定认识。后 5 章属于进阶和高级内容，适合有一定基础的人阅读。通读完后 5 章的内容之后，无论是初学者或是有一定基础的读者都能从中获益良多。

本书专注于 SQL 优化技巧，因此书中不会涉及太多数据库系统优化的内容。

虽然本书是基于 Oracle 编写的，但是关系型数据库的优化方法都殊途同归，因此无论是 DB2 从业者、SQL SERVER 从业者、MYSQL 从业者，亦或是 PostGre SQL 从业者等，都能从本书中学到所需要的 SQL 优化知识。

因水平有限，本书在编写过程中难免有错漏之处，恳请读者批评、指正。联系我们的方式如下：692162374@qq.com（QQ 好友数已达上限）或者 327165427@qq.com（新开 QQ 账号）。

如果有读者想进一步学习 SQL 优化技能或者一些公司或机构需要开展 SQL 优化方面的培训，都可以联系作者。另外，作者还开设了实体培训班，可以实现零基础学习，结业后可以顺利就业，欢迎联系罗老师。

本 书 约 定

在阅读本书之前请读者安装好 Oracle 数据库并且配置好示例账户 Scott，因为本书均以 Scott 账户进行讲解。推荐读者安装与本书相同版本的数据库进行测试，具有专研精神的读者请安装好 Oracle12c 进行对比实验，这样一来，你将发现 Oracle12c CBO 的一些新特征。本书使用的版本是 Oracle11gR2。

```
SQL> select * from v$version where rownum=1;

BANNER
--------------------------------------------------------------------------------
Oracle Database 11g Enterprise Edition Release 11.2.0.1.0 - Production

SQL> show user
USER is "SYS"

SQL> grant dba to scott;

Grant succeeded.

SQL> alter user scott account unlock;

User altered.

SQL> alter user scott identified by tiger;

User altered.

SQL> conn scott/tiger
Connected.

SQL> create table test as select * from dba_objects;

Table created.
```

目 录

第 1 章 SQL 优化必懂概念 ... 1
1.1 基数（CARDINALITY） ... 1
1.2 选择性（SELECTIVITY） ... 3
1.3 直方图（HISTOGRAM） ... 7
1.4 回表（TABLE ACCESS BY INDEX ROWID） ... 13
1.5 集群因子（CLUSTERING FACTOR） ... 15
1.6 表与表之间关系 ... 19

第 2 章 统计信息 ... 21
2.1 什么是统计信息 ... 21
2.2 统计信息重要参数设置 ... 24
2.3 检查统计信息是否过期 ... 32
2.4 扩展统计信息 ... 37
2.5 动态采样 ... 42
2.6 定制统计信息收集策略 ... 47

第 3 章 执行计划 ... 49
3.1 获取执行计划常用方法 ... 49
3.1.1 使用 AUTOTRACE 查看执行计划 ... 49
3.1.2 使用 EXPLAIN PLAN FOR 查看执行计划 ... 52
3.1.3 查看带有 A-TIME 的执行计划 ... 54
3.1.4 查看正在执行的 SQL 的执行计划 ... 56
3.2 定制执行计划 ... 57
3.3 怎么通过查看执行计划建立索引 ... 59
3.4 运用光标移动大法阅读执行计划 ... 63

第 4 章 访问路径（ACCESS PATH） ... 67
4.1 常见访问路径 ... 67
4.1.1 TABLE ACCESS FULL ... 67
4.1.2 TABLE ACCESS BY USER ROWID ... 71
4.1.3 TABLE ACCESS BY ROWID RANGE ... 71
4.1.4 TABLE ACCESS BY INDEX ROWID ... 72
4.1.5 INDEX UNIQUE SCAN ... 72
4.1.6 INDEX RANGE SCAN ... 73
4.1.7 INDEX SKIP SCAN ... 74
4.1.8 INDEX FULL SCAN ... 75
4.1.9 INDEX FAST FULL SCAN ... 77
4.1.10 INDEX FULL SCAN（MIN/MAX） ... 80
4.1.11 MAT_VIEW REWRITE ACCESS FULL ... 83
4.2 单块读与多块读 ... 83
4.3 为什么有时候索引扫描比全表扫描更慢 ... 84
4.4 DML 对于索引维护的影响 ... 84

第 5 章 表连接方式 ... 86
5.1 嵌套循环（NESTED LOOPS） ... 86
5.2 HASH 连接（HASH JOIN） ... 90
5.3 排序合并连接（SORT MERGE JOIN） ... 93
5.4 笛卡儿连接（CARTESIAN JOIN） ... 95
5.5 标量子查询（SCALAR SUBQUERY） ... 98
5.6 半连接（SEMI JOIN） ... 100
5.6.1 半连接等价改写 ... 100
5.6.2 控制半连接执行计划 ... 101
5.6.3 读者思考 ... 103
5.7 反连接（ANTI JOIN） ... 104
5.7.1 反连接等价改写 ... 104

目录

　　5.7.2　控制反连接执行计划 ············ 105
　　5.7.3　读者思考 ·························· 108
5.8　FILTER ····································· 108
5.9　IN 与 EXISTS 谁快谁慢 ············· 111
5.10　SQL 语句的本质 ····················· 111

第 6 章　成本计算 ························· 112
6.1　优化 SQL 需要看 COST 吗 ······· 112
6.2　全表扫描成本计算 ···················· 112
6.3　索引范围扫描成本计算 ············· 116
6.4　SQL 优化核心思想 ··················· 119

第 7 章　必须掌握的查询变换 ········ 120
7.1　子查询非嵌套 ·························· 120
7.2　视图合并 ································ 125
7.3　谓词推入 ································ 129

第 8 章　调优技巧 ························· 133
8.1　查看真实的基数（Rows） ········· 133
8.2　使用 UNION 代替 OR ··············· 134
8.3　分页语句优化思路 ···················· 135
　　8.3.1　单表分页优化思路 ·············· 135
　　8.3.2　多表关联分页优化思路 ········ 150
8.4　使用分析函数优化自连接 ·········· 153
8.5　超大表与超小表关联优化方法 ··· 154
8.6　超大表与超大表关联优化方法 ··· 155
8.7　LIKE 语句优化方法 ·················· 159
8.8　DBLINK 优化 ·························· 161
8.9　对表进行 ROWID 切片 ············· 167
8.10　SQL 三段分拆法 ····················· 169

第 9 章　SQL 优化案例赏析 ············ 170
9.1　组合索引优化案例 ···················· 170
9.2　直方图优化案例 ······················· 173
9.3　NL 被驱动表不能走 INDEX SKIP
　　　SCAN ····································· 177
9.4　优化 SQL 需要注意表与表之间
　　　关系 ······································· 178
9.5　INDEX FAST FULL SCAN 优化
　　　案例 ······································· 179
9.6　分页语句优化案例 ···················· 181
9.7　ORDER BY 取别名列优化案例 ···· 183
9.8　半连接反向驱动主表案例一 ······· 185
9.9　半连接反向驱动主表案例二 ······· 187
9.10　连接列数据分布不均衡导致性能
　　　问题 ······································· 192
9.11　Filter 优化经典案例 ················ 198
9.12　树形查询优化案例 ·················· 202
9.13　本地索引优化案例 ·················· 204
9.14　标量子查询优化案例 ··············· 206
　　9.14.1　案例一 ···························· 206
　　9.14.2　案例二 ···························· 207
9.15　关联更新优化案例 ·················· 211
9.16　外连接有 OR 关联条件只能走 NL ··· 213
9.17　把你脑袋当 CBO ···················· 217
9.18　扩展统计信息优化案例 ··········· 221
9.19　使用 LISGAGG 分析函数优化
　　　WMSYS.WM_CONCAT ··········· 227
9.20　INSTR 非等值关联优化案例 ···· 230
9.21　REGEXP_LIKE 非等值关联优化
　　　案例 ······································· 233
9.22　ROW LEVEL SECURITY 优化
　　　案例 ······································· 237
9.23　子查询非嵌套优化案例一 ········ 240
9.24　子查询非嵌套优化案例二 ········ 247
9.25　烂用外连接导致无法谓词推入 ··· 252
9.26　谓词推入优化案例 ·················· 262
9.27　使用 CARDINALITY 优化 SQL ···· 268
9.28　利用等待事件优化 SQL ··········· 272

第 10 章　全自动 SQL 审核 ············ 281
10.1　抓出外键没创建索引的表 ········ 281
10.2　抓出需要收集直方图的列 ········ 282
10.3　抓出必须创建索引的列 ··········· 283
10.4　抓出 SELECT * 的 SQL ··········· 284
10.5　抓出有标量子查询的 SQL ········ 285
10.6　抓出带有自定义函数的 SQL ···· 286
10.7　抓出表被多次反复调用 SQL ···· 287
10.8　抓出走了 FILTER 的 SQL ········ 288
10.9　抓出返回行数较多的嵌套
　　　循环 SQL ······························· 290

10.10 抓出 NL 被驱动表走了全表扫描的 SQL ············ 292
10.11 抓出走了 TABLE ACCESS FULL 的 SQL ············ 293
10.12 抓出走了 INDEX FULL SCAN 的 SQL ············ 294
10.13 抓出走了 INDEX SKIP SCAN 的 SQL ············ 295
10.14 抓出索引被哪些 SQL 引用 ············ 297
10.15 抓出走了笛卡儿积的 SQL ············ 298
10.16 抓出走了错误的排序合并连接的 SQL ············ 299
10.17 抓出 LOOP 套 LOOP 的 PSQL ············ 301
10.18 抓出走了低选择性索引的 SQL ············ 302
10.19 抓出可以创建组合索引的 SQL（回表再过滤选择性高的列）············ 304
10.20 抓出可以创建组合索引的 SQL（回表只访问少数字段）············ 306

第 1 章 SQL 优化必懂概念

1.1 基数（CARDINALITY）

某个列唯一键（Distinct_Keys）的数量叫作基数。比如性别列，该列只有男女之分，所以这一列基数是 2。主键列的基数等于表的总行数。基数的高低影响列的数据分布。

以测试表 test 为例，owner 列和 object_id 列的基数分别如下所示。

```
SQL> select count(distinct owner),count(distinct object_id),count(*) from test;
COUNT(DISTINCTOWNER) COUNT(DISTINCTOBJECT_ID)   COUNT(*)
-------------------- ------------------------ ----------
                  29                    72462      72462
```

TEST 表的总行数为 72 462，owner 列的基数为 29，说明 owner 列里面有大量重复值，object_id 列的基数等于总行数，说明 object_id 列没有重复值，相当于主键。owner 列的数据分布如下。

```
SQL> select owner,count(*) from test group by owner order by 2 desc;
OWNER                     COUNT(*)
------------------------- ----------
SYS                          30808
PUBLIC                       27699
SYSMAN                        3491
ORDSYS                        2532
APEX_030200                   2406
MDSYS                         1509
XDB                            844
OLAPSYS                        719
SYSTEM                         529
CTXSYS                         366
WMSYS                          316
EXFSYS                         310
SH                             306
ORDDATA                        248
OE                             127
DBSNMP                          57
IX                              55
HR                              34
PM                              27
FLOWS_FILES                     12
OWBSYS_AUDIT                    12
ORDPLUGINS                      10
OUTLN                            9
BI                               8
SI_INFORMTN_SCHEMA               8
ORACLE_OCM                       8
SCOTT                            7
```

```
APPQOSSYS                              3
OWBSYS                                 2
```

owner 列的数据分布极不均衡，我们运行如下 SQL。

```
select * from test where owner='SYS';
```

SYS 有 30 808 条数据，从 72 462 条数据里面查询 30 808 条数据，也就是说要返回表中 42.5%的数据。

```
SQL> select 30808/72462*100 "Percent" from dual;

   Percent
----------
42.5160774
```

那么请思考，你认为以上查询应该使用索引吗？现在我们换一种查询语句。

```
select * from test where owner='SCOTT';
```

SCOTT 有 7 条数据，从 72 462 条数据里面查询 7 条数据，也就是说要返回表中 0.009%的数据。

```
SQL> select 7/72462*100 "Percent" from dual;

   Percent
----------
.009660236
```

请思考，返回表中 0.009%的数据应不应该走索引？

如果你还不懂索引，没关系，后面的章节我们会详细介绍。如果你回答不了上面的问题，我们先提醒一下。当查询结果是返回表中 5%以内的数据时，应该走索引；当查询结果返回的是超过表中 5%的数据时，应该走全表扫描。

当然了，返回表中 5%以内的数据走索引，返回超过 5%的数据就使用全表扫描，这个结论太绝对了，因为你还没掌握后面章节的知识，这里暂且记住 5%这个界限就行。我们之所以在这里讲 5%，是怕一些初学者不知道上面问题的答案而纠结。

现在有如下查询语句。

```
select * from test where owner=:B1;
```

语句中，":B1" 是绑定变量，可以传入任意值，该查询可能走索引也可能走全表扫描。

现在得到一个结论：如果某个列基数很低，该列数据分布就会非常不均衡，由于该列数据分布不均衡，会导致 SQL 查询可能走索引，也可能走全表扫描。在做 SQL 优化的时候，如果怀疑列数据分布不均衡，我们可以使用 select 列, count(*) from 表 group by 列 order by 2 desc 来查看列的数据分布。

如果 SQL 语句是单表访问，那么可能走索引，可能走全表扫描，也可能走物化视图扫描。在不考虑有物化视图的情况下，单表访问要么走索引，要么走全表扫描。现在，回忆一下走索引的条件：返回表中 5%以内的数据走索引，超过 5%的时候走全表扫描。相信大家读到这里，已经搞懂了单表访问的优化方法。

我们来看如下查询。

```
select * from test where object_id=:B1;
```

不管 object_id 传入任何值，都应该走索引。

我们再思考如下查询语句。

```
select * from test where object_name=:B1;
```

不管给 object_name 传入任何值，请问该查询应该走索引吗？

请你去查看 object_name 的数据分布。写到这里，其实有点想把本节名称改为"数据分布"。大家在以后的工作中一定要注意列的数据分布！

1.2 选择性（SELECTIVITY）

基数与总行数的比值再乘以 100%就是某个列的选择性。

在进行 SQL 优化的时候，单独看列的基数是没有意义的，基数必须对比总行数才有实际意义，正是因为这个原因，我们才引出了选择性这个概念。

下面我们查看 test 表各个列的基数与选择性，为了查看选择性，必须先收集统计信息。关于统计信息，我们在第 2 章会详细介绍。下面的脚本用于收集 test 表的统计信息。

```
SQL> BEGIN
  2    DBMS_STATS.GATHER_TABLE_STATS(ownname          => 'SCOTT',
  3                                  tabname          => 'TEST',
  4                                  estimate_percent => 100,
  5                                  method_opt       => 'for all columns size 1',
  6                                  no_invalidate    => FALSE,
  7                                  degree           => 1,
  8                                  cascade          => TRUE);
  9  END;
 10  /

PL/SQL procedure successfully completed.
```

下面的脚本用于查看 test 表中每个列的基数与选择性。

```
SQL> select a.column_name,
  2         b.num_rows,
  3         a.num_distinct Cardinality,
  4         round(a.num_distinct / b.num_rows * 100, 2) selectivity,
  5         a.histogram,
  6         a.num_buckets
  7    from dba_tab_col_statistics a, dba_tables b
  8   where a.owner = b.owner
  9     and a.table_name = b.table_name
 10     and a.owner = 'SCOTT'
 11     and a.table_name = 'TEST';

COLUMN_NAME        NUM_ROWS   CARDINALITY  SELECTIVITY HISTOGRAM NUM_BUCKETS
---------------    --------   -----------  ----------- --------- -----------
OWNER              72462             29          .04   NONE               1
OBJECT_NAME        72462          44236        61.05   NONE               1
SUBOBJECT_NAME     72462            106          .15   NONE               1
OBJECT_ID          72462          72462          100   NONE               1
```

```
DATA_OBJECT_ID        72462         7608        10.5 NONE              1
OBJECT_TYPE           72462           44         .06 NONE              1
CREATED               72462         1366        1.89 NONE              1
LAST_DDL_TIME         72462         1412        1.95 NONE              1
TIMESTAMP             72462         1480        2.04 NONE              1
STATUS                72462            1           0 NONE              1
TEMPORARY             72462            2           0 NONE              1
GENERATED             72462            2           0 NONE              1
SECONDARY             72462            2           0 NONE              1
NAMESPACE             72462           21         .03 NONE              1
EDITION_NAME          72462            0           0 NONE              0

15 rows selected.
```

请思考：什么样的列必须建立索引呢？

有人说基数高的列，有人说在 where 条件中的列。这些答案并不完美。基数高究竟是多高？没有和总行数对比，始终不知道有多高。比如某个列的基数有几万行，但是总行数有几十亿行，那么这个列的基数还高吗？这就是要引出选择性的根本原因。

当一个列选择性大于 20%，说明该列的数据分布就比较均衡了。测试表 test 中 object_name、object_id 的选择性均大于 20%，其中 object_name 列的选择性为 61.05%。现在我们查看该列数据分布（为了方便展示，只输出前 10 行数据的分布情况）。

```
SQL> select *
  2    from (select object_name, count(*)
  3            from test
  4           group by object_name
  5           order by 2 desc)
  6   where rownum <= 10;

OBJECT_NAME              COUNT(*)
-------------------- ----------
COSTS                        30
SALES                        30
SALES_CHANNEL_BIX            29
COSTS_TIME_BIX               29
COSTS_PROD_BIX               29
SALES_TIME_BIX               29
SALES_PROMO_BIX              29
SALES_PROD_BIX               29
SALES_CUST_BIX               29
DBMS_REPCAT_AUTH              5

10 rows selected.
```

由上面的查询结果我们可知，object_name 列的数据分布非常均衡。我们查询以下 SQL。

```
select * from test where object_name=:B1;
```

不管 object_name 传入任何值，最多返回 30 行数据。

什么样的列必须要创建索引呢？当一个列出现在 where 条件中，该列没有创建索引并且选择性大于 20%，那么该列就必须创建索引，从而提升 SQL 查询性能。当然了，如果表只有几百条数据，那我们就不用创建索引了。

下面抛出 SQL 优化核心思想第一个观点：**只有大表才会产生性能问题**。

也许有人会说："我有个表很小，只有几百条，但是该表经常进行 DML，会产生热点块，

1.2 选择性（SELECTIVITY）

也会出性能问题。"对此我们并不想过多地讨论此问题，这属于应用程序设计问题，不属于 SQL 优化的范畴。

下面我们将通过实验为大家分享本书第一个全自动优化脚本。

抓出必须创建索引的列（请读者对该脚本适当修改，以便用于生产环境）。

首先，该列必须出现在 where 条件中，怎么抓出表的哪个列出现在 where 条件中呢？有两种方法，一种是可以通过 V$SQL_PLAN 抓取，另一种是通过下面的脚本抓取。

先执行下面的存储过程，刷新数据库监控信息。

```
begin
  dbms_stats.flush_database_monitoring_info;
end;
```

运行完上面的命令之后，再运行下面的查询语句就可以查询出哪个表的哪个列出现在 where 条件中。

```
select r.name owner,
       o.name table_name,
       c.name column_name,
       equality_preds, ---等值过滤
       equijoin_preds, ---等值 JOIN 比如 where a.id=b.id
       nonequijoin_preds, ----不等 JOIN
       range_preds, ----范围过滤次数 > >= < <= between and
       like_preds, ----LIKE 过滤
       null_preds, ----NULL 过滤
       timestamp
  from sys.col_usage$ u, sys.obj$ o, sys.col$ c, sys.user$ r
 where o.obj# = u.obj#
   and c.obj# = u.obj#
   and c.col# = u.intcol#
   and r.name = 'SCOTT'
   and o.name = 'TEST';
```

下面是实验步骤。

我们首先运行一个查询语句，让 owner 与 object_id 列出现在 where 条件中。

```
SQL> select object_id, owner, object_type
  2    from test
  3   where owner = 'SYS'
  4     and object_id < 100
  5     and rownum <= 10;

OBJECT_ID OWNER                OBJECT_TYPE
---------- -------------------- -----------
        20 SYS                  TABLE
        46 SYS                  INDEX
        28 SYS                  TABLE
        15 SYS                  TABLE
        29 SYS                  CLUSTER
         3 SYS                  INDEX
        25 SYS                  TABLE
        41 SYS                  INDEX
        54 SYS                  INDEX
        40 SYS                  INDEX

10 rows selected.
```

其次刷新数据库监控信息。

```
SQL> begin
  2     dbms_stats.flush_database_monitoring_info;
  3  end;
  4  /

PL/SQL procedure successfully completed.
```

然后我们查看 test 表有哪些列出现在 where 条件中。

```
SQL> select r.name owner, o.name table_name, c.name column_name
  2    from sys.col_usage$ u, sys.obj$ o, sys.col$ c, sys.user$ r
  3   where o.obj# = u.obj#
  4     and c.obj# = u.obj#
  5     and c.col# = u.intcol#
  6     and r.name = 'SCOTT'
  7     and o.name = 'TEST';

OWNER      TABLE_NAME COLUMN_NAME
---------- ---------- ------------------------------
SCOTT      TEST       OWNER
SCOTT      TEST       OBJECT_ID
```

接下来我们查询出选择性大于等于20%的列。

```
SQL> select a.owner,
  2         a.table_name,
  3         a.column_name,
  4         round(a.num_distinct / b.num_rows * 100, 2) selectivity
  5    from dba_tab_col_statistics a, dba_tables b
  6   where a.owner = b.owner
  7     and a.table_name = b.table_name
  8     and a.owner = 'SCOTT'
  9     and a.table_name = 'TEST'
 10     and a.num_distinct / b.num_rows >= 0.2;

OWNER      TABLE_NAME COLUMN_NAME   SELECTIVITY
---------- ---------- ------------- -----------
SCOTT      TEST       OBJECT_NAME         61.05
SCOTT      TEST       OBJECT_ID             100
```

最后，确保这些列没有创建索引。

```
SQL> select table_owner, table_name, column_name, index_name
  2    from dba_ind_columns
  3   where table_owner = 'SCOTT'
  4     and table_name = 'TEST';
未选定行
```

把上面的脚本组合起来，我们就可以得到全自动的优化脚本了。

```
SQL> select owner,
  2         column_name,
  3         num_rows,
  4         Cardinality,
  5         selectivity,
  6         'Need index' as notice
  7    from (select b.owner,
  8                 a.column_name,
  9                 b.num_rows,
```

```
 10                   a.num_distinct Cardinality,
 11                   round(a.num_distinct / b.num_rows * 100, 2) selectivity
 12              from dba_tab_col_statistics a, dba_tables b
 13             where a.owner = b.owner
 14               and a.table_name = b.table_name
 15               and a.owner = 'SCOTT'
 16               and a.table_name = 'TEST')
 17   where selectivity >= 20
 18     and column_name not in (select column_name
 19                               from dba_ind_columns
 20                              where table_owner = 'SCOTT'
 21                                and table_name = 'TEST')
 22     and column_name in
 23         (select c.name
 24            from sys.col_usage$ u, sys.obj$ o, sys.col$ c, sys.user$ r
 25           where o.obj# = u.obj#
 26             and c.obj# = u.obj#
 27             and c.col# = u.intcol#
 28             and r.name = 'SCOTT'
 29             and o.name = 'TEST');

OWNER        COLUMN_NAME     NUM_ROWS    CARDINALITY   SELECTIVITY   NOTICE
----------   -------------   ---------   -----------   -----------   ----------
SCOTT        OBJECT_ID          72462         72462           100    Need index
```

1.3 直方图（HISTOGRAM）

前面提到，当某个列基数很低，该列数据分布就会不均衡。数据分布不均衡会导致在查询该列的时候，要么走全表扫描，要么走索引扫描，这个时候很容易走错执行计划。

如果没有对基数低的列收集直方图统计信息，基于成本的优化器（CBO）会认为该列数据分布是均衡的。

下面我们还是以测试表 test 为例，用实验讲解直方图。

首先我们对测试表 test 收集统计信息，在收集统计信息的时候，不收集列的直方图，语句 `for all columns size 1` 表示对所有列都不收集直方图。

```
SQL> BEGIN
  2    DBMS_STATS.GATHER_TABLE_STATS(ownname          => 'SCOTT',
  3                                  tabname          => 'TEST',
  4                                  estimate_percent => 100,
  5                                  method_opt       => 'for all columns size 1',
  6                                  no_invalidate    => FALSE,
  7                                  degree           => 1,
  8                                  cascade          => TRUE);
  9  END;
 10  /

PL/SQL procedure successfully completed.
```

Histogram 为 none 表示没有收集直方图。

```
SQL> select a.column_name,
  2         b.num_rows,
  3         a.num_distinct Cardinality,
  4         round(a.num_distinct / b.num_rows * 100, 2) selectivity,
  5         a.histogram,
```

```
  6         a.num_buckets
  7    from dba_tab_col_statistics a, dba_tables b
  8    where a.owner = b.owner
  9      and a.table_name = b.table_name
 10      and a.owner = 'SCOTT'
 11      and a.table_name = 'TEST';

COLUMN_NAME         NUM_ROWS CARDINALITY SELECTIVITY HISTOGRAM NUM_BUCKETS
------------------- -------- ----------- ----------- --------- -----------
OWNER                  72462          29         .04 NONE                1
OBJECT_NAME            72462       44236       61.05 NONE                1
SUBOBJECT_NAME         72462         106         .15 NONE                1
OBJECT_ID              72462       72462         100 NONE                1
DATA_OBJECT_ID         72462        7608        10.5 NONE                1
OBJECT_TYPE            72462          44         .06 NONE                1
CREATED                72462        1366        1.89 NONE                1
LAST_DDL_TIME          72462        1412        1.95 NONE                1
TIMESTAMP              72462        1480        2.04 NONE                1
STATUS                 72462           1           0 NONE                1
TEMPORARY              72462           2           0 NONE                1
GENERATED              72462           2           0 NONE                1
SECONDARY              72462           2           0 NONE                1
NAMESPACE              72462          21         .03 NONE                1
EDITION_NAME           72462           0           0 NONE                0

15 rows selected.
```

owner 列基数很低，现在我们对 owner 列进行查询。

```
SQL> set autot trace
SQL> select * from test where owner='SCOTT';

7 rows selected.

Execution Plan
----------------------------------------------------------
Plan hash value: 1357081020

--------------------------------------------------------------------------
| Id  | Operation         | Name | Rows  | Bytes | Cost (%CPU)| Time     |
--------------------------------------------------------------------------
|   0 | SELECT STATEMENT  |      |  2499 |  236K |   289   (1)| 00:00:04 |
|*  1 |  TABLE ACCESS FULL| TEST |  2499 |  236K |   289   (1)| 00:00:04 |
--------------------------------------------------------------------------

Predicate Information (identified by operation id):
---------------------------------------------------

   1 - filter("OWNER"='SCOTT')
```

请注意看粗体字部分，查询 owner='SCOTT'返回了 7 条数据，但是 CBO 在计算 **Rows** 的时候认为 owner='SCOTT'返回 **2 499** 条数据，Rows 估算得不是特别准确。从 72 462 条数据里面查询出 7 条数据，应该走索引，所以现在我们对 owner 列创建索引。

```
SQL> create index idx_owner on test(owner);

Index created.
```

我们再来查询一下。

1.3 直方图（HISTOGRAM）

```
SQL> select * from test where owner='SCOTT';

7 rows selected.

Execution Plan
----------------------------------------------------------
Plan hash value: 3932013684

--------------------------------------------------------------------------------
| Id |Operation                     |Name     | Rows  | Bytes | Cost(%CPU)| Time     |
--------------------------------------------------------------------------------
|  0 | SELECT STATEMENT             |         | 2499  | 236K  |  73   (0)| 00:00:01 |
|  1 |  TABLE ACCESS BY INDEX ROWID |TEST     | 2499  | 236K  |  73   (0)| 00:00:01 |
|* 2 |   INDEX RANGE SCAN           |IDX_OWNER| 2499  |       |   6   (0)| 00:00:01 |
--------------------------------------------------------------------------------

Predicate Information (identified by operation id):
---------------------------------------------------

   2 - access("OWNER"='SCOTT')
```

现在我们查询 owner='SYS'。

```
SQL> select * from test where owner='SYS';

30808 rows selected.

Execution Plan
----------------------------------------------------------
Plan hash value: 3932013684

--------------------------------------------------------------------------------
| Id |Operation                     |Name     | Rows  | Bytes | Cost(%CPU)| Time     |
--------------------------------------------------------------------------------
|  0 | SELECT STATEMENT             |         | 2499  | 236K  |  73   (0)| 00:00:01 |
|  1 |  TABLE ACCESS BY INDEX ROWID |TEST     | 2499  | 236K  |  73   (0)| 00:00:01 |
|* 2 |   INDEX RANGE SCAN           |IDX_OWNER| 2499  |       |   6   (0)| 00:00:01 |
--------------------------------------------------------------------------------

Predicate Information (identified by operation id):
---------------------------------------------------

   2 - access("OWNER"='SYS')
```

注意粗字体部分，查询 owner='SYS' 返回了 30 808 条数据。从 72 462 条数据里面返回 30 808 条数据能走索引吗？很明显应该走全表扫描。也就是说该执行计划是错误的。

为什么查询 owner='SYS' 的执行计划会用错呢？因为 owner 这个列基数很低，只有 29，而表的总行数是 72 462。**前文着重强调过，当列没有收集直方图统计信息的时候，CBO 会认为该列数据分布是均衡的**。正是因为 CBO 认为 owner 列数据分布是均衡的，不管 owner 等于任何值，CBO 估算的 Rows 永远都是 **2 499**。而这 **2 499** 是怎么来的呢？答案如下。

```
SQL> select round(72462/29) from dual;

round(72462/29)
---------------
           2499
```

现在大家也知道了，**执行计划里面的 Rows 是假的**。执行计划中的 Rows 是根据统计信息以及一些数学公式计算出来的。很多 DBA 到现在还不知道执行计划中 Rows 是假的这个真相，真是令人遗憾。

在做 SQL 优化的时候，经常需要做的工作就是帮助 CBO 计算出比较准确的 Rows。注意：我们说的是比较准确的 Rows。CBO 是无法得到精确的 Rows 的，因为对表收集统计信息的时候，统计信息一般都不会按照 100%的标准采样收集，即使按照 100%的标准采样收集了表的统计信息，表中的数据也随时在发生变更。另外计算 Rows 的数学公式目前也是有缺陷的，CBO 永远不可能计算得到精确的 Rows。

如果 CBO 每次都能计算得到精确的 Rows，那么相信我们这个时候只需要关心业务逻辑、表设计、SQL 写法以及如何建立索引了，**再也不用担心 SQL 会走错执行计划了**。

Oracle12c 的新功能 SQL Plan Directives 在一定程度上解决了 Rows 估算不准而引发的 SQL 性能问题。关于 SQL Plan Directives，本书不做过多讨论。

为了让 CBO 选择正确的执行计划，我们需要对 owner 列收集直方图信息，从而告知 CBO 该列数据分布不均衡，让 CBO 在计算 Rows 的时候参考直方图统计。现在我们对 owner 列收集直方图。

```
SQL> BEGIN
  2     DBMS_STATS.GATHER_TABLE_STATS(ownname          => 'SCOTT',
  3                                   tabname          => 'TEST',
  4                                   estimate_percent => 100,
  5                                   method_opt       => 'for columns owner size skewonly',
  6                                   no_invalidate    => FALSE,
  7                                   degree           => 1,
  8                                   cascade          => TRUE);
  9  END;
 10  /

PL/SQL procedure successfully completed.
```

查看一下 owner 列的直方图信息。

```
SQL> select a.column_name,
  2         b.num_rows,
  3         a.num_distinct Cardinality,
  4         round(a.num_distinct / b.num_rows * 100, 2) selectivity,
  5         a.histogram,
  6         a.num_buckets
  7    from dba_tab_col_statistics a, dba_tables b
  8   where a.owner = b.owner
  9     and a.table_name = b.table_name
 10     and a.owner = 'SCOTT'
 11     and a.table_name = 'TEST';

COLUMN_NAME        NUM_ROWS CARDINALITY SELECTIVITY HISTOGRAM  NUM_BUCKETS
---------------    -------- ----------- ----------- ---------- -----------
OWNER                 72462          29         .04 FREQUENCY           29
OBJECT_NAME           72462       44236       61.05 NONE                 1
SUBOBJECT_NAME        72462         106         .15 NONE                 1
OBJECT_ID             72462       72462         100 NONE                 1
DATA_OBJECT_ID        72462        7608        10.5 NONE                 1
OBJECT_TYPE           72462          44         .06 NONE                 1
CREATED               72462        1366        1.89 NONE                 1
LAST_DDL_TIME         72462        1412        1.95 NONE                 1
```

1.3 直方图（HISTOGRAM）

```
TIMESTAMP          72462       1480      2.04 NONE                1
STATUS             72462          1         0 NONE                1
TEMPORARY          72462          2         0 NONE                1
GENERATED          72462          2         0 NONE                1
SECONDARY          72462          2         0 NONE                1
NAMESPACE          72462         21       .03 NONE                1
EDITION_NAME       72462          0         0 NONE                0

15 rows selected.
```

现在我们再来查询上面的 SQL，看执行计划是否还会走错并且验证 Rows 是否还会算错。

```
SQL> select * from test where owner='SCOTT';

7 rows selected.

Execution Plan
----------------------------------------------------------
Plan hash value: 3932013684

---------------------------------------------------------------------------------------
| Id  | Operation                   | Name      | Rows | Bytes | Cost (%CPU)| Time     |
---------------------------------------------------------------------------------------
|   0 | SELECT STATEMENT            |           |    7 |   679 |     2   (0)| 00:00:01 |
|   1 |  TABLE ACCESS BY INDEX ROWID| TEST      |    7 |   679 |     2   (0)| 00:00:01 |
|*  2 |   INDEX RANGE SCAN          | IDX_OWNER |    7 |       |     1   (0)| 00:00:01 |
---------------------------------------------------------------------------------------

Predicate Information (identified by operation id):
---------------------------------------------------

   2 - access("OWNER"='SCOTT')

SQL> select * from test where owner='SYS';

30808 rows selected.

Execution Plan
----------------------------------------------------------
Plan hash value: 1357081020

--------------------------------------------------------------------------
| Id  | Operation         | Name | Rows  | Bytes | Cost (%CPU)| Time     |
--------------------------------------------------------------------------
|   0 | SELECT STATEMENT  |      | 30808 | 2918K |   290   (1)| 00:00:04 |
|*  1 |  TABLE ACCESS FULL| TEST | 30808 | 2918K |   290   (1)| 00:00:04 |
--------------------------------------------------------------------------

Predicate Information (identified by operation id):
---------------------------------------------------

   1 - filter("OWNER"='SYS')
```

对 owner 列收集完直方图之后，CBO 估算的 Rows 就基本准确了，一旦 Rows 估算对了，那么执行计划也就不会出错了。

大家是不是很好奇，为什么收集完直方图之后，Rows 计算得那么精确，收集直方图究竟完成了什么操作呢？对 owner 列收集直方图其实就相当于运行了以下 SQL。

```
select owner,count(*) from test group by owner;
```

直方图信息就是以上 SQL 的查询结果，这些查询结果会保存在数据字典中。这样当我们查询 owner 为任意值的时候，CBO 总会算出正确的 Rows，因为直方图已经知道每个值有多少行数据。

如果 SQL 使用了绑定变量，绑定变量的列收集了直方图，那么该 SQL 就会引起绑定变量窥探。绑定变量窥探是一个老生常谈的问题，这里不多做讨论。Oracle11g 引入了自适应游标共享（Adaptive Cursor Sharing），基本上解决了绑定变量窥探问题，但是自适应游标共享也会引起一些新问题，对此也不做过多讨论。

当我们遇到一个 SQL 有绑定变量怎么办？其实很简单，我们只需要运行以下语句。

```
select 列, count(*) from test group by 列 order by 2 desc;
```

如果列数据分布均衡，基本上 SQL 不会出现问题；如果列数据分布不均衡，我们需要对列收集直方图统计。

关于直方图，其实还有非常多的话题，比如直方图的种类、直方图的桶数等，本书在此不做过多讨论。在我们看来，**读者只需要知道直方图是用来帮助 CBO 在对基数很低、数据分布不均衡的列进行 Rows 估算的时候，可以得到更精确的 Rows 就够了**。

什么样的列需要收集直方图呢？当列出现在 where 条件中，列的选择性小于 1%并且该列没有收集过直方图，这样的列就应该收集直方图。注意：千万不能对没有出现在 where 条件中的列收集直方图。对没有出现在 where 条件中的列收集直方图完全是做无用功，浪费数据库资源。

下面我们为大家分享本书第二个全自动化优化脚本。

抓出必须创建直方图的列（大家可以对该脚本进行适当修改，以便用于生产环境）。

```
SQL> select a.owner,
  2         a.table_name,
  3         a.column_name,
  4         b.num_rows,
  5         a.num_distinct,
  6         trunc(num_distinct / num_rows * 100,2) selectivity,
  7         'Need Gather Histogram' notice
  8    from dba_tab_col_statistics a, dba_tables b
  9   where a.owner = 'SCOTT'
 10     and a.table_name = 'TEST'
 11     and a.owner = b.owner
 12     and a.table_name = b.table_name
 13     and num_distinct / num_rows<0.01
 14     and (a.owner, a.table_name, a.column_name) in
 15         (select r.name owner, o.name table_name, c.name column_name
 16            from sys.col_usage$ u, sys.obj$ o, sys.col$ c, sys.user$ r
 17           where o.obj# = u.obj#
 18             and c.obj# = u.obj#
 19             and c.col# = u.intcol#
 20             and r.name = 'SCOTT'
 21             and o.name = 'TEST')
 22     and a.histogram ='NONE';

OWNER TABLE COLUM    NUM_ROWS NUM_DISTINCT SELECTIVITY NOTICE
----- ----- -----    -------- ------------ ----------- ---------------------
SCOTT TEST  OWNER       72462           29         .04 Need Gather Histogram
```

1.4 回表（TABLE ACCESS BY INDEX ROWID）

当对一个列创建索引之后，索引会包含该列的键值以及键值对应行所在的 rowid。**通过索引中记录的 rowid 访问表中的数据就叫回表。回表一般是单块读，回表次数太多会严重影响 SQL 性能，如果回表次数太多，就不应该走索引扫描了，应该直接走全表扫描。**

在进行 SQL 优化的时候，一定要注意回表次数！特别是要注意回表的物理 I/O 次数！

大家还记得 1.3 节中错误的执行计划吗？

```
SQL> select * from test where owner='SYS';

30808 rows selected.

Execution Plan
----------------------------------------------------------
Plan hash value: 3932013684

--------------------------------------------------------------------------------
| Id  | Operation                   | Name      | Rows  | Bytes | Cost(%CPU)| Time     |
--------------------------------------------------------------------------------
|   0 | SELECT STATEMENT            |           |  2499 |  236K|    73   (0)| 00:00:01 |
|   1 |  TABLE ACCESS BY INDEX ROWID| TEST      |  2499 |  236K|    73   (0)| 00:00:01 |
|*  2 |   INDEX RANGE SCAN          | IDX_OWNER |  2499 |       |     6   (0)| 00:00:01 |
--------------------------------------------------------------------------------

Predicate Information (identified by operation id):
---------------------------------------------------

   2 - access("OWNER"='SYS')
```

执行计划中加粗部分（TABLE ACCESS BY INDEX ROWID）就是回表。索引返回多少行数据，回表就要回多少次，每次回表都是单块读（因为一个 rowid 对应一个数据块）。该 SQL 返回了 30 808 行数据，那么回表一共就需要 30 808 次。

请思考：上面执行计划的性能是耗费在索引扫描中还是耗费在回表中？

为了得到答案，请大家在 SQLPLUS 中进行实验。为了消除 arraysize 参数对逻辑读的影响，设置 `arraysize=5000`。arraysize 表示 Oracle 服务器每次传输多少行数据到客户端，默认为 15。如果一个块有 150 行数据，那么这个块就会被读 10 次，因为每次只传输 15 行数据到客户端，逻辑读会被放大。设置了 `arraysize=5000` 之后，就不会发生一个块被读 n 次的问题了。

```
SQL> set arraysize 5000
SQL> set autot trace
SQL> select owner from test where owner='SYS';

30808 rows selected.

Execution Plan
----------------------------------------------------------
Plan hash value: 373050211

--------------------------------------------------------------------------------
| Id  | Operation          | Name      | Rows  | Bytes | Cost (%CPU)| Time     |
--------------------------------------------------------------------------------
```

```
|   0 | SELECT STATEMENT    |            | 2499 | 14994 |     6  (0)| 00:00:01 |
|*  1 |  INDEX RANGE SCAN| IDX_OWNER | 2499 | 14994 |     6  (0)| 00:00:01 |
---------------------------------------------------------------------------

Predicate Information (identified by operation id):
---------------------------------------------------

   1 - access("OWNER"='SYS')


Statistics
----------------------------------------------------------
          0  recursive calls
          0  db block gets
         74  consistent gets
          0  physical reads
          0  redo size
     155251  bytes sent via SQL*Net to client
        486  bytes received via SQL*Net from client
          8  SQL*Net roundtrips to/from client
          0  sorts (memory)
          0  sorts (disk)
      30808  rows processed
```

从上面的实验可见，索引扫描只耗费了 74 个逻辑读。

```
SQL> select * from test where owner='SYS';

30808 rows selected.

Execution Plan
----------------------------------------------------------
Plan hash value: 3932013684

--------------------------------------------------------------------------------
| Id |Operation                     | Name      | Rows | Bytes | Cost(%CPU)| Time     |
--------------------------------------------------------------------------------
|  0 | SELECT STATEMENT             |           | 2499 | 236K|    73  (0)| 00:00:01 |
|  1 |  TABLE ACCESS BY INDEX ROWID| TEST       | 2499 | 236K|    73  (0)| 00:00:01 |
|* 2 |   INDEX RANGE SCAN           | IDX_OWNER | 2499 |     |     6  (0)| 00:00:01 |
--------------------------------------------------------------------------------

Predicate Information (identified by operation id):
---------------------------------------------------

   2 - access("OWNER"='SYS')


Statistics
----------------------------------------------------------
          0  recursive calls
          0  db block gets
        877  consistent gets
          0  physical reads
          0  redo size
    3120934  bytes sent via SQL*Net to client
        486  bytes received via SQL*Net from client
          8  SQL*Net roundtrips to/from client
          0  sorts (memory)
          0  sorts (disk)
      30808  rows processed
```

1.5 集群因子（CLUSTERING FACTOR）

```
SQL> set autot off
SQL> select count(distinct dbms_rowid.rowid_block_number(rowid)) blocks
  2    from test
  3   where owner = 'SYS';

    BLOCKS
----------
       796
```

SQL 在有回表的情况下，一共耗费了 877 个逻辑读，那么这 877 个逻辑读是怎么来的呢？

SQL 返回的 30 808 条数据一共存储在 796 个数据块中，访问这 796 个数据块就需要消耗 796 个逻辑读，加上索引扫描的 74 个逻辑读，再加上 7 个逻辑读[其中 7=ROUND（30808/5000）]，这样累计起来刚好就是 877 个逻辑读。

因此我们可以判断，该 SQL 的性能确实绝大部分损失在回表中！

更糟糕的是：假设 30 808 条数据都在不同的数据块中，表也没有被缓存在 buffer cache 中，那么回表一共需要耗费 30 808 个物理 I/O，这太可怕了。

大家看到这里，是否能回答为什么返回表中 5%以内的数据走索引、超过表中 5%的数据走全表扫描？根本原因就在于回表。

在无法避免回表的情况下，走索引如果返回数据量太多，必然会导致回表次数太多，从而导致性能严重下降。

Oracle12c 的新功能批量回表（TABLE ACCESS BY INDEX ROWID BATCHED）在一定程度上改善了单行回表（TABLE ACCESS BY INDEX ROWID）的性能。关于批量回表本书不做讨论。

什么样的 SQL 必须要回表？

```
Select * from table where ...
```

这样的 SQL 就必须回表，所以我们必须严禁使用 Select *。那什么样的 SQL 不需要回表？

```
Select count(*) from table
```

这样的 SQL 就不需要回表。

当要查询的列也包含在索引中，这个时候就不需要回表了，所以我们往往会建立组合索引来消除回表，从而提升查询性能。

当一个 SQL 有多个过滤条件但是只在一个列或者部分列建立了索引，这个时候会发生回表再过滤（TABLE ACCESS BY INDEX ROWID 前面有 "*"），也需要创建组合索引，进而消除回表再过滤，从而提升查询性能。

关于如何创建组合索引，这问题太复杂了，我们在本书 8.3 节、9.1 节以及第 10 章都会反复提及如何创建组合索引。

1.5 集群因子（CLUSTERING FACTOR）

集群因子用于判断**索引回表**需要消耗的**物理 I/O 次数**。

我们先对测试表 test 的 object_id 列创建一个索引 idx_id。

```
SQL> create index idx_id on test(object_id);

Index created.
```

然后我们查看该索引的集群因子。

```
SQL> select owner, index_name, clustering_factor
  2    from dba_indexes
  3   where owner = 'SCOTT'
  4     and index_name = 'IDX_ID';

OWNER        INDEX_NAME    CLUSTERING_FACTOR
----------   ----------    -----------------
SCOTT        IDX_ID                     1094
```

索引 idx_id 的叶子块中有**序地存储**了索引的键值以及键值对应行所在的 ROWID。

```
SQL> select * from (
  2   select object_id, rowid
  3     from test
  4    where object_id is not null
  5    order by object_id) where rownum<=5;

 OBJECT_ID ROWID
---------- ------------------
         2 AAASNJAAEAAAAITAAw
         3 AAASNJAAEAAAAITAAF
         4 AAASNJAAEAAAAITAAx
         5 AAASNJAAEAAAAITAAa
         6 AAASNJAAEAAAAITAAV
```

集群因子的算法如下。

首先我们比较 2、3 对应的 ROWID 是否在同一个数据块，如果在同一个数据块，Clustering Factor +0；如果不在同一个数据块，那么 Clustering Factor 值加 1。

然后我们比较 3、4 对应的 ROWID 是否在同一个数据块，如果在同一个数据块，Clustering Factor 值不变；如果不在同一个数据块，那么 Clustering Factor 值加 1。

接下来我们比较 4、5 对应的 ROWID 是否在同一个数据块，如果在同一个数据块，Clustering Factor +0；如果不在同一个数据块，那么 Clustering Factor 值加 1。

像上面步骤一样，一直这样**有序地比较**下去，直到比较完索引中最后一个键值。

根据算法我们知道集群因子介于表的块数和表行数之间。

如果集群因子与块数接近，说明表的数据基本上是有序的，而且其顺序基本与索引顺序一样。这样在进行索引范围或者索引全扫描的时候，回表只需要读取少量的数据块就能完成。

如果集群因子与表记录数接近，说明表的数据和索引顺序差异很大，在进行索引范围扫描或者索引全扫描的时候，回表会读取更多的数据块。

集群因子只会影响索引范围扫描（INDEX RANGE SCAN）以及索引全扫描（INDEX FULL SCAN），因为只有这两种索引扫描方式会有大量数据回表。

集群因子不会影响索引唯一扫描（INDEX UNIQUE SCAN），因为索引唯一扫描只返回一条数据。集群因子更不会影响索引快速全扫描（INDEX FAST FULL SCAN），因为索引快速全扫描不回表。

1.5 集群因子（CLUSTERING FACTOR）

下面是根据集群因子算法人工计算集群因子的 SQL 脚本。

```
SQL> select sum(case
  2              when block#1 = block#2 and file#1 = file#2 then
  3                 0
  4              else
  5                 1
  6              end) CLUSTERING_FACTOR
  7     from (select dbms_rowid.rowid_relative_fno(rowid) file#1,
  8           lead(dbms_rowid.rowid_relative_fno(rowid), 1, null) over(order by object_id) file#2,
  9                  dbms_rowid.rowid_block_number(rowid) block#1,
 10           lead(dbms_rowid.rowid_block_number(rowid), 1, null) over(order by object_id) block#2
 11             from test
 12            where object_id is not null);

CLUSTERING_FACTOR
-----------------
             1094
```

我们来查看索引 idx_id 的集群因子接近表的总行数还是表的总块数。

通过前面的章节我们知道，表的总行数为 72 462 行。

表的总块数如下可知。

```
SQL> select count(distinct dbms_rowid.rowid_block_number(rowid)) blocks
  2    from test;

    BLOCKS
----------
      1032
```

集群因子非常接近表的总块数。现在，我们来查看下面 SQL 语句的执行计划。

```
SQL> set arraysize 5000
SQL> set autot trace
SQL> select * from test where object_id < 1000;

942 rows selected.

Execution Plan
----------------------------------------------------------
Plan hash value: 3946039639

--------------------------------------------------------------------------------------
| Id  | Operation                   | Name   | Rows  | Bytes | Cost (%CPU)| Time     |
--------------------------------------------------------------------------------------
|   0 | SELECT STATEMENT            |        |   970 | 94090 |    19   (0)| 00:00:01 |
|   1 |  TABLE ACCESS BY INDEX ROWID| TEST   |   970 | 94090 |    19   (0)| 00:00:01 |
|*  2 |   INDEX RANGE SCAN          | IDX_ID |   970 |       |     4   (0)| 00:00:01 |
--------------------------------------------------------------------------------------

Predicate Information (identified by operation id):
---------------------------------------------------

   2 - access("OBJECT_ID"<1000)


Statistics
----------------------------------------------------------
```

```
         0  recursive calls
         0  db block gets
        17  consistent gets
         0  physical reads
         0  redo size
     86510  bytes sent via SQL*Net to client
       420  bytes received via SQL*Net from client
         2  SQL*Net roundtrips to/from client
         0  sorts (memory)
         0  sorts (disk)
       942  rows processed
```

该 SQL 耗费了 17 个逻辑读。

现在我们新建一个测试表 test2 并且对数据进行**随机排序**。

```
SQL> create table test2 as select * from test order by dbms_random.value;

Table created.
```

我们在 object_id 列创建一个索引 idx_id2。

```
SQL> create index idx_id2 on test2(object_id);

Index created.
```

我们查看索引 idx_id2 的集群因子。

```
SQL> select owner, index_name, clustering_factor
  2    from dba_indexes
  3   where owner = 'SCOTT'
  4     and index_name = 'IDX_ID2';

OWNER        INDEX_NAME  CLUSTERING_FACTOR
----------   ----------  -----------------
SCOTT        IDX_ID2                 72393
```

索引 idx_id2 的集群因子接近于表的总行数，回表的时候会读取更多的数据块，现在我们来看一下 SQL 的执行计划。

```
SQL> set arraysize 5000
SQL> set autot trace
SQL> select /*+ index(test2) */ * from test2 where object_id <1000;

942 rows selected.

Execution Plan
----------------------------------------------------------
Plan hash value: 3711990673

--------------------------------------------------------------------------------------------
| Id  | Operation                    | Name    | Rows | Bytes | Cost (%CPU)| Time     |
--------------------------------------------------------------------------------------------
|   0 | SELECT STATEMENT             |         |  942 |  190K |   855   (0)| 00:00:11 |
|   1 |  TABLE ACCESS BY INDEX ROWID | TEST2   |  942 |  190K |   855   (0)| 00:00:11 |
|*  2 |   INDEX RANGE SCAN           | IDX_ID2 |  942 |       |     4   (0)| 00:00:01 |
--------------------------------------------------------------------------------------------

Predicate Information (identified by operation id):
---------------------------------------------------
```

```
    2 - access("OBJECT_ID"<1000)
Note
-----
   - dynamic sampling used for this statement (level=2)


Statistics
----------------------------------------------------------
          0  recursive calls
          0  db block gets
        943  consistent gets
          0  physical reads
          0  redo size
      86510  bytes sent via SQL*Net to client
        420  bytes received via SQL*Net from client
          2  SQL*Net roundtrips to/from client
          0  sorts (memory)
          0  sorts (disk)
        942  rows processed
```

通过上面实验我们得知，集群因子太大会严重影响索引回表的性能。

集群因子究竟影响的是什么性能呢？**集群因子影响的是索引回表的物理 I/O 次数**。我们假设索引范围扫描返回了 1 000 行数据，如果 buffer cache 中没有缓存表的数据块，假设这 1000 行数据都在同一个数据块中，那么回表需要耗费的物理 I/O 就只需要一个；假设这 1000 行数据都在不同的数据块中，那么回表就需要耗费 1 000 个物理 I/O。因此，**集群因子影响索引回表的物理 I/O 次数**。

请注意，不要尝试重建索引来降低集群因子，这根本没用，因为表中的数据顺序始终没变。唯一能降低集群因子的办法就是根据索引列排序对表进行重建（create table new_table as select * from old_table order by 索引列），但是这在实际操作中是不可取的，因为我们无法照顾到每一个索引。

怎么才能避免集群因子对 SQL 查询性能产生影响呢？其实前文已经有了答案，集群因子只影响索引范围扫描和索引全扫描。**当索引范围扫描，索引全扫描不回表或者返回数据量很少的时候，不管集群因子多大，对 SQL 查询性能几乎没有任何影响**。

再次强调一遍，在进行 SQL 优化的时候，往往会建立合适的组合索引消除回表，或者建立组合索引尽量减少回表次数。

如果无法避免回表，怎么做才能消除回表对 SQL 查询性能产生影响呢？当我们把表中所有的数据块缓存在 buffer cache 中，这个时候不管集群因子多大，对 SQL 查询性能也没有多大影响，因为这时不需要物理 I/O，数据块全在内存中访问速度是非常快的。

在本书第 6 章中我们还会进一步讨论集群因子。

1.6 表与表之间关系

关系型数据库中，表与表之间会进行关联，在进行关联的时候，我们一定要理清楚表与表之间的关系。表与表之间存在 3 种关系。一种是 1∶1 关系，一种是 1∶N 关系，最后一种是 N∶N

关系。搞懂表与表之间关系，对于 SQL 优化、SQL 等价改写、表设计优化以及分表分库都有巨大帮助。

两表在进行关联的时候，如果两表属于 1：1 关系，关联之后返回的结果也是属于 1 的关系，数据不会重复。如果两表属于 1：N 关系，关联之后返回的结果集属于 N 的关系。如果两表属于 N：N 关系，关联之后返回的结果集会产生局部范围的笛卡儿积，N：N 关系一般不存在内/外连接中，只能存在于半连接或者反连接中。

如果我们不知道业务，不知道数据字典，怎么判断两表是什么关系呢？我们以下面 SQL 为例子。

```sql
select * from emp e, dept d where e.deptno = d.deptno;
```

我们只需要对两表关联列进行汇总统计就能知道两表是什么关系。

```
SQL> select deptno, count(*) from emp group by deptno order by 2 desc;

    DEPTNO   COUNT(*)
---------- ----------
        30          6
        20          5
        10          3

SQL> select deptno, count(*) from dept group by deptno order by 2 desc;

    DEPTNO   COUNT(*)
---------- ----------
        10          1
        40          1
        30          1
        20          1
```

从上面查询我们可以知道两表 emp 与 dept 是 N：1 关系。搞清楚表与表之间关系对于 SQL 优化很有帮助。

2013 年，我们曾遇到一个案例，SQL 运行了 12 秒，SQL 文本如下。

```sql
select count(*) from a left join b on a.id=b.id;
```

案例中 a 与 b 是 1：1 关系，a 与 b 都是上千万数据量。因为 a 与 b 是使用外连接进行关联，不管 a 与 b 是否关联上，始终都会返回 a 的数据，SQL 语句中求的是两表关联后的总行数，因为两表是 1：1 关系，关联之后数据不会翻番，那么该 SQL 等价于如下文本。

```sql
select count(*) from a;
```

我们将 SQL 改写之后，查询可以秒出。如果 a 与 b 是 n：1 关系，我们也可以将 b 表去掉，因为两表关联之后数据不会翻倍。如果 b 表属于 n 的关系，这时我们不能去掉 b 表，因为这时关联之后数据量会翻番。

在本书后面的标量子查询等价改写、半连接等价改写以及 SQL 优化案例章节中我们就会用到表与表之间关系这个重要的概念。

第 2 章　统计信息

2.1 什么是统计信息

前面提到，只有大表才会产生性能问题，那么怎么才能让优化器知道某个表多大呢？这就需要对表收集统计信息。我们在第一章提到的基数、直方图、集群因子等概念都需要事先收集统计信息才能得到。

统计信息类似于战争中的侦察兵，如果情报工作没有做好，打仗就会输掉战争。同样的道理，如果没有正确地收集表的统计信息，或者没有及时地更新表的统计信息，SQL 的执行计划就会跑偏，SQL 也就会出现性能问题。收集统计信息是为了让优化器选择最佳执行计划，以最少的代价（成本）查询出表中的数据。

统计信息主要分为表的统计信息、列的统计信息、索引的统计信息、系统的统计信息、数据字典的统计信息以及动态性能视图基表的统计信息。

关于系统的统计信息、数据字典的统计信息以及动态性能视图基表的统计信息本书不做讨论，本书重点讨论表的统计信息、列的统计信息以及索引的统计信息。

表的统计信息主要包含表的总行数（num_rows）、表的块数（blocks）以及行平均长度（avg_row_len），我们可以通过查询数据字典 DBA_TABLES 获取表的统计信息。

现在我们创建一个测试表 T_STATS。

```
SQL> create table t_stats as select * from dba_objects;

Table created.
```

我们查看表 T_STATS 常用的表的统计信息。

```
SQL> select owner, table_name, num_rows, blocks, avg_row_len
  2    from dba_tables
  3   where owner = 'SCOTT'
  4     and table_name = 'T_STATS';

OWNER            TABLE_NAME         NUM_ROWS    BLOCKS AVG_ROW_LEN
---------------- ---------------- ---------- ---------- -----------
SCOTT            T_STATS
```

因为 T_STATS 是新创建的表，没有收集过统计信息，所以从 DBA_TABLES 查询数据是空的。

现在我们来收集表 T_STATS 的统计信息。

```
SQL> BEGIN
  2    DBMS_STATS.GATHER_TABLE_STATS(ownname          => 'SCOTT',
  3                                  tabname          => 'T_STATS',
  4                                  estimate_percent => 100,
```

```
    5                                    method_opt        => 'for all columns size auto',
    6                                    no_invalidate     => FALSE,
    7                                    degree            => 1,
    8                                    cascade           => TRUE);
    9  END;
   10  /

PL/SQL procedure successfully completed.
```

我们再次查看表的统计信息。

```
SQL> select owner, table_name, num_rows, blocks, avg_row_len
  2    from dba_tables
  3   where owner = 'SCOTT'
  4     and table_name = 'T_STATS';

OWNER           TABLE_NAME        NUM_ROWS     BLOCKS AVG_ROW_LEN
--------------- ---------------- ---------- ---------- -----------
SCOTT           T_STATS              72674       1061          97
```

从查询中我们可以看到，表 T_STATS 一共有 72 674 行数据，1 061 个数据块，平均行长度为 97 字节。

列的统计信息主要包含列的基数、列中的空值数量以及列的数据分布情况（直方图）。我们可以通过数据字典 DBA_TAB_COL_STATISTICS 查看列的统计信息。

现在我们查看表 T_STATS 常用的列统计信息。

```
SQL> select column_name, num_distinct, num_nulls, num_buckets, histogram
  2    from dba_tab_col_statistics
  3   where owner = 'SCOTT'
  4     and table_name = 'T_STATS';

COLUMN_NAME     NUM_DISTINCT  NUM_NULLS NUM_BUCKETS HISTOGRAM
--------------- ------------ ---------- ----------- -----------
EDITION_NAME               0      72674           0 NONE
NAMESPACE                 21          1           1 NONE
SECONDARY                  2          0           1 NONE
GENERATED                  2          0           1 NONE
TEMPORARY                  2          0           1 NONE
STATUS                     2          0           1 NONE
TIMESTAMP               1592          1           1 NONE
LAST_DDL_TIME           1521          1           1 NONE
CREATED                 1472          0           1 NONE
OBJECT_TYPE               45          0           1 NONE
DATA_OBJECT_ID          7796      64833           1 NONE
OBJECT_ID              72673          1           1 NONE
SUBOBJECT_NAME           140      72145           1 NONE
OBJECT_NAME            44333          0           1 NONE
OWNER                     31          0           1 NONE

15 rows selected.
```

上面查询中，第一个列表示列名字，第二个列表示列的基数，第三个列表示列中 NULL 值的数量，第四个列表示直方图的桶数，最后一个列表示直方图类型。

在工作中，我们经常使用下面脚本查看表和列的统计信息。

```
SQL> select a.column_name,
  2         b.num_rows,
```

```
  3            a.num_nulls,
  4            a.num_distinct Cardinality,
  5            round(a.num_distinct / b.num_rows * 100, 2) selectivity,
  6            a.histogram,
  7            a.num_buckets
  8     from dba_tab_col_statistics a, dba_tables b
  9    where a.owner = b.owner
 10      and a.table_name = b.table_name
 11      and a.owner = 'SCOTT'
 12      and a.table_name = 'T_STATS';

COLUMN_NAME          NUM_ROWS   NUM_NULLS   CARDINALITY   SELECTIVITY   HISTOGRAM   NUM_BUCKETS
-------------------- ---------- ----------- ------------- ------------- ----------- -----------
EDITION_NAME         72674      72674       0             0             NONE        0
NAMESPACE            72674      1           21            .03           NONE        1
SECONDARY            72674      0           2             0             NONE        1
GENERATED            72674      0           2             0             NONE        1
TEMPORARY            72674      0           2             0             NONE        1
STATUS               72674      0           2             0             NONE        1
TIMESTAMP            72674      1           1592          2.19          NONE        1
LAST_DDL_TIME        72674      1           1521          2.09          NONE        1
CREATED              72674      0           1472          2.03          NONE        1
OBJECT_TYPE          72674      0           45            .06           NONE        1
DATA_OBJECT_ID       72674      64833       7796          10.73         NONE        1
OBJECT_ID            72674      1           72673         100           NONE        1
SUBOBJECT_NAME       72674      72145       140           .19           NONE        1
OBJECT_NAME          72674      0           44333         61            NONE        1
OWNER                72674      0           31            .04           NONE        1

15 rows selected.
```

索引的统计信息主要包含索引 blevel（索引高度-1）、叶子块的个数（leaf_blocks）以及集群因子（clustering_factor）。我们可以通过数据字典 DBA_INDEXES 查看索引的统计信息。

我们在 OBJECT_ID 列上创建一个索引。

```
SQL> create index idx_t_stats_id on t_stats(object_id);

Index created.
```

创建索引的时候会自动收集索引的统计信息，运行下面脚本查看索引的统计信息。

```
SQL> select blevel, leaf_blocks, clustering_factor,status
  2    from dba_indexes
  3   where owner = 'SCOTT'
  4     and index_name = 'IDX_T_STATS_ID';

    BLEVEL LEAF_BLOCKS CLUSTERING_FACTOR STATUS
---------- ----------- ----------------- ----------------
         1         161              1127 VALID
```

如果要单独对索引收集统计信息，可以使用下面脚本收集。

```
SQL> BEGIN
  2    DBMS_STATS.GATHER_INDEX_STATS(ownname => 'SCOTT',
  3                                  indname => 'IDX_T_STATS_ID');
  4  END;
  5  /

PL/SQL procedure successfully completed.
```

在本书第 6 章中，我们会详细介绍表的统计信息、列的统计信息以及索引的统计信息是如何被应用于成本计算的。

2.2 统计信息重要参数设置

我们通常使用下面脚本收集表和索引的统计信息。

```
BEGIN
  DBMS_STATS.GATHER_TABLE_STATS(ownname          => 'TAB_OWNER',
                                tabname          => 'TAB_NAME',
                                estimate_percent => 根据表大小设置,
                                method_opt       => 'for all columns size repeat',
                                no_invalidate    => FALSE,
                                degree           => 根据表大小, CPU 资源和负载设置,
                                granularity      => 'AUTO',
                                cascade          => TRUE);
END;
/
```

ownname 表示表的拥有者，不区分大小写。

tabname 表示表名字，不区分大小写。

granularity 表示收集统计信息的粒度，该选项只对分区表生效，默认为 AUTO，表示让 Oracle 根据表的分区类型自己判断如何收集分区表的统计信息。对于该选项，我们一般采用 AUTO 方式，也就是数据库默认方式，因此，在后面的脚本中，省略该选项。

estimate_percent 表示采样率，范围是 0.000 001～100。

我们一般对小于 1GB 的表进行 100%采样，因为表很小，即使 100%采样速度也比较快。有时候小表有可能数据分布不均衡，如果没有 100%采样，可能会导致统计信息不准。因此我们建议对小表 100%采样。

我们一般对表大小在 1GB～5GB 的表采样 50%，对大于 5GB 的表采样 30%。如果表特别大，有几十甚至上百 GB，我们建议应该先对表进行分区，然后分别对每个分区收集统计信息。

一般情况下，为了确保统计信息比较准确，我们建议采样率不要低于 30%。

我们可以使用下面脚本查看表的采样率。

```
SQL> SELECT owner,
  2         table_name,
  3         num_rows,
  4         sample_size,
  5         round(sample_size / num_rows * 100) estimate_percent
  6    FROM DBA_TAB_STATISTICS
  7   WHERE owner='SCOTT' AND table_name='T_STATS';

OWNER             TABLE_NAME         NUM_ROWS SAMPLE_SIZE ESTIMATE_PERCENT
----------------- ---------------- ---------- ----------- ----------------
SCOTT             T_STATS               72674       72674              100
```

从上面查询我们可以看到,对表 T_STATS 是 100%采样的。现在我们将采样率设置为 30%。

```
SQL> BEGIN
  2    DBMS_STATS.GATHER_TABLE_STATS(ownname          => 'SCOTT',
```

```
  3                                       tabname          => 'T_STATS',
  4                                       estimate_percent => 30,
  5                                       method_opt       => 'for all columns size auto',
  6                                       no_invalidate    => FALSE,
  7                                       degree           => 1,
  8                                       cascade          => TRUE);
  9    END;
 10   /

PL/SQL procedure successfully completed.

SQL> SELECT owner,
  2         table_name,
  3         num_rows,
  4         sample_size,
  5         round(sample_size / num_rows * 100) estimate_percent
  6    FROM DBA_TAB_STATISTICS
  7   WHERE owner='SCOTT' AND table_name='T_STATS';

OWNER           TABLE_NAME        NUM_ROWS  SAMPLE_SIZE ESTIMATE_PERCENT
--------------- ---------------- ---------- ------------ ----------------
SCOTT           T_STATS               73067        21920               30
```

从上面查询我们可以看到采样率为 30%，表的总行数被估算为 73 067，而实际上表的总行数为 72 674。设置采样率 30%的时候，一共分析了 21 920 条数据，表的总行数等于 round(21 920*100/30)，也就是 73 067。

除非一个表是小表，否则没有必要对一个表 100%采样。因为表一直都会进行 DML 操作，表中的数据始终是变化的。

method_opt 用于控制收集直方图策略。

method_opt => 'for all columns size 1' 表示所有列都不收集直方图，如下所示。

```
SQL> BEGIN
  2    DBMS_STATS.GATHER_TABLE_STATS(ownname          => 'SCOTT',
  3                                  tabname          => 'T_STATS',
  4                                  estimate_percent => 100,
  5                                  method_opt       => 'for all columns size 1',
  6                                  no_invalidate    => FALSE,
  7                                  degree           => 1,
  8                                  cascade          => TRUE);
  9  END;
 10  /

PL/SQL procedure successfully completed.
```

我们查看直方图信息。

```
SQL> select a.column_name,
  2         b.num_rows,
  3         a.num_nulls,
  4         a.num_distinct Cardinality,
  5         round(a.num_distinct / b.num_rows * 100, 2) selectivity,
  6         a.histogram,
  7         a.num_buckets
  8    from dba_tab_col_statistics a, dba_tables b
  9   where a.owner = b.owner
 10     and a.table_name = b.table_name
```

第 2 章 统计信息

```
 11      and a.owner = 'SCOTT'
 12      and a.table_name = 'T_STATS';

COLUMN_NAME       NUM_ROWS  NUM_NULLS  CARDINALITY  SELECTIVITY  HISTOGRAM  NUM_BUCKETS
---------------   --------  ---------  -----------  -----------  ---------  -----------
EDITION_NAME        72674      72674            0            0   NONE                 0
NAMESPACE           72674          1           21          .03   NONE                 1
SECONDARY           72674          0            2            0   NONE                 1
GENERATED           72674          0            2            0   NONE                 1
TEMPORARY           72674          0            2            0   NONE                 1
STATUS              72674          0            2            0   NONE                 1
TIMESTAMP           72674          1         1592         2.19   NONE                 1
LAST_DDL_TIME       72674          1         1521         2.09   NONE                 1
CREATED             72674          0         1472         2.03   NONE                 1
OBJECT_TYPE         72674          0           45          .06   NONE                 1
DATA_OBJECT_ID      72674      64833         7796        10.73   NONE                 1
OBJECT_ID           72674          1        72673          100   NONE                 1
SUBOBJECT_NAME      72674      72145          140          .19   NONE                 1
OBJECT_NAME         72674          0        44333           61   NONE                 1
OWNER               72674          0           31          .04   NONE                 1

15 rows selected.
```

从上面查询我们看到，所有列都没有收集直方图。

method_opt => 'for all columns size skewonly' 表示对表中所有列收集自动判断是否收集直方图，如下所示。

```
SQL> BEGIN
  2      DBMS_STATS.GATHER_TABLE_STATS(ownname          => 'SCOTT',
  3                                    tabname          => 'T_STATS',
  4                                    estimate_percent => 100,
  5                                    method_opt       => 'for all columns size skewonly',
  6                                    no_invalidate    => FALSE,
  7                                    degree           => 1,
  8                                    cascade          => TRUE);
  9  END;
 10  /

PL/SQL procedure successfully completed.
```

我们查看直方图信息，如下所示。

```
SQL> select a.column_name,
  2         b.num_rows,
  3         a.num_nulls,
  4         a.num_distinct Cardinality,
  5         round(a.num_distinct / b.num_rows * 100, 2) selectivity,
  6         a.histogram,
  7         a.num_buckets
  8    from dba_tab_col_statistics a, dba_tables b
  9   where a.owner = b.owner
 10     and a.table_name = b.table_name
 11     and a.owner = 'SCOTT'
 12     and a.table_name = 'T_STATS';

COLUMN_NAME       NUM_ROWS  NUM_NULLS  CARDINALITY  SELECTIVITY  HISTOGRAM   NUM_BUCKETS
---------------   --------  ---------  -----------  -----------  ----------  -----------
EDITION_NAME        72674      72674            0            0   NONE                  0
NAMESPACE           72674          1           21          .03   FREQUENCY            21
SECONDARY           72674          0            2            0   FREQUENCY             2
```

```
GENERATED          72674        0         2         0 FREQUENCY           2
TEMPORARY          72674        0         2         0 FREQUENCY           2
STATUS             72674        0         2         0 FREQUENCY           2
TIMESTAMP          72674        1      1592      2.19 HEIGHT BALANCED    254
LAST_DDL_TIME      72674        1      1521      2.09 HEIGHT BALANCED    254
CREATED            72674        0      1472      2.03 HEIGHT BALANCED    254
OBJECT_TYPE        72674        0        45       .06 FREQUENCY          45
DATA_OBJECT_ID     72674    64833      7796     10.73 HEIGHT BALANCED    254
OBJECT_ID          72674        1     72673       100 NONE                 1
SUBOBJECT_NAME     72674    72145       140       .19 FREQUENCY         140
OBJECT_NAME        72674        0     44333        61 HEIGHT BALANCED    254
OWNER              72674        0        31       .04 FREQUENCY          31

15 rows selected.
```

从上面查询我们可以看到，除了 OBJECT_ID 列和 EDITION_NAME 列，其余所有列都收集了直方图。因为 EDITION_NAME 列全是 NULL，所以没必要收集直方图。OBJECT_ID 列选择性为 100%，没必要收集直方图。

在实际工作中千万不要使用 method_opt => 'for all columns size skewonly' 收集直方图信息，因为并不是表中所有的列都会出现在 where 条件中，对没有出现在 where 条件中的列收集直方图没有意义。

method_opt => 'for all columns size auto' 表示对出现在 where 条件中的列自动判断是否收集直方图。

现在我们删除表中所有列的直方图。

```
SQL> BEGIN
  2    DBMS_STATS.GATHER_TABLE_STATS(ownname          => 'SCOTT',
  3                                  tabname          => 'T_STATS',
  4                                  estimate_percent => 100,
  5                                  method_opt       => 'for all columns size 1',
  6                                  no_invalidate    => FALSE,
  7                                  degree           => 1,
  8                                  cascade          => TRUE);
  9  END;
 10  /

PL/SQL procedure successfully completed.
```

我们执行下面 SQL，以便将 owner 列放入 where 条件中。

```
SQL> select count(*) from t_stats where owner='SYS';

  COUNT(*)
----------
     30850
```

接下来我们刷新数据库监控信息。

```
SQL> begin
  2    dbms_stats.flush_database_monitoring_info;
  3  end;
  4  /

PL/SQL procedure successfully completed.
```

我们使用 method_opt => 'for all columns size auto' 方式对表收集统计信息。

```
SQL> BEGIN
  2     DBMS_STATS.GATHER_TABLE_STATS(ownname          => 'SCOTT',
  3                                   tabname          => 'T_STATS',
  4                                   estimate_percent => 100,
  5                                   method_opt       => 'for all columns size auto',
  6                                   no_invalidate    => FALSE,
  7                                   degree           => 1,
  8                                   cascade          => TRUE);
  9  END;
 10  /

PL/SQL procedure successfully completed.
```

然后我们查看直方图信息。

```
SQL> select a.column_name,
  2         b.num_rows,
  3         a.num_nulls,
  4         a.num_distinct Cardinality,
  5         round(a.num_distinct / b.num_rows * 100, 2) selectivity,
  6         a.histogram,
  7         a.num_buckets
  8    from dba_tab_col_statistics a, dba_tables b
  9   where a.owner = b.owner
 10     and a.table_name = b.table_name
 11     and a.owner = 'SCOTT'
 12     and a.table_name = 'T_STATS';

COLUMN_NAME       NUM_ROWS  NUM_NULLS  CARDINALITY  SELECTIVITY  HISTOGRAM  NUM_BUCKETS
---------------   --------  ---------  -----------  -----------  ---------  -----------
EDITION_NAME         72674      72674            0            0  NONE                 0
NAMESPACE            72674          1           21          .03  NONE                 1
SECONDARY            72674          0            2            0  NONE                 1
GENERATED            72674          0            2            0  NONE                 1
TEMPORARY            72674          0            2            0  NONE                 1
STATUS               72674          0            2            0  NONE                 1
TIMESTAMP            72674          1         1592         2.19  NONE                 1
LAST_DDL_TIME        72674          1         1521         2.09  NONE                 1
CREATED              72674          0         1472         2.03  NONE                 1
OBJECT_TYPE          72674          0           45          .06  NONE                 1
DATA_OBJECT_ID       72674      64833         7796        10.73  NONE                 1
OBJECT_ID            72674          1        72673          100  NONE                 1
SUBOBJECT_NAME       72674      72145          140          .19  NONE                 1
OBJECT_NAME          72674          0        44333           61  NONE                 1
OWNER                72674          0           31          .04  FREQUENCY           31

15 rows selected.
```

从上面查询我们可以看到，Oracle 自动地对 owner 列收集了直方图。

思考，如果将选择性比较高的列放入 where 条件中，会不会自动收集直方图？现在我们将 OBJECT_NAME 列放入 where 条件中。

```
SQL> select count(*) from t_stats where object_name='EMP';

  COUNT(*)
----------
         3
```

然后我们刷新数据库监控信息。

```
SQL> begin
  2    dbms_stats.flush_database_monitoring_info;
  3  end;
  4  /

PL/SQL procedure successfully completed.
```

我们收集统计信息。

```
SQL> BEGIN
  2    DBMS_STATS.GATHER_TABLE_STATS(ownname          => 'SCOTT',
  3                                  tabname          => 'T_STATS',
  4                                  estimate_percent => 100,
  5                                  method_opt       => 'for all columns size auto',
  6                                  no_invalidate    => FALSE,
  7                                  degree           => 1,
  8                                  cascade          => TRUE);
  9  END;
 10  /

PL/SQL procedure successfully completed.
```

我们查看 OBJECT_NAME 列是否收集了直方图。

```
SQL> select a.column_name,
  2         b.num_rows,
  3         a.num_nulls,
  4         a.num_distinct Cardinality,
  5         round(a.num_distinct / b.num_rows * 100, 2) selectivity,
  6         a.histogram,
  7         a.num_buckets
  8    from dba_tab_col_statistics a, dba_tables b
  9   where a.owner = b.owner
 10     and a.table_name = b.table_name
 11     and a.owner = 'SCOTT'
 12     and a.table_name = 'T_STATS';

COLUMN_NAME          NUM_ROWS   NUM_NULLS  CARDINALITY  SELECTIVITY  HISTOGRAM        NUM_BUCKETS
-------------------- ---------- ---------- ------------ ------------ ---------------- -----------
EDITION_NAME              72674      72674            0            0 NONE                       0
NAMESPACE                 72674          1           21          .03 NONE                       1
SECONDARY                 72674          0            2            0 NONE                       1
GENERATED                 72674          0            2            0 NONE                       1
TEMPORARY                 72674          0            2            0 NONE                       1
STATUS                    72674          0            2            0 NONE                       1
TIMESTAMP                 72674          1         1592         2.19 NONE                       1
LAST_DDL_TIME             72674          1         1521         2.09 NONE                       1
CREATED                   72674          0         1472         2.03 NONE                       1
OBJECT_TYPE               72674          0           45          .06 NONE                       1
DATA_OBJECT_ID            72674      64833         7796        10.73 NONE                       1
OBJECT_ID                 72674          1        72673          100 NONE                       1
SUBOBJECT_NAME            72674      72145          140          .19 NONE                       1
OBJECT_NAME               72674          0        44333           61 NONE                       1
OWNER                     72674          0           31          .04 FREQUENCY                 31

15 rows selected.
```

从上面查询我们可以看到，OBJECT_NAME 列没有收集直方图。由此可见，使用 AUTO 方式收集直方图很智能。mothod_opt 默认的参数就是 for all columns size auto。

method_opt => 'for all columns size repeat' 表示当前有哪些列收集了直

方图,现在就对哪些列收集直方图。

当前只对 OWNER 列收集了直方图,现在我们使用 REPEAT 方式收集直方图。

```
SQL> BEGIN
  2      DBMS_STATS.GATHER_TABLE_STATS(ownname          => 'SCOTT',
  3                                    tabname          => 'T_STATS',
  4                                    estimate_percent => 100,
  5                                    method_opt       => 'for all columns size repeat',
  6                                    no_invalidate    => FALSE,
  7                                    degree           => 1,
  8                                    cascade          => TRUE);
  9  END;
 10  /

PL/SQL procedure successfully completed.
```

我们查看直方图信息。

```
SQL> select a.column_name,
  2         b.num_rows,
  3         a.num_nulls,
  4         a.num_distinct Cardinality,
  5         round(a.num_distinct / b.num_rows * 100, 2) selectivity,
  6         a.histogram,
  7         a.num_buckets
  8    from dba_tab_col_statistics a, dba_tables b
  9   where a.owner = b.owner
 10     and a.table_name = b.table_name
 11     and a.owner = 'SCOTT'
 12     and a.table_name = 'T_STATS';
```

COLUMN_NAME	NUM_ROWS	NUM_NULLS	CARDINALITY	SELECTIVITY	HISTOGRAM	NUM_BUCKETS
EDITION_NAME	72674	72674	0	0	NONE	0
NAMESPACE	72674	1	21	.03	NONE	1
SECONDARY	72674	0	2	0	NONE	1
GENERATED	72674	0	2	0	NONE	1
TEMPORARY	72674	0	2	0	NONE	1
STATUS	72674	0	2	0	NONE	1
TIMESTAMP	72674	1	1592	2.19	NONE	1
LAST_DDL_TIME	72674	1	1521	2.09	NONE	1
CREATED	72674	0	1472	2.03	NONE	1
OBJECT_TYPE	72674	0	45	.06	NONE	1
DATA_OBJECT_ID	72674	64833	7796	10.73	NONE	1
OBJECT_ID	72674	1	72673	100	NONE	1
SUBOBJECT_NAME	72674	72145	140	.19	NONE	1
OBJECT_NAME	72674	0	44333	61	NONE	1
OWNER	72674	0	31	.04	FREQUENCY	31

15 rows selected.

从查询中我们可以看到,使用 REPEAT 方式延续了上次收集直方图的策略。对一个运行稳定的系统,我们应该采用 REPEAT 方式收集直方图。

method_opt => 'for columns object_type size skewonly' 表示单独对 OBJECT_TYPE 列收集直方图,对于其余列,如果之前收集过直方图,现在也收集直方图。

```
SQL> BEGIN
  2      DBMS_STATS.GATHER_TABLE_STATS(ownname          => 'SCOTT',
```

```
  3                                         tabname          => 'T_STATS',
  4                                         estimate_percent => 100,
  5                    method_opt           => 'for columns object_type size skewonly',
  6                                         no_invalidate    => FALSE,
  7                                         degree           => 1,
  8                                         cascade          => TRUE);
  9  END;
 10  /

PL/SQL procedure successfully completed.
```

我们查看直方图信息。

```
SQL> select a.column_name,
  2         b.num_rows,
  3         a.num_nulls,
  4         a.num_distinct Cardinality,
  5         round(a.num_distinct / b.num_rows * 100, 2) selectivity,
  6         a.histogram,
  7         a.num_buckets
  8    from dba_tab_col_statistics a, dba_tables b
  9   where a.owner = b.owner
 10     and a.table_name = b.table_name
 11     and a.owner = 'SCOTT'
 12     and a.table_name = 'T_STATS';
```

COLUMN_NAME	NUM_ROWS	NUM_NULLS	CARDINALITY	SELECTIVITY	HISTOGRAM	NUM_BUCKETS
EDITION_NAME	72674	72674	0	0	NONE	0
NAMESPACE	72674	1	21	.03	NONE	1
SECONDARY	72674	0	2	0	NONE	1
GENERATED	72674	0	2	0	NONE	1
TEMPORARY	72674	0	2	0	NONE	1
STATUS	72674	0	2	0	NONE	1
TIMESTAMP	72674	1	1592	2.19	NONE	1
LAST_DDL_TIME	72674	1	1521	2.09	NONE	1
CREATED	72674	0	1472	2.03	NONE	1
OBJECT_TYPE	72674	0	45	.06	FREQUENCY	45
DATA_OBJECT_ID	72674	64833	7796	10.73	NONE	1
OBJECT_ID	72674	1	72673	100	NONE	1
SUBOBJECT_NAME	72674	72145	140	.19	NONE	1
OBJECT_NAME	72674	0	44333	61	NONE	1
OWNER	72674	0	31	.04	FREQUENCY	31

15 rows selected.

从查询中我们可以看到，OBJECT_TYPE 列收集了直方图，因为之前收集过 owner 列直方图，现在也跟着收集了 owner 列的直方图。

在实际工作中，我们需要对列收集直方图就收集直方图，需要删除某列直方图就删除其直方图，当系统趋于稳定之后，使用 REPEAT 方式收集直方图。

no_invalidate 表示共享池中涉及到该表的游标是否立即失效，默认值为 DBMS_STATS.AUTO_INVALIDATE，表示让 Oracle 自己决定是否立即失效。我们建议将 no_invalidate 参数设置为 FALSE，立即失效。因为我们发现有时候 SQL 执行缓慢是因为统计信息过期导致，重新收集了统计信息之后执行计划还是没有更改，原因就在于没有将这个参数设置为 false。

degree 表示收集统计信息的并行度，默认为 NULL。如果表没有设置 degree，收集统计信息的时候后就不开并行；如果表设置了 degree，收集统计信息的时候就按照表的 degree 来开并行。可以查询 DBA_TABLES.degree 来查看表的 degree，一般情况下，表的 degree 都为 1。我们建议可以根据当时系统的负载、系统中 CPU 的个数以及表大小来综合判断设置并行度。

cascade 表示在收集表的统计信息的时候，是否级联收集索引的统计信息，默认值为 DBMS_STATS.AUTO_CASCADE，表示让 Oracle 自己判断是否级联收集索引的统计信息。我们一般将其设置为 TRUE，在收集表的统计信息的时候，级联收集索引的统计信息。

2.3 检查统计信息是否过期

收集完表的统计信息之后，如果表中有大量数据发生变化，这时表的统计信息就过期了，我们需要重新收集表的统计信息，如果不重新收集，可能会导致执行计划走偏。

以 T_STATS 为例，我们先在 owner 列上创建一个索引。

```
SQL> create index idx_t_stats_owner on t_stats(owner);

Index created.
```

我们收集 owner 列的直方图信息。

```
SQL> BEGIN
  2     DBMS_STATS.GATHER_TABLE_STATS(ownname          => 'SCOTT',
  3                                   tabname          => 'T_STATS',
  4                                   estimate_percent => 100,
  5                                   method_opt       => 'for columns owner size skewonly',
  6                                   no_invalidate    => FALSE,
  7                                   degree           => 1,
  8                                   cascade          => TRUE);
  9  END;
 10  /

PL/SQL procedure successfully completed.
```

我们执行下面 SQL 并且查看执行计划（为了方便排版，省略了执行计划中的 Time 列）。

```
SQL> select * from t_stats where owner='SCOTT';

122 rows selected.

Execution Plan
----------------------------------------------------------
Plan hash value: 3912915053

--------------------------------------------------------------------------------
| Id  | Operation                   | Name              | Rows  | Bytes | Cost (%CPU)|
--------------------------------------------------------------------------------
|   0 | SELECT STATEMENT            |                   |   122 | 11834 |     5   (0)|
|   1 |  TABLE ACCESS BY INDEX ROWID| T_STATS           |   122 | 11834 |     5   (0)|
|*  2 |   INDEX RANGE SCAN          | IDX_T_STATS_OWNER |   122 |       |     1   (0)|
--------------------------------------------------------------------------------

Predicate Information (identified by operation id):
---------------------------------------------------
```

```
   2 - access("OWNER"='SCOTT')


Statistics
----------------------------------------------------------
          0  recursive calls
          0  db block gets
         26  consistent gets
          0  physical reads
          0  redo size
      13440  bytes sent via SQL*Net to client
        508  bytes received via SQL*Net from client
         10  SQL*Net roundtrips to/from client
          0  sorts (memory)
          0  sorts (disk)
        122  rows processed
```

SQL 的过滤条件是 where owner='SCOTT'，因为收集了 owner 列的直方图统计，优化器能准确地估算出 SQL 返回 122 行数据，该 SQL 走的是索引范围扫描，执行计划是正确的。

现在我们更新表中的数据，将 object_id<=10000 的 owner 更新为'SCOTT'。

```
SQL> update t_stats set owner='SCOTT' where object_id<=10000;

9709 rows updated.

SQL> commit;

Commit complete.
```

我们再次执行 SQL 并且查看执行计划。

```
SQL> select * from t_stats where owner='SCOTT';

9831 rows selected.


Execution Plan
----------------------------------------------------------
Plan hash value: 3912915053

--------------------------------------------------------------------------------
| Id  | Operation                   | Name              | Rows  | Bytes | Cost (%CPU)|
--------------------------------------------------------------------------------
|   0 | SELECT STATEMENT            |                   |   122 | 11834 |     5   (0)|
|   1 |  TABLE ACCESS BY INDEX ROWID| T_STATS           |   122 | 11834 |     5   (0)|
|*  2 |   INDEX RANGE SCAN          | IDX_T_STATS_OWNER |   122 |       |     1   (0)|
--------------------------------------------------------------------------------

Predicate Information (identified by operation id):
---------------------------------------------------

   2 - access("OWNER"='SCOTT')


Statistics
----------------------------------------------------------
          0  recursive calls
          0  db block gets
       1502  consistent gets
          0  physical reads
```

```
      3236  redo size
   1005607  bytes sent via SQL*Net to client
      7625  bytes received via SQL*Net from client
       657  SQL*Net roundtrips to/from client
         0  sorts (memory)
         0  sorts (disk)
      9831  rows processed
```

从执行计划中可以看到，SQL 一共返回了 9 831 行数据，但是优化器评估只返回 122 行数据，因为优化器评估 where owner='SCOTT' 只返回 122 行数据，所以执行计划走了索引，但是实际上应该走全表扫描。

为什么优化器会评估 where owner='SCOTT' 只返回 122 行数据呢？原因在于表中有大量数据发生了变化，但是统计信息没有得到及时更新，优化器还是采用的老的（过期的）统计信息来估算返回行数。

我们可以使用下面方法检查表统计信息是否过期，先刷新数据库监控信息。

```
SQL> begin
  2    dbms_stats.flush_database_monitoring_info;
  3  end;
  4  /

PL/SQL procedure successfully completed.
```

然后我们执行下面查询。

```
SQL> select owner, table_name , object_type, stale_stats, last_analyzed
  2    from dba_tab_statistics
  3   where owner = 'SCOTT'
  4     and table_name = 'T_STATS';

OWNER      TABLE_NAME      OBJECT_TYPE     STALE_STATS     LAST_ANALYZED
---------- --------------- --------------- --------------- -------------
SCOTT      T_STATS         TABLE           YES             24-MAY-17
```

STALE_STATS 显示为 YES 表示表的统计信息过期了。如果 STALE_STATS 显示为 NO，表示表的统计信息没有过期。

我们可以通过下面查询找出统计信息过期的原因。

```
SQL> select table_owner, table_name, inserts, updates, deletes, timestamp
  2    from all_tab_modifications
  3   where table_owner = 'SCOTT'
  4     and table_name = 'T_STATS';

TABLE_OWNER     TABLE_NAME      INSERTS    UPDATES    DELETES  TIMESTAMP
--------------- --------------- ---------- ---------- -------- ---------
SCOTT           T_STATS                  0       9709        0  24-MAY-17
```

从查询结果我们可以看到，从上一次收集统计信息到现在，表被更新了 9 709 行数据，所以表的统计信息过期了。

现在我们重新收集表的统计信息。

```
SQL> BEGIN
  2    DBMS_STATS.GATHER_TABLE_STATS(ownname         => 'SCOTT',
  3                                  tabname         => 'T_STATS',
```

```
  4                                      estimate_percent => 100,
  5                                      method_opt       => 'for columns owner size skewonly',
  6                                      no_invalidate    => FALSE,
  7                                      degree           => 1,
  8                                      cascade          => TRUE);
  9  END;
 10  /

PL/SQL procedure successfully completed.
```

我们再次查看 SQL 的执行计划。

```
SQL> select * from t_stats where owner='SCOTT';

9831 rows selected.


Execution Plan
----------------------------------------------------------
Plan hash value: 1525972472

--------------------------------------------------------------------------------
| Id  | Operation         | Name    | Rows  | Bytes | Cost (%CPU)| Time     |
--------------------------------------------------------------------------------
|   0 | SELECT STATEMENT  |         |  9831 |   931K|   187   (2)| 00:00:03 |
|*  1 |  TABLE ACCESS FULL| T_STATS |  9831 |   931K|   187   (2)| 00:00:03 |
--------------------------------------------------------------------------------

Predicate Information (identified by operation id):
---------------------------------------------------

   1 - filter("OWNER"='SCOTT')


Statistics
----------------------------------------------------------
          0  recursive calls
          0  db block gets
       1690  consistent gets
          0  physical reads
          0  redo size
     418062  bytes sent via SQL*Net to client
       7625  bytes received via SQL*Net from client
        657  SQL*Net roundtrips to/from client
          0  sorts (memory)
          0  sorts (disk)
       9831  rows processed
```

重新收集完统计信息之后，优化器估算返回 9 831 行数据，这次 SQL 没走索引扫描而是走的全表扫描，SQL 走了正确的执行计划。

细心的读者可能会认为走索引扫描的性能高于全表扫描，因为索引扫描逻辑读为 1 502，而全表扫描逻辑读为 1 690，所以索引扫描性能高。其实这是不对的，衡量一个 SQL 的性能不能只看逻辑读，还要结合 SQL 的物理 I/O 次数综合判断。本书第 4 章会就为什么这里全表扫描性能比索引扫描性能更高给出详细解释。

Oracle 是怎么判断一个表的统计信息过期了呢？当表中有超过 10%的数据发生变化（INSERT，UPDATE，DELETE），就会引起统计信息过期。

现在我们查看表一共有多少行数据。

```
SQL> select count(*) from t_stats;

  COUNT(*)
----------
     72674
```

删除表中 10%的数据,然后我们查看表的统计信息是否过期。

```
SQL> delete t_stats where rownum<=72674*0.1+1;

7268 rows deleted.

SQL> commit;
```

我们刷新数据库监控信息。

```
SQL> begin
  2    dbms_stats.flush_database_monitoring_info;
  3  end;
  4  /

PL/SQL procedure successfully completed.
```

我们检查表统计信息是否过期。

```
SQL> select owner, table_name, object_type, stale_stats, last_analyzed
  2    from dba_tab_statistics
  3   where owner = 'SCOTT'
  4     and table_name = 'T_STATS';

OWNER      TABLE_NAME OBJECT_TYP STALE_STATS     LAST_ANALYZED
---------- ---------- ---------- --------------- -----------------
SCOTT      T_STATS    TABLE      YES             24-MAY-17
```

STALE_STATS 显示为 YES,说明表的统计信息过期了。

我们查看统计信息过期原因。

```
SQL> select table_owner, table_name, inserts, updates, deletes, timestamp
  2    from all_tab_modifications
  3   where table_owner = 'SCOTT'
  4     and table_name = 'T_STATS';

TABLE_OWNE TABLE_NAME    INSERTS    UPDATES    DELETES TIMESTAMP
---------- ---------- ---------- ---------- ---------- -----------------
SCOTT      T_STATS             0          0       7268 24-MAY-17
```

从上面查询我们可以看到表被删除了 7 268 行数据,从而导致表的统计信息过期。

在进行 SQL 优化的时候,我们需要检查表的统计信息是否过期,如果表的统计信息过期了,要及时更新表的统计信息。

数据字典 all_tab_modifications 还可以用来判断哪些表需要定期降低高水位,比如一个表经常进行 insert、delete,那么这个表应该定期降低高水位,这个表的索引也应该定期重建。除此之外,all_tab_modifications 还可以用来判断系统中哪些表是业务核心表、表的数据每天增长量等。

如果一个 SQL 有七八个表关联或者有视图套视图等，怎么快速检查 SQL 语句中所有的表统计信息是否过期呢？

现有如下 SQL。

```
select * from emp e,dept d where e.deptno=d.deptno;
```

我们可以先用 explain plan for 命令，在 plan_table 中生成 SQL 的执行计划。

```
SQL> explain plan for select * from emp e,dept d where e.deptno=d.deptno;
Explained.
```

然后我们使用下面脚本检查 SQL 语句中所有的表的统计信息是否过期。

```
SQL> select owner, table_name, object_type, stale_stats, last_analyzed
  2    from dba_tab_statistics
  3   where (owner, table_name) in
  4         (select object_owner, object_name
  5            from plan_table
  6           where object_type like '%TABLE%'
  7          union
  8          select table_owner, table_name
  9            from dba_indexes
 10           where (owner, index_name) in
 11                 (select object_owner, object_name
 12                    from plan_table
 13                   where object_type like '%INDEX%'));

OWNER      TABLE_NAME OBJECT_TYP STALE_STATS     LAST_ANALYZED
---------- ---------- ---------- --------------- -----------------
SCOTT      DEPT       TABLE      NO              05-DEC-16
SCOTT      EMP        TABLE      YES             22-OCT-16
```

最后我们可以使用下面脚本检查 SQL 语句中表统计信息的过期原因。

```
select *
  from all_tab_modifications
 where (table_owner, table_name) in
       (select object_owner, object_name
          from plan_table
         where object_type like '%TABLE%'
        union
        select table_owner, table_name
          from dba_indexes
         where (owner, index_name) in
               (select object_owner, object_name
                  from plan_table
                 where object_type like '%INDEX%'));
```

2.4 扩展统计信息

当 where 条件中有多个谓词过滤条件，但是这些谓词过滤条件彼此是有关系的而不是相互独立的，这时我们可能需要收集扩展统计信息以便优化器能够估算出较为准确的行数（Rows）。

我们创建一个表 T。

第 2 章 统计信息

```
SQL> create table t as
  2    select level as id, level || 'a' as a, level || level || 'b' as b
  3      from dual
  4   connect by level < 100;

Table created.
```

在 T 表中，知道 A 列的值就知道 B 列的值，A 和 B 这样的列就叫作相关列。

我们一直执行 `insert into t select * from t;` 直到 T 表中有 3244032 行数据。

我们对 T 表收集统计信息。

```
SQL> BEGIN
  2    DBMS_STATS.GATHER_TABLE_STATS(ownname          => 'SCOTT',
  3                                  tabname          => 'T',
  4                                  estimate_percent => 100,
  5                                  method_opt       => 'for all columns size skewonly',
  6                                  no_invalidate    => FALSE,
  7                                  degree           => 1,
  8                                  cascade          => TRUE);
  9  END;
 10  /

PL/SQL procedure successfully completed.
```

我们查看 T 表的统计信息。

```
SQL> select a.column_name,
  2         b.num_rows,
  3         a.num_distinct Cardinality,
  4         round(a.num_distinct / b.num_rows * 100, 2) selectivity,
  5         a.histogram,
  6         a.num_buckets
  7    from dba_tab_col_statistics a, dba_tables b
  8   where a.owner = b.owner
  9     and a.table_name = b.table_name
 10     and a.owner = 'SCOTT'
 11     and a.table_name = 'T';

COLUMN_NAME        NUM_ROWS CARDINALITY SELECTIVITY HISTOGRAM       NUM_BUCKETS
---------------- ---------- ----------- ----------- --------------- -----------
ID                  3244032          99           0 FREQUENCY                99
A                   3244032          99           0 FREQUENCY                99
B                   3244032          99           0 FREQUENCY                99
```

我们创建两个索引。

```
SQL> create index idx1 on t(a);

Index created.

SQL> create index idx2 on t(a,b);

Index created.
```

现有如下 SQL 及其执行计划。

```
SQL> select * from t where a='1a' and b='11b';

32768 rows selected.
```

```
Execution Plan
----------------------------------------------------------
Plan hash value: 2303463401
----------------------------------------------------------
| Id  | Operation                    | Name | Rows | Bytes | Cost (%CPU)| Time     |
----------------------------------------------------------
|   0 | SELECT STATEMENT             |      |  331 |  4303 |    84   (0)| 00:00:02 |
|   1 |  TABLE ACCESS BY INDEX ROWID | T    |  331 |  4303 |    84   (0)| 00:00:02 |
|*  2 |   INDEX RANGE SCAN           | IDX2 |  331 |       |     3   (0)| 00:00:01 |
----------------------------------------------------------

Predicate Information (identified by operation id):
---------------------------------------------------

   2 - access("A"='1a' AND "B"='11b')

Statistics
----------------------------------------------------------
          0  recursive calls
          0  db block gets
      11854  consistent gets
         78  physical reads
          0  redo size
     775996  bytes sent via SQL*Net to client
      24444  bytes received via SQL*Net from client
       2186  SQL*Net roundtrips to/from client
          0  sorts (memory)
          0  sorts (disk)
      32768  rows processed
```

优化器估算返回 331 行数据，但是实际上返回了 32 768 行数据。为什么优化器估算返回的行数与真实返回的行数有这么大差异呢？这是因为优化器不知道 A 与 B 的关系，所以在估算返回行数的时候采用的是总行数*A 的选择性*B 的选择性。

```
SQL> select round(1/99/99*3244032) from dual;

round(1/99/99*3244032)
----------------------
                   331
```

因为 A 列的值可以决定 B 列的值，所以上述 SQL 可以去掉 B 列的过滤条件。

```
SQL> select * from t where a='1a';

32768 rows selected.

Execution Plan
----------------------------------------------------------
Plan hash value: 1601196873
----------------------------------------------------------
| Id  | Operation         | Name | Rows  | Bytes | Cost (%CPU)| Time     |
----------------------------------------------------------
|   0 | SELECT STATEMENT  |      | 32768 |  416K |  1775   (3)| 00:00:22 |
|*  1 |  TABLE ACCESS FULL| T    | 32768 |  416K |  1775   (3)| 00:00:22 |
----------------------------------------------------------

Predicate Information (identified by operation id):
---------------------------------------------------

   1 - filter("A"='1a')
```

```
Statistics
----------------------------------------------------------
          0  recursive calls
          0  db block gets
      10118  consistent gets
          0  physical reads
          0  redo size
     441776  bytes sent via SQL*Net to client
      24444  bytes received via SQL*Net from client
       2186  SQL*Net roundtrips to/from client
          0  sorts (memory)
          0  sorts (disk)
      32768  rows processed
```

这时优化器能正确地估算返回的 Rows。如果不想改写 SQL，怎么才能让优化器得到比较准确的 Rows 呢？在 Oracle11g 之前可以使用动态采样（至少 Level 4）。

```
SQL> alter session set optimizer_dynamic_sampling=4;

Session altered.

SQL> select * from t where a='1a' and b='11b';

32768 rows selected.

Execution Plan
----------------------------------------------------------
Plan hash value: 1601196873

---------------------------------------------------------------------------
| Id  | Operation         | Name | Rows  | Bytes | Cost (%CPU)| Time     |
---------------------------------------------------------------------------
|   0 | SELECT STATEMENT  |      | 33845 |  429K|  1778   (3)| 00:00:22 |
|*  1 |  TABLE ACCESS FULL| T    | 33845 |  429K|  1778   (3)| 00:00:22 |
---------------------------------------------------------------------------

Predicate Information (identified by operation id):
---------------------------------------------------
   1 - filter("A"='1a' AND "B"='11b')

Note
-----
   - dynamic sampling used for this statement (level=4)


Statistics
----------------------------------------------------------
          0  recursive calls
          0  db block gets
      10118  consistent gets
          0  physical reads
          0  redo size
     441776  bytes sent via SQL*Net to client
      24444  bytes received via SQL*Net from client
       2186  SQL*Net roundtrips to/from client
          0  sorts (memory)
          0  sorts (disk)
      32768  rows processed
```

使用动态采样 Level4 采样之后，优化器估算返回 33 845 行数据，实际返回了 32 768 行数据，这已经比较精确了。在 Oracle11g 以后，我们可以使用扩展统计信息将相关的列组合成一个列。

```
SQL> SELECT DBMS_STATS.CREATE_EXTENDED_STATS(USER, 'T', '(A, B)') FROM DUAL;

DBMS_STATS.CREATE_EXTENDED_STATS(USER,'T','(A,B)')
--------------------------------------------------------------------------------
SYS_STUNA$6DVXJXTP05EH56DTIR0X
```

现在我们对表重新收集统计信息。

```
SQL> BEGIN
  2    DBMS_STATS.GATHER_TABLE_STATS(ownname          => 'SCOTT',
  3                                  tabname          => 'T',
  4                                  estimate_percent => 100,
  5      method_opt => 'for columns SYS_STUNA$6DVXJXTP05EH56DTIR0X size skewonly',
  6                                  no_invalidate    => FALSE,
  7                                  degree           => 1,
  8                                  cascade          => TRUE);
  9  END;
 10  /

PL/SQL procedure successfully completed.
```

我们查看 T 表的统计信息。

```
SQL> select a.column_name,
  2         b.num_rows,
  3         a.num_distinct Cardinality,
  4         round(a.num_distinct / b.num_rows * 100, 2) selectivity,
  5         a.histogram,
  6         a.num_buckets
  7    from dba_tab_col_statistics a, dba_tables b
  8   where a.owner = b.owner
  9     and a.table_name = b.table_name
 10     and a.owner = 'SCOTT'
 11     and a.table_name = 'T';

COLUMN_NAME                    NUM_ROWS CARDINALITY SELECTIVITY HISTOGRAM  NUM_BUCKETS
------------------------------ -------- ----------- ----------- ---------- -----------
ID                              3244032          99           0 FREQUENCY           99
A                               3244032          99           0 FREQUENCY           99
B                               3244032          99           0 FREQUENCY           99
SYS_STUNA$6DVXJXTP05EH56DTIR0X  3244032          99           0 FREQUENCY           99
```

重新收集统计信息之后，扩展列 SYS_STUNA$6DVXJXTP05EH56DTIR0X 也收集了直方图。

我们再次执行 SQL。

```
SQL> select * from t where a='1a' and b='11b';

32768 rows selected.


Execution Plan
----------------------------------------------------------
Plan hash value: 1601196873
```

```
---------------------------------------------------------------------
| Id  | Operation          | Name | Rows  | Bytes | Cost (%CPU)| Time     |
---------------------------------------------------------------------
|   0 | SELECT STATEMENT   |      | 32768 |  416K |  1778   (3)| 00:00:22 |
|*  1 |  TABLE ACCESS FULL | T    | 32768 |  416K |  1778   (3)| 00:00:22 |
---------------------------------------------------------------------

Predicate Information (identified by operation id):
---------------------------------------------------

   1 - filter("A"='1a' AND "B"='11b')


Statistics
----------------------------------------------------------
          1  recursive calls
          0  db block gets
      10118  consistent gets
          0  physical reads
          0  redo size
     441776  bytes sent via SQL*Net to client
      24444  bytes received via SQL*Net from client
       2186  SQL*Net roundtrips to/from client
          0  sorts (memory)
          0  sorts (disk)
      32768  rows processed
```

收集完扩展统计信息之后，优化器就能估算出较为准确的 Rows。

需要注意的是，扩展统计信息只能用于等值查询，不能用于非等值查询。

在本书的 SQL 优化案例赏析章节中，我们将会为各位读者分享一个经典的扩展统计信息优化案例。

2.5 动态采样

如果一个表从来没收集过统计信息，默认情况下 Oracle 会对表进行动态采样（Level=2）以便优化器估算出较为准确的 Rows，**动态采样的最终目的就是为了让优化器能够评估出较为准确的 Rows。**

现在我们创建一个测试表 T_DYNA。

```
SQL> create table t_dyna as select * from dba_objects;

Table created.
```

我们执行下面 SQL 并且查看执行计划。

```
SQL> select count(*) from t_dyna;

Execution Plan
----------------------------------------------------------
Plan hash value: 3809964769

---------------------------------------------------------------------
| Id  | Operation          | Name | Rows  | Cost (%CPU)| Time     |
```

```
-------------------------------------------------------------------
| 0 | SELECT STATEMENT   |       |    1 |   187  (1)| 00:00:03 |
| 1 |  SORT AGGREGATE    |       |    1 |           |          |
| 2 |   TABLE ACCESS FULL| T_DYNA| 65305|   187  (1)| 00:00:03 |
-------------------------------------------------------------------

Note
-----
   - dynamic sampling used for this statement (level=2)
```

因为表 T_DYNA 是才创建的新表，没有收集过统计信息，所以会启用动态采样。执行计划中 dynamic sampling used for this statement (level=2) 表示启用了动态采样，level 表示采样级别，默认情况下采样级别为 2。

动态采样的级别分为 11 级。

level 0：不启用动态采样。

level 1：当表（非分区表）没有收集过统计信息并且这个表要与另外的表进行关联（不能是单表访问），同时该表没有索引，表的数据块必须大于 32 个，满足这些条件的时候，Oracle 会随机扫描表中 32 个数据块，然后评估返回的 Rows。

level 2：对没有收集过统计信息的表启用动态采样，采样的块数为 64 个，如果表的块数小于 64 个，表有多少个块就会采样多少个块。

level 3：对没有收集过统计信息的表启用动态采样，采样的块数为 64 个。如果表已经收集过统计信息，但是优化器不能准确地估算出返回的 Rows，而是靠猜，比如 WHERE SUBSTR(owner,1,3)，这时会随机扫描 64 个数据块进行采样。

level 4：对没有收集过统计信息的表启用动态采样，采样的块数为 64 个。如果表已经收集过统计信息，但是表有两个或者两个以上过滤条件（AND/OR），这时会随机扫描 64 个数据块进行采样，相关列问题就必须启用至少 level 4 进行动态采样。level4 采样包含了 level 3 的采样数据。

level 5：收集满足 level 4 采样条件的数据，采样的块数为 128 个。

level 6：收集满足 level 4 采样条件的数据，采样的块数为 256 个。

level 7：收集满足 level 4 采样条件的数据，采样的块数为 512 个。

level 8：收集满足 level 4 采样条件的数据，采样的块数为 1 024 个。

level 9：收集满足 level 4 采样条件的数据，采样的块数为 4 086 个。

level 10：收集满足 level 4 采样条件的数据，采样表中所有的数据块。

level 11：Oracle 自动判断如何采样，采样的块数由 Oracle 自动决定。

在 2.4 节中我们已经演示过动态采样 level 4 的用途,现在将为各位读者演示动态采样 level 3 的用途。

我们执行下面 SQL 并且查看执行计划。

```
SQL> select * from t_dyna where substr(owner,4,3)='LIC';

27699 rows selected.
```

```
Execution Plan
----------------------------------------------------------
Plan hash value: 1744410282

--------------------------------------------------------------------------
| Id  | Operation         | Name   | Rows  | Bytes  | Cost (%CPU)| Time     |
--------------------------------------------------------------------------
|   0 | SELECT STATEMENT  |        | 23044 |  4658K |   190   (3)| 00:00:03 |
|*  1 |  TABLE ACCESS FULL| T_DYNA | 23044 |  4658K |   190   (3)| 00:00:03 |
--------------------------------------------------------------------------

Predicate Information (identified by operation id):
---------------------------------------------------

   1 - filter(SUBSTR("OWNER",4,3)='LIC')

Note
-----
   - dynamic sampling used for this statement (level=2)
```

因为 T_DYNA 没有收集过统计信息，启用了动态采样，采样级别默认为 level 2，动态采样估算的 Rows(23 044)与真实的 Rows(27 699)比较接近。

现在我们对表 T_DYNA 收集统计信息。

```
SQL> BEGIN
  2     DBMS_STATS.GATHER_TABLE_STATS(ownname          => 'SCOTT',
  3                                   tabname          => 'T_DYNA',
  4                                   estimate_percent => 100,
  5                                   method_opt       => 'for all columns size skewonly',
  6                                   no_invalidate    => FALSE,
  7                                   degree           => 1,
  8                                   cascade          => TRUE);
  9  END;
 10  /

PL/SQL procedure successfully completed.
```

我们再次查看执行计划。

```
SQL> select * from t_dyna where substr(owner,4,3)='LIC';

27699 rows selected.


Execution Plan
----------------------------------------------------------
Plan hash value: 1744410282

--------------------------------------------------------------------------
| Id  | Operation         | Name   | Rows  | Bytes  | Cost (%CPU)| Time     |
--------------------------------------------------------------------------
|   0 | SELECT STATEMENT  |        |   728 | 70616  |   190   (3)| 00:00:03 |
|*  1 |  TABLE ACCESS FULL| T_DYNA |   728 | 70616  |   190   (3)| 00:00:03 |
--------------------------------------------------------------------------

Predicate Information (identified by operation id):
---------------------------------------------------

   1 - filter(SUBSTR("OWNER",4,3)='LIC')
```

2.5 动态采样

对表 T_DYNA 收集了统计信息之后，因为统计信息中没有包含 substr(owner,4,3) 的统计，所以优化器无法估算出较为准确的 Rows，优化器估算返回了 728 行数据，而实际上返回了 27 699 行数据。现在我们将动态采样 level 设置为 3。

```
SQL> alter session set optimizer_dynamic_sampling=3;

Session altered.
```

我们执行 SQL 并且查看执行计划。

```
SQL> select * from t_dyna where substr(owner,4,3)='LIC';

27699 rows selected.

Execution Plan
----------------------------------------------------------
Plan hash value: 1744410282

--------------------------------------------------------------------------
| Id  | Operation         | Name   | Rows  | Bytes | Cost (%CPU)| Time     |
--------------------------------------------------------------------------
|   0 | SELECT STATEMENT  |        | 28795 | 2727K |   191   (3)| 00:00:03 |
|*  1 |  TABLE ACCESS FULL| T_DYNA | 28795 | 2727K |   191   (3)| 00:00:03 |
--------------------------------------------------------------------------

Predicate Information (identified by operation id):
---------------------------------------------------

   1 - filter(SUBSTR("OWNER",4,3)='LIC')

Note
-----
   - dynamic sampling used for this statement (level=3)
```

将动态采样设置为 level 3 之后，优化器发现 where 条件中有 substr(owner,4,3)，无法估算出准确的 Rows，因此对 SQL 启用了动态采样，动态采样估算返回了 28 795 行数据，接近于真实的行数 27 699。

除了设置参数 optimizer_dynamic_sampling 启用动态采样外，我们还可以添加 HINT 启用动态采样。

```
SQL> alter session set optimizer_dynamic_sampling=2;

Session altered.

SQL> select /*+ dynamic_sampling(3) */ * from t_dyna where substr(owner,4,3)='LIC';

27699 rows selected.

Execution Plan
----------------------------------------------------------
Plan hash value: 1744410282

--------------------------------------------------------------------------
| Id  | Operation         | Name   | Rows  | Bytes | Cost (%CPU)| Time     |
--------------------------------------------------------------------------
```

```
|   0 | SELECT STATEMENT    |        | 28795 | 2727K|   191   (3)| 00:00:03 |
|*  1 |  TABLE ACCESS FULL  | T_DYNA | 28795 | 2727K|   191   (3)| 00:00:03 |
---------------------------------------------------------------------------

Predicate Information (identified by operation id):
---------------------------------------------------

   1 - filter(SUBSTR("OWNER",4,3)='LIC')

Note
-----
   - dynamic sampling used for this statement (level=3)
```

如果表已经收集过统计信息并且优化器能够准确地估算出返回的 Rows，即使添加了动态采样的 HINT 或者是设置了动态采样的参数为 level 3，也不会启用动态采样。

```
SQL> select /*+ dynamic_sampling(3) */ * from t_dyna where owner='SYS';

30928 rows selected.

Execution Plan
----------------------------------------------------------
Plan hash value: 1744410282

---------------------------------------------------------------------------
| Id  | Operation          | Name   | Rows  | Bytes | Cost (%CPU)| Time     |
---------------------------------------------------------------------------
|   0 | SELECT STATEMENT   |        | 30928 | 2929K|   188   (2)| 00:00:03 |
|*  1 |  TABLE ACCESS FULL | T_DYNA | 30928 | 2929K|   188   (2)| 00:00:03 |
---------------------------------------------------------------------------

Predicate Information (identified by operation id):
---------------------------------------------------

   1 - filter("OWNER"='SYS')
```

因为表 T_DYNA 收集过统计信息，优化器能够直接根据统计信息估算出较为准确的 Rows，所以，即使添加了 HINT：/*+ dynamic_sampling(3) */，也没有启用动态采样。

什么时候需要启用动态采样呢？

当系统中有全局临时表，就需要使用动态采样，因为全局临时表无法收集统计信息，我们建议对全局临时表至少启用 level 4 进行采样。

当执行计划中表的 Rows 估算有严重偏差的时候，例如相关列问题，或者两表关联有多个连接列，关联之后 Rows 算少，或者是 where 过滤条件中对列使用了 substr、instr、like，又或者是 where 过滤条件中有非等值过滤，或者 group by 之后导致 Rows 估算错误，此时我们可以考虑使用动态采样，同样，我们建议动态采样至少设置为 level 4。

在数据仓库系统中，有些报表 SQL 是采用 Obiee/SAP BO/Congnos 自动生成的，此类 SQL 一般都有几十行甚至几百行，SQL 的过滤条件一般也比较复杂，有大量的 AND 和 OR 过滤条件，同时也可能有大量的 where 子查询过滤条件，SQL 最终返回的数据量其实并不多。对于此类 SQL，如果 SQL 执行缓慢，有可能是因为 SQL 的过滤条件太复杂，从而导致优化器不能估算出较为准确的 Rows 而产生了错误的执行计划。我们可以考虑启用动态采样 level 6 观察性能是否有所改善，我们曾利用该方法优化了大量的报表 SQL。

最后，需要注意的是，不要在系统级更改动态采样级别，默认为 2 就行，如果某个表需要启用动态采样，直接在 SQL 语句中添加 HINT 即可。

2.6 定制统计信息收集策略

优化器在计算执行计划的成本时依赖于统计信息，如果没有收集统计信息，或者是统计信息过期了，那么优化器就会出现严重偏差，从而导致性能问题。因此要确保统计信息准确性。虽然数据库自带有 JOB 每天晚上会定时收集数据库中所有表的统计信息，但是如果数据库特别大，自带的 JOB 无法完成全库统计信息收集。一些资深的 DBA 会关闭数据库自带的统计信息收集 JOB，根据实际情况自己定制收集统计信息策略。

下面脚本用于收集 SCOTT 账户下统计信息过期了或者是从没收集过统计信息的表的统计信息，采样率也根据表的段大小做出了相应调整。

```
declare
  cursor stale_table is
    select owner,
           segment_name,
           case
             when segment_size < 1 then
              100
             when segment_size >= 1 and segment_size <= 5 then
              50
             when segment_size > 5 then
              30
           end as percent,
           6 as degree
      from (select owner,
                   segment_name,
                   sum(bytes / 1024 / 1024 / 1024) segment_size
              from DBA_SEGMENTS
             where owner = 'SCOTT'
               and segment_name in
                   (select table_name
                      from DBA_TAB_STATISTICS
                     where (last_analyzed is null or stale_stats = 'YES')
                       and owner = 'SCOTT')
             group by owner, segment_name);
begin
  dbms_stats.flush_database_monitoring_info;
  for stale in stale_table loop
    dbms_stats.gather_table_stats(ownname          => stale.owner,
                                  tabname          => stale.segment_name,
                                  estimate_percent => stale.percent,
                                  method_opt       => 'for all columns size repeat',
                                  degree           => stale.degree,
                                  cascade          => true);
  end loop;
end;
/
```

在实际工作中，我们可以根据自身数据库中实际情况，对以上脚本进行修改。

全局临时表无法收集统计信息,我们可以抓出系统中的全局临时表,抓出系统中使用到全局临时表的 SQL,然后根据实际情况,对全局临时表进行动态采样,或者是人工对全局临时表设置统计信息(DBMS_STATS.SET_TABLE_STATS)。

下面脚本抓出系统中使用到全局临时表的 SQL。

```sql
select b.object_owner, b.object_name, a.temporary, sql_text
  from dba_tables a, v$sql_plan b, v$sql c
 where a.owner = b.object_owner
   and a.temporary = 'Y'
   and a.table_name = b.object_name
   and b.sql_id = c.sql_id;
```

第 3 章 执行计划

SQL 执行缓慢有很多原因，有时候是数据库本身原因，比如 LATCH 争用，或者某些参数设置不合理。有时候是 SQL 写法有问题，有时候是缺乏索引，可能是因为统计信息过期或者没收集直方图，也可能是优化器本身并不完善或者优化器自身 BUG 而导致的性能问题，还有可能是业务原因，比如要访问一年的数据，然而一年累计有数亿条数据，数据量太大导致 SQL 性能缓慢。

如果是数据库自身原因导致 SQL 缓慢，我们需要通过分析等待事件，做出相应处理。本书侧重讨论单纯的 SQL 优化，因此更侧重于分析 SQL 写法，分析 SQL 的执行计划。

SQL 调优就是通过各种手段和方法使优化器选择最佳执行计划，以最小的资源消耗获取到想要的数据。

3.1 获取执行计划常用方法

3.1.1 使用 AUTOTRACE 查看执行计划

我们利用 SQLPLUS 中自带的 AUTOTRACE 工具查看执行计划。AUTOTRACE 用法如下。

```
SQL> set autot
Usage: SET AUTOT[RACE] {OFF | ON | TRACE[ONLY]} [EXP[LAIN]] [STAT[ISTICS]]
```

方括号内的字符可以省略。

set autot on：该命令会运行 SQL 并且显示运行结果，执行计划和统计信息。

set autot trace：该命令会运行 SQL，但不显示运行结果，会显示执行计划和统计信息。

set autot trace exp：运行该命令查询语句不执行，DML 语句会执行，只显示执行计划。

set autot trace stat：该命令会运行 SQL，只显示统计信息。

set autot off：关闭 AUTOTRACE。

我们使用 set autot on 查看执行计划（基于 Oracle11gR2，Scott 账户）。

```
SQL> conn scott/tiger
```

显示已连接。

```
SQL> set lines 200 pages 200
SQL> set autot on
SQL> select count(*) from emp;

  COUNT(*)
----------
        14
```

执行计划

Plan hash value: 1006289799

| Id | Operation | Name | Rows | Cost (%CPU)| Time |

0	SELECT STATEMENT		1	2 (0)	00:00:01
1	SORT AGGREGATE		1		
2	INDEX FAST FULL SCAN	PK_EMP	14	2 (0)	00:00:01

Note

   - dynamic sampling used for this statement (level=2)

统计信息
----
        233 recursive calls
          0 db block gets
         51 consistent gets
         10 physical reads
          0 redo size
        430 bytes sent via SQL*Net to client
        419 bytes received via SQL*Net from client
          2 SQL*Net roundtrips to/from client
          4 sorts (memory)
          0 sorts (disk)
          1 rows processed

使用 set autot on 查看执行计划会输出 SQL 运行结果，如果 SQL 要返回大量结果，我们可以使用 set autot trace 查看执行计划，set autot trace 不会输出 SQL 运行结果。

```
SQL> set autot trace
SQL> select count(*) from emp;
```
执行计划

Plan hash value: 1006289799

| Id | Operation | Name | Rows | Cost (%CPU)| Time |

0	SELECT STATEMENT		1	2 (0)	00:00:01
1	SORT AGGREGATE		1		
2	INDEX FAST FULL SCAN	PK_EMP	14	2 (0)	00:00:01

Note

   - dynamic sampling used for this statement (level=2)
统计信息
----
          0 recursive calls
          0 db block gets
          4 consistent gets
          0 physical reads
          0 redo size
        430 bytes sent via SQL*Net to client
        419 bytes received via SQL*Net from client
          2 SQL*Net roundtrips to/from client
          0 sorts (memory)

```
        0  sorts (disk)
        1  rows processed
```

笔者经常使用 set autot trace 命令查看执行计划。

利用 AUTOTRACE 查看执行计划会带来一个额外的好处,当 SQL 执行完毕之后,会在执行计划的末尾显示 SQL 在运行过程中耗费的一些统计信息。

recursive calls 表示递归调用的次数。一个 SQL 第一次执行就会发生硬解析,在硬解析的时候,优化器会隐含地调用一些内部 SQL,因此当一个 SQL 第一次执行,recursive calls 会大于 0;第二次执行的时候不需要递归调用,recursive calls 会等于 0。

如果 SQL 语句中有自定义函数,recursive calls 永远不会等于 0,自定义函数被调用了多少次,recursive calls 就会显示为多少次。

```
SQL> create or replace function f_getdname(v_deptno in number) return varchar2 as
  2    v_dname dept.dname%type;
  3  begin
  4    select dname into v_dname from dept where deptno = v_deptno;
  5    return v_dname;
  6  end f_getdname;
  7  /

Function created.
```

SQL 多次执行后的执行计划如下。

```
SQL> select ename,f_getdname(deptno) from emp;

14 rows selected.

Execution Plan
----------------------------------------------------------
Plan hash value: 3956160932

--------------------------------------------------------------------------
| Id  | Operation         | Name | Rows | Bytes | Cost (%CPU)| Time     |
--------------------------------------------------------------------------
|   0 | SELECT STATEMENT  |      |   14 |   126 |     3   (0)| 00:00:01 |
|   1 |  TABLE ACCESS FULL| EMP  |   14 |   126 |     3   (0)| 00:00:01 |
--------------------------------------------------------------------------


Statistics
----------------------------------------------------------
         14  recursive calls
          0  db block gets
         36  consistent gets
          0  physical reads
          0  redo size
        769  bytes sent via SQL*Net to client
        419  bytes received via SQL*Net from client
          2  SQL*Net roundtrips to/from client
          0  sorts (memory)
          0  sorts (disk)
         14  rows processed
```

SQL 一共返回了 14 行数据,每返回一行数据,就会调用一次自定义函数,所以执行计划中 recursive calls 为 14。

db block gets 表示有多少个块发生变化，一般情况下，只有 DML 语句才会导致块发生变化，所以查询语句中 db block gets 一般为 0。如果有延迟块清除，或者 SQL 语句中调用了返回 CLOB 的函数，db block gets 也有可能会大于 0，不要觉得奇怪。

consistent gets 表示逻辑读，单位是块。在进行 SQL 优化的时候，我们应该想方设法减少逻辑读个数。通常情况下逻辑读越小，性能也就越好。需要注意的是，逻辑读并不是衡量 SQL 执行快慢的唯一标准，需要结合 I/O 等其他综合因素共同判断。

怎么通过逻辑读判断一个 SQL 还存在较大优化空间呢？如果 SQL 的逻辑读远远大于 SQL 语句中所有表的段大小之和（假设所有表都走全表扫描，表关联方式为 HASH JOIN），那么该 SQL 就存在较大优化空间。动手能力强的读者可以据此编写一个 SQL，抓出 SQL 逻辑读远远大于语句中所有表段大小之和的 SQL 语句。

physical reads 表示从磁盘读取了多少个数据块，如果表已经被缓存在 buffer cache 中，没有物理读，physical reads 等于 0。

redo size 表示产生了多少字节的重做日志，一般情况下只有 DML 语句才会产生 redo，查询语句一般情况下不会产生 redo，所以这里 redo size 为 0。如果有延迟块清除，查询语句也会产生 redo。

bytes sent via SQL*Net to client 表示从数据库服务器发送了多少字节到客户端。

bytes received via SQL*Net from client 表示从客户端发送了多少字节到服务端。

SQL*Net roundtrips to/from client 表示客户端与数据库服务端交互次数，我们可以通过设置 arraysize 减少交互次数。

sorts (memory)和 sorts (disk)分别表示内存排序和磁盘排序的次数。

rows processed 表示 SQL 一共返回多少行数据。我们在做 SQL 优化的时候最关心这部分数据，因为可以根据 SQL 返回的行数判断整个 SQL 应该是走 HASH 连接还是走嵌套循环。如果 **rows processed** 很大，一般走 HASH 连接；如果 **rows processed** 很小，一般走嵌套循环。

3.1.2 使用 EXPLAIN PLAN FOR 查看执行计划

使用 explain plan for 查看执行计划，用法如下。

```
explain plan for SQL 语句;
select * from table(dbms_xplan.display);
```

示例（Oracle11gR2，Scott 账户）如下。

```
SQL> explain plan for select ename, deptno
  2    from emp
  3   where deptno in (select deptno from dept where dname = 'CHICAGO');

Explained.

SQL> select * from table(dbms_xplan.display);

PLAN_TABLE_OUTPUT
```

```
------------------------------------------------------------------------------------
Plan hash value: 844388907

------------------------------------------------------------------------------------
| Id  | Operation                    | Name    | Rows | Bytes | Cost(%CPU)| Time     |
------------------------------------------------------------------------------------
|   0 | SELECT STATEMENT             |         |    5 |   110 |     6 (17)| 00:00:01 |
|   1 |  MERGE JOIN                  |         |    5 |   110 |     6 (17)| 00:00:01 |
|*  2 |   TABLE ACCESS BY INDEX ROWID| DEPT    |    1 |    13 |     2  (0)| 00:00:01 |
|   3 |    INDEX FULL SCAN           | PK_DEPT |    4 |       |     1  (0)| 00:00:01 |
|*  4 |   SORT JOIN                  |         |   14 |   126 |     4 (25)| 00:00:01 |
|   5 |    TABLE ACCESS FULL         | EMP     |   14 |   126 |     3  (0)| 00:00:01 |
------------------------------------------------------------------------------------

Predicate Information (identified by operation id):
---------------------------------------------------

   2 - filter("DNAME"='CHICAGO')
   4 - access("DEPTNO"="DEPTNO")
       filter("DEPTNO"="DEPTNO")

19 rows selected.
```

查看高级（ADVANCED）执行计划，用法如下。

explain plan for SQL 语句;
select * from table(dbms_xplan.display(NULL, NULL, 'advanced -projection'));

示例（Oracle11gR2，Scott 账户）如下。

```
SQL> explain plan for select ename, deptno
  2  from emp
  3  where deptno in (select deptno from dept where dname = 'CHICAGO');

Explained.

SQL> select * from table(dbms_xplan.display(NULL, NULL, 'advanced -projection'));

PLAN_TABLE_OUTPUT
------------------------------------------------------------------------------------

Plan hash value: 844388907

------------------------------------------------------------------------------------
| Id  | Operation                    | Name    | Rows | Bytes | Cost(%CPU)| Time     |
------------------------------------------------------------------------------------
|   0 | SELECT STATEMENT             |         |    5 |   110 |     6 (17)| 00:00:01 |
|   1 |  MERGE JOIN                  |         |    5 |   110 |     6 (17)| 00:00:01 |
|*  2 |   TABLE ACCESS BY INDEX ROWID| DEPT    |    1 |    13 |     2  (0)| 00:00:01 |
|   3 |    INDEX FULL SCAN           | PK_DEPT |    4 |       |     1  (0)| 00:00:01 |
|*  4 |   SORT JOIN                  |         |   14 |   126 |     4 (25)| 00:00:01 |
|   5 |    TABLE ACCESS FULL         | EMP     |   14 |   126 |     3  (0)| 00:00:01 |
------------------------------------------------------------------------------------

Query Block Name / Object Alias (identified by operation id):
-------------------------------------------------------------

   1 - SEL$5DA710D3
   2 - SEL$5DA710D3 / DEPT@SEL$2
   3 - SEL$5DA710D3 / DEPT@SEL$2
   5 - SEL$5DA710D3 / EMP@SEL$1
```

```
Outline Data
-------------

  /*+
      BEGIN_OUTLINE_DATA
      PX_JOIN_FILTER(@"SEL$5DA710D3" "EMP"@"SEL$1")
      USE_MERGE(@"SEL$5DA710D3" "EMP"@"SEL$1")
      LEADING(@"SEL$5DA710D3" "DEPT"@"SEL$2" "EMP"@"SEL$1")
      FULL(@"SEL$5DA710D3" "EMP"@"SEL$1")
      INDEX(@"SEL$5DA710D3" "DEPT"@"SEL$2" ("DEPT"."DEPTNO"))
      OUTLINE(@"SEL$2")
      OUTLINE(@"SEL$1")
      UNNEST(@"SEL$2")
      OUTLINE_LEAF(@"SEL$5DA710D3")
      ALL_ROWS
      DB_VERSION('11.2.0.1')
      OPTIMIZER_FEATURES_ENABLE('10.2.0.3')
      IGNORE_OPTIM_EMBEDDED_HINTS
      END_OUTLINE_DATA
  */

Predicate Information (identified by operation id):
---------------------------------------------------

   2 - filter("DNAME"='CHICAGO')
   4 - access("DEPTNO"="DEPTNO")
       filter("DEPTNO"="DEPTNO")

48 rows selected.
```

高级执行计划比普通执行计划多了 Query Block Name /Object Alias 和 Outline Data。

当需要控制半连接/反连接执行计划的时候，我们就可能需要查看高级执行计划。有时候我们需要使用 SQL PROFILE 固定执行计划，也可能需要查看高级执行计划。

Query Block Name 表示查询块名称，Object Alias 表示对象别名。Outline Data 表示 SQL 内部的 HINT。一条 SQL 语句可能会包含多个子查询，每个子查询在执行计划内部就是一个 Query Block。为什么会有 Query Block 呢？比如一个 SQL 语句包含有多个子查询，假如每个子查询都要访问同一个表，不给表取别名，这个时候我们怎么区分表属于哪个子查询呢？所以 Oracle 会给同一个 SQL 语句中的子查询取别名，这个名字就是 Query Block Name，以此来区分子查询中的表。Query Block Name 默认会命名为 SEL$1，SEL$2，SEL$3 等，我们可以使用 HINT：qb_name（别名）给子查询取别名。

关于高级执行计划更为详细的内容，请阅读本书 5.6.2 节。

3.1.3 查看带有 A-TIME 的执行计划

查看带有 A-TIME 的执行计划的用法如下。

```
alter session set statistics_level=all;
或者在 SQL 语句中添加 hint:/*+ gather_plan_statistics */
```

运行完 SQL 语句，然后执行下面的查询语句就可以获取带有 A-TIME 的执行计划。

```
select * from table(dbms_xplan.display_cursor(null,null,'allstats last'));
```

示例（Oracle11gR2，Scott 账户）如下。

```
SQL> select /*+ gather_plan_statistics full(test) */ count(*) from test where owner='SYS';

  COUNT(*)
----------
     30808

SQL> select * from table(dbms_xplan.display_cursor(null,null,'allstats last'));

PLAN_TABLE_OUTPUT
---------------------------------------------------------------------------------
SQL_ID  fswg73p1zmvqu, child number 0
-------------------------------------
select /*+ gather_plan_statistics full(test) */ count(*) from test
where owner='SYS'

Plan hash value: 1950795681

---------------------------------------------------------------------------------
| Id |Operation           | Name | Starts | E-Rows | A-Rows |   A-Time   | Buffers | Reads |
---------------------------------------------------------------------------------
|  0 |SELECT STATEMENT    |      |      1 |        |      1 |00:00:00.03 |    1037 |  1033 |
|  1 | SORT AGGREGATE     |      |      1 |      1 |      1 |00:00:00.03 |    1037 |  1033 |
|* 2 |  TABLE ACCESS FULL | TEST |      1 |   2518 |  30808 |00:00:00.01 |    1037 |  1033 |
---------------------------------------------------------------------------------

Predicate Information (identified by operation id):
---------------------------------------------------

   2 - filter("OWNER"='SYS')

20 rows selected.
```

Starts 表示这个操作执行的次数。

E-Rows 表示优化器估算的行数，就是普通执行计划中的 Rows。

A-Rows 表示真实的行数。

A-Time 表示累加的总时间。与普通执行计划不同的是，普通执行计划中的 Time 是假的，而 A-Time 是真实的。

Buffers 表示累加的逻辑读。

Reads 表示累加的物理读。

上面介绍了 3 种方法查看执行计划。使用 AUTOTRACE 或者 EXPLAIN PLAN FOR 获取的执行计划来自于 PLAN_TABLE。PLAN_TABLE 是一个会话级的临时表，里面的执行计划并不是 SQL 真实的执行计划，它只是优化器估算出来的。真实的执行计划不应该是估算的，应该是真正执行过的。SQL 执行过的执行计划存在于共享池中，具体存在于数据字典 V$SQL_PLAN 中，带有 A-Time 的执行计划来自于 V$SQL_PLAN，是真实的执行计划，而通过 AUTOTRACE、通过 EXPLAIN PLAN FOR 获取的执行计划只是优化器估算获得的执行计划。有读者会有疑问，使用 AUTOTRACE 查看执行计划，SQL 是真正运行过的，怎么得到的执行计划不是真实的呢？原因在于 AUTOTRACE 获取的执行计划来自于 PLAN_TABLE，而非来自于共享池中的 V$SQL_PLAN。

3.1.4 查看正在执行的 SQL 的执行计划

有时需要抓取正在运行的 SQL 的执行计划，这时我们需要获取 SQL 的 SQL_ID 以及 SQL 的 CHILD_NUMEBR，然后将其代入下面 SQL，就能获取正在运行的 SQL 的执行计划。

```
select * from table(dbms_xplan.display_cursor('sql_id',child_number));
```

示例（Oracle11gR2，Scott 账户）如下。

先创建两个测试表 a，b。

```
SQL> create table a as select * from dba_objects;
Table created.
SQL> create table b as select * from dba_objects;
Table created.
```

然后在一个会话中执行如下 SQL。

```
select count(*) from a,b where a.owner=b.owner;
```

在另外一个会话中执行如下 SQL，结果如图 3-1 所示。

```
select a.sid, a.event, a.sql_id, a.sql_child_number, b.sql_text
  from v$session a, v$sql b
 where a.sql_address = b.address
   and a.sql_hash_value = b.hash_value
   and a.sql_child_number = b.child_number
 order by 1 desc;
```

SID	EVENT	SQL_ID	SQL_CHILD_NUMBER	SQL_TEXT
98	SQL*Net message from client	ach0j2bvtabtu	0	select a.sid, a.event, a.sql_id, a.sql_child_number, b...
33	db file scattered read	czr9jwxv0xra6	0	select count(*) from a,b where a.owner=b.owner

图 3-1

接下来我们将 SQL_ID 和 CHILD_NUMBER 代入以下 SQL。

```
SQL> select * from table(dbms_xplan.display_cursor('czr9jwxv0xra6',0));

PLAN_TABLE_OUTPUT
--------------------------------------------------------------------------------
SQL_ID  czr9jwxv0xra6, child number 0
-------------------------------------
select count(*) from a,b where a.owner=b.owner

Plan hash value: 319234518

--------------------------------------------------------------------------------
| Id  | Operation       | Name | Rows | Bytes |TempSpc| Cost (%CPU)| Time     |
--------------------------------------------------------------------------------
|   0 | SELECT STATEMENT|      |      |       |       |  2556 (100)|          |
|   1 |  SORT AGGREGATE |      |    1 |    34 |       |            |          |
```

```
|*  2 |   HASH JOIN          |       |   400M|   12G| 1920K| 2556  (78)| 00:00:31 |
|   3 |    TABLE ACCESS FULL | B     | 67547 | 1121K|      |  187   (1)| 00:00:03 |
|   4 |    TABLE ACCESS FULL | A     | 77054 | 1279K|      |  187   (1)| 00:00:03 |
---------------------------------------------------------------------------------

Predicate Information (identified by operation id):
---------------------------------------------------

   2 - access("A"."OWNER"="B"."OWNER")

Note
-----
   - dynamic sampling used for this statement (level=2)
```

3.2 定制执行计划

在 Oracle 数据库中，执行计划是树形结构，因此我们可以利用树形查询来定制执行计划。我们打开 PLSQL dev SQL 窗口，登录示例账户 Scott 并且运行如下 SQL。

```
explain plan for select /*+ use_hash(a,dept) */ *
  from emp a, dept
 where a.deptno = dept.deptno
   and a.sal > 3000;
```

然后执行下面的脚本，结果如图 3-2 所示。

```
select case
         when (filter_predicates is not null or
               access_predicates is not null) then
          '*'
         else
          ' '
       end || id as "Id",
       lpad(' ', level) || operation || ' ' || options "Operation",
       object_name "Name",
       cardinality as "Rows",
       filter_predicates "Filter",
       access_predicates "Access"
  from plan_table
 start with id = 0
connect by prior id = parent_id;
```

Id	Operation	Name	Rows	Filter	Access
0	SELECT STATEMENT		1		
*1	HASH JOIN		1		"A"."DEPTNO"="DEPT"."DEPTNO"
*2	TABLE ACCESS FULL	EMP	1	"A"."SAL">3000	
3	TABLE ACCESS FULL	DEPT	4		

图 3-2

我们曾在 1.2 节中提到，只有大表才会产生性能问题，因此可以将表的段大小添加到定制执行计划中，这样我们在用定制执行计划优化 SQL 的时候，可以很方便地知道表大小，从而更快地判断该步骤是否可能是性能瓶颈。下面脚本添加表的段大小以及索引段大小到定制执行计划中，结果如图 3-3 所示。

```
select case
         when (filter_predicates is not null or
```

```
                   access_predicates is not null) then
            '*'
          else
            ' '
        end || id as "Id",
        lpad(' ', level) || operation || ' ' || options "Operation",
        object_name "Name",
        cardinality as "Rows",
        b.size_mb "Size_Mb",
        filter_predicates "Filter",
        access_predicates "Access"
  from plan_table a,
       (select owner, segment_name, sum(bytes / 1024 / 1024) size_mb
          from dba_segments
         group by owner, segment_name) b
 where a.object_owner = b.owner(+)
   and a.object_name = b.segment_name(+)
 start with id = 0
connect by prior id = parent_id;
```

如图 3-3 所示，Size_Mb 显示表的段大小，单位是 MB。

Id	Operation	Name	Rows	Size_Mb	Filter	Access
0	SELECT STATEMENT		1			
*1	HASH JOIN		1			"A"."DEPTNO"="DEPT"."DEPTNO"
*2	TABLE ACCESS FULL	EMP	1	0.0625	"A"."SAL">3000	
3	TABLE ACCESS FULL	DEPT	4	0.0625		

图 3-3

我们曾在 1.4 节中提到建立组合索引避免回表或者建立合适的组合索引减少回表次数。如果一个 SQL 只访问了某个表的极少部分列，那么我们可以将这些被访问的列联合在一起，从而建立组合索引。下面脚本将添加表的总字段数以及被访问字段数量到定制执行计划中，结果如图 3-4 所示。

```
select case
         when access_predicates is not null or filter_predicates is not null then
           '*' || id
         else
           ' ' || id
       end as "Id",
       lpad(' ', level) || operation || ' ' || options "Operation",
       object_name "Name",
       cardinality "Rows",
       b.size_mb "Mb",
       case
         when object_type like '%TABLE%' then
           REGEXP_COUNT(a.projection, ']') || '/' || c.column_cnt
       end as "Column",
       access_predicates "Access",
       filter_predicates "Filter",
       case
         when object_type like '%TABLE%' then
           projection
       end as "Projection"
  from plan_table a,
       (select owner, segment_name, sum(bytes / 1024 / 1024) size_mb
          from dba_segments
         group by owner, segment_name) b,
```

```
            (select owner, table_name, count(*) column_cnt
               from dba_tab_cols
             group by owner, table_name) c
 where a.object_owner = b.owner(+)
    and a.object_name = b.segment_name(+)
    and a.object_owner = c.owner(+)
    and a.object_name = c.table_name(+)
 start with id = 0
 connect by prior id = parent_id;
```

Id	Operation	Name	Rows	Mb	Column	Access	Filter
0	SELECT STATEMENT		1				
*1	HASH JOIN		1			"A"."DEPTNO"="DEPT"."DEPTNO"	
*2	TABLE ACCESS FULL	EMP	1	0.0625	8/8		"A"."SAL">3000
3	TABLE ACCESS FULL	DEPT	4	0.0625	3/3		

图 3-4

如图 3-4 中所示，Column 表示访问了表多少列/表一共有多少列。Projection 显示了具体的访问列信息，限于书本宽度，图中没有显示 Projection 列信息。

限于书本限制，定制执行计划本书不做进一步讨论，有兴趣的读者请自行添加其余定制信息到定制执行计划中。

3.3 怎么通过查看执行计划建立索引

我们利用如下 SQL 讲解（基于 Oracle11gR2 scott）。

```
SQL> explain plan for select e.ename,e.job,d.dname from emp e,dept d  where e.deptno=
d.deptno and e.sal<2000;

Explained.

SQL> select * from table(dbms_xplan.display);

PLAN_TABLE_OUTPUT
--------------------------------------------------------------------------------

Plan hash value: 615168685

--------------------------------------------------------------------------------
| Id  | Operation          | Name | Rows  | Bytes | Cost (%CPU)| Time     |
--------------------------------------------------------------------------------
|   0 | SELECT STATEMENT   |      |     8 |   488 |     7  (15)| 00:00:01 |
|*  1 |  HASH JOIN         |      |     8 |   488 |     7  (15)| 00:00:01 |
|   2 |   TABLE ACCESS FULL| DEPT |     4 |    88 |     3   (0)| 00:00:01 |
|*  3 |   TABLE ACCESS FULL| EMP  |     8 |   312 |     3   (0)| 00:00:01 |
--------------------------------------------------------------------------------

Predicate Information (identified by operation id):
---------------------------------------------------

   1 - access("E"."DEPTNO"="D"."DEPTNO")
   3 - filter("E"."SAL"<2000)

Note
-----
   - dynamic sampling used for this statement (level=2)
```

执行计划分为两部分，Plan hash value 和 Predicate Information 之间这部分主要是表的访问路径以及表的连接方式。关于访问路径以及表连接方式会在之后章节详细解释。另外一部分是谓词过滤信息，这部分信息位于 Predicate Information 下面，谓词过滤信息非常重要。一些老 DBA 因为之前接触的是 Oracle8i 或者 Oracle9i，那个时候执行计划还没有谓词信息，所以就遗留了一个传统，看执行计划只看访问路径和表连接方式了，而不关心谓词过滤信息。还有些人做 SQL 优化喜欢用 10 046 trace 或者 10 053 trace，如果仅仅是优化一个 SQL，根本就不需要使用上面两个工具，直接分析 SQL 语句以及执行计划即可。当然，如果是为了深入研究为什么不走索引，为什么走了嵌套循环而没走 HASH 连接等，这个时候我们可以用 10 053 trace；如果想研究访问路径是单块读或者是多块读，可以使用 10 046 trace。

我们这里先不讲怎么阅读执行计划，后面会讲利用光标移动大法阅读执行计划。

注意观察 Id 这列，有些 Id 前面有"*"号，这表示发生了谓词过滤，或者发生了 HASH 连接，或者是走了索引。Id=1 前面有"*"号，它是 HASH 连接的"*"号，我们观察对应的谓词过滤信息就能知道是哪两个表进行的 HASH 连接，而且能知道是对哪些列进行的 HASH 连接，这里是 e 表（emp 表的别名）的 deptno 列与 d 表（dept 的别名）deptno 列进行 HASH 连接的。Id=3 前面有"*"号，这里表示表 emp 有谓词过滤，它的过滤条件就是 Id=3 对应的谓词过滤信息，也就是 e.sal<2000。Id=2 前面没有"*"号，那么说明 dept 表没有谓词过滤条件。

提问：TABLE ACCESS FULL 前面没有"*"号怎么办？

回答：如果表很小，那么不需理会，小表不会产生性能问题。如果表很大，那么我们要询问开发人员是不是忘了写过滤条件，当然了一般也不会遇到这种情况。如果真的是没过滤条件呢？比如一个表有 10GB，但是没有过滤条件，那么它就会成为整个 SQL 的性能瓶颈。这个时候我们需要查看 SQL 语句中该表访问了多少列，如果访问的列不多，就可以把这些列组合起来，建立一个组合索引，索引的大小可能就只有 1GB 左右。我们利用 INDEX FAST FULL SCAN 代替 TABLE ACCESS FULL。在访问列不多的情况，索引的大小（Segment Size）肯定比表的大小（Segment Size）小，那么就不需要扫描 10GB 了，只需要扫描 1GB，从而达到优化目的。如果 SQL 语句里面要访问表中大部分列，这时就不应该建立组合索引了，因为此时索引大小比表更大，可以通过其他方法优化，比如开启并行查询，或者更改表连接方式，让大表作为嵌套循环的被驱动表，同时在大表的连接列上建立索引。关于表连接方式，我们会在后面章节详细介绍。

提问：TABLE ACCESS FULL 前面有"*"号怎么办？

回答：如果表很小，那么我们不需理会；如果表很大，可以使用"select count(*) from 表"，查看有多少行数据，然后通过"select count(*) from 表 where *"对应的谓词过滤条件，查看返回多少行数据。如果返回的行数在表总行数的 5%以内，我们可以在过滤列上建立索引。如果已经存在索引，但是没走索引，这时我们要检查统计信息，特别是直方图信息。如果统计信息

3.3 怎么通过查看执行计划建立索引

已经收集过了，我们可以用 HINT 强制走索引。如果有多个谓词过滤条件，我们需要建立组合索引并且要将选择性高的列放在前面，选择性低的列在后面。如果返回的行数超过表总行数的 5%，这个时候我们要查看 SQL 语句中该表访问了多少列，如果访问的列少，同样可以把这些列组合起来，建立组合索引，建立组合索引的时候，谓词过滤列在前面，连接列在中间，select 部分的列在最后。如果访问的列多，这个时候就只能走全表扫描了。

提问：TABLE ACCESS BY INDEX ROWID 前面有 "*" 号怎么办？

回答：我们利用如下 SQL 讲解（基于 Oracle11gR2 scott）。

```
SQL> grant dba to scott;
```

授权成功。

```
SQL> create table test as select * from dba_objects;
```

表已创建。

```
SQL> create index idx_name on test(object_name);
```

索引已创建。

```
SQL> set autot trace
SQL> select /*+ index(test) */ * from test where object_name like 'V_$%' and owner='S
COTT' ;
未选定行
```

执行计划

Plan hash value: 461797767

```
--------------------------------------------------------------------------------
| Id  | Operation                    | Name     | Rows  | Bytes | Cost(%CPU)| Time     |
--------------------------------------------------------------------------------
|   0 | SELECT STATEMENT             |          |    38 |  7866 |   334   (0)| 00:00:05 |
|*  1 |  TABLE ACCESS BY INDEX ROWID | TEST     |    38 |  7866 |   334   (0)| 00:00:05 |
|*  2 |   INDEX RANGE SCAN           | IDX_NAME |   672 |       |     6   (0)| 00:00:01 |
--------------------------------------------------------------------------------

Predicate Information (identified by operation id):
---------------------------------------------------

   1 - filter("OWNER"='SCOTT')
   2 - access("OBJECT_NAME" LIKE 'V_$%')
       filter("OBJECT_NAME" LIKE 'V_$%')

Note
-----
   - dynamic sampling used for this statement (level=2)
```

统计信息

```
          0  recursive calls
          0  db block gets
        332  consistent gets
```

```
        0  physical reads
        0  redo size
     1191  bytes sent via SQL*Net to client
      409  bytes received via SQL*Net from client
        1  SQL*Net roundtrips to/from client
        0  sorts (memory)
        0  sorts (disk)
        0  rows processed
```

TABLE ACCESS BY INDEX ROWID 前面有 "*" 号，表示回表再过滤。回表再过滤说明数据没有在索引中过滤干净。当 TABLE ACCESS BY INDEX ROWID 前面有 "*" 号时，可以将 "*" 号下面的过滤条件包含在索引中，这样可以减少回表次数，提升查询性能。

```
SQL> create index idx_ownername on test(owner,object_name);
索引已创建
SQL> select /*+ index(test) */ * from test where  object_name like 'V_$%' and owner='SCOTT' ;
未选定行

执行计划
----------------------------------------------------------
Plan hash value: 3756723214

--------------------------------------------------------------------------------------
| Id |Operation                    |Name          |Rows | Bytes | Cost(%CPU)| Time     |
--------------------------------------------------------------------------------------
|  0 |SELECT STATEMENT             |              |  38 |  7866 |    5  (0) | 00:00:01 |
|  1 | TABLE ACCESS BY INDEX ROWID |TEST          |  38 |  7866 |    5  (0) | 00:00:01 |
|* 2 |  INDEX RANGE SCAN           |IDX_OWNERNAME |   3 |       |    3  (0) | 00:00:01 |
--------------------------------------------------------------------------------------

Predicate Information (identified by operation id):
---------------------------------------------------

   2 - access("OWNER"='SCOTT' AND "OBJECT_NAME" LIKE 'V_$%')
       filter("OBJECT_NAME" LIKE 'V_$%')

Note
-----
   - dynamic sampling used for this statement (level=2)
统计信息
----------------------------------------------------------
        0  recursive calls
        0  db block gets
        3  consistent gets
        0  physical reads
        0  redo size
     1191  bytes sent via SQL*Net to client
      409  bytes received via SQL*Net from client
        1  SQL*Net roundtrips to/from client
        0  sorts (memory)
        0  sorts (disk)
        0  rows processed
```

如果索引返回的数据本身很少，即使 TABLE ACCESS BY INDEX ROWID 前面有 "*" 号，也可以不用理会，因为索引本身返回的数据少，回表也没有多少次，因此可以不用再创建组合索引。

通过上面的讲解，相信大家也明白了为什么我们不推荐使用工具查看执行计划，因为有些工具看不到"*"号，看不到谓词过滤信息。

3.4 运用光标移动大法阅读执行计划

执行计划中，最需要关心的有 Id，Operation，Name，Rows。

看 Id 是为了观察 Id 前面是否有 "*" 号。

Operation 表示表的访问路径或者连接方式。第 4 章我们会详细介绍常见访问路径，第 5 章会详细介绍表连接方式。

Name 是 SQL 语句中对象的名字，可以是表名、索引名、视图名、物化视图名或者 CBO 自动生成的名字。

Rows 是 CBO 根据统计信息以及数学公式计算出来的，也就是说 Rows 是假的，不是真实的。这里的 Rows 也被称作执行计划中返回的基数。再一次强调，Rows 是假的，别被它骗了。前面介绍过带有 A-Time 的执行计划，带有 A-Time 的执行计划中 E-Rows 就是普通执行计划中的 Rows，A-Rows 才是真实的。在进行 SQL 优化的时候，我们经常需要手工计算某个访问路径的真实 Rows，然后对比执行计划中的 Rows。如果手工计算的 Rows 与执行计划中的 Rows 相差很大，执行计划往往就出错了。

有些人可能还会特意查看执行计划中的 Cost，在进行 SQL 优化的时候，千万别看 Cost！如果一个 SQL 语句都需要优化了，那么它的 Cost 还是准确的吗？有很大概率算错了！既然算错了，你还去看错误的 Cost 干什么呢？关于 Cost，我们会在第 6 章详细介绍，同时由此引出 SQL 优化核心思想。

下面我们将为大家介绍如何利用光标移动大法阅读执行计划。

现有如下执行计划。

```
-------------------------------------------------------------------------------
| Id  | Operation                           | Name                    | Rows  |
-------------------------------------------------------------------------------
|   0 | SELECT STATEMENT                    |                         |     1 |
|   1 |  TABLE ACCESS BY INDEX ROWID        | INTRC_PROD_DIM          |     1 |
|   2 |   NESTED LOOPS                      |                         |     1 |
|   3 |    NESTED LOOPS                     |                         |     1 |
|   4 |     NESTED LOOPS                    |                         |   330 |
|   5 |      NESTED LOOPS                   |                         | 1312K |
|*  6 |       HASH JOIN                     |                         |  6558 |
|   7 |        TABLE ACCESS FULL            | INTRC_GEO_DIM           |  2532 |
|*  8 |        HASH JOIN                    |                         |  6558 |
|*  9 |         TABLE ACCESS FULL           | INTRC_INITV_DIM         |   833 |
|* 10 |         HASH JOIN                   |                         |  6558 |
|  11 |          PARTITION RANGE SINGLE     |                         |   171 |
|* 12 |           TABLE ACCESS FULL         | INTRC_TIME_DIM          |   171 |
|* 13 |          HASH JOIN                  |                         |  6558 |
|  14 |           PARTITION RANGE SINGLE    |                         |   171 |
|* 15 |            TABLE ACCESS FULL        | INTRC_TIME_DIM          |   171 |
|  16 |           PARTITION RANGE SINGLE    |                         |  6558 |
|* 17 |            TABLE ACCESS FULL        | INTRC_INITV_TIME_BRDG_DIM |  6558 |
|  18 |       PARTITION RANGE SINGLE        |                         |   200 |
```

```
|*  19 |          TABLE ACCESS BY LOCAL INDEX ROWID| INTRC_INBR_FCT          |   200 |
|   20 |           BITMAP CONVERSION TO ROWIDS     |                         |       |
|   21 |            BITMAP INDEX FULL SCAN         | INTRC_INBR_FCT_BX1      |       |
|   22 |        PARTITION RANGE SINGLE             |                         |     1 |
|   23 |         BITMAP CONVERSION TO ROWIDS       |                         |     1 |
|   24 |          BITMAP AND                       |                         |       |
|*  25 |           BITMAP INDEX SINGLE VALUE       | INTRC_TIME_DIM_BX1      |       |
|   26 |           BITMAP CONVERSION FROM ROWIDS   |                         |       |
|   27 |            SORT ORDER BY                  |                         |       |
|*  28 |             INDEX RANGE SCAN              | INTRC_TIME_DIM_PK       |     1 |
|   29 |           BITMAP CONVERSION FROM ROWIDS   |                         |       |
|*  30 |            INDEX RANGE SCAN               | INTRC_TIME_DIM_NX1      |     1 |
|   31 |         BITMAP CONVERSION TO ROWIDS       |                         |     1 |
|   32 |          BITMAP AND                       |                         |       |
|   33 |           BITMAP CONVERSION FROM ROWIDS   |                         |       |
|*  34 |            INDEX RANGE SCAN               | INTRC_INPR_BRDG_DIM_PK  |     1 |
|*  35 |           BITMAP INDEX SINGLE VALUE       | INTRC_INPR_BRDG_DIM_BX1 |       |
|*  36 |      INDEX RANGE SCAN                     | INTRC_PROD_DIM_PK       |     1 |
------------------------------------------------------------------------------------------
```

有些读者可能会认为 Id=15 最先执行，因为 Id=15 的缩进最大，其实这是错误的。

现在给大家介绍一种方法：光标移动大法。光标就是我们打字的时候，鼠标点到某个地方，闪烁的光标。**阅读执行计划的时候，一般从上往下看，找到执行计划的入口之后，再往上看。**

阅读执行计划的时候，我们将光标移动到 Id=0 SELECT 的 S 前面，然后按住键盘的向下移动的箭头，向下移动，然后向右移动，然后再向下，再向右……Id=0 和 Id=1 相差一个空格（缩进），上下相差一个空格（缩进）就是父子关系，上面的是父亲，下面的是儿子，儿子比父亲先执行。那么这里 Id=1 是 Id=0 的儿子，Id=1 先执行。Id=2 是 Id=1 的儿子，Id=2 先执行。Id=3 是 Id=2 的儿子，Id=3 先执行。这样我们一直将光标移动到 Id=7（向下，向右移动），Id=7 与 Id=8 对齐，表示 Id=7 与 Id=8 是兄弟关系，上面的是兄，下面的是弟，兄优先于弟先执行，也就是说 Id=7 先于 Id=8 执行。Id=7 也跟 Id=19、Id=24、Id=34 对齐，将光标移动到 Id=7 前面，向下移动光标，Id=19 在 Id=18 的下面，光标移动大法是不能"穿墙"的，从 Id=7 移动到 Id=19 会穿过 Id=18，同理 Id=24、Id=34 也"穿墙"了，因此 Id=7 只是和 Id=8 对齐。因为 Id=7 下面没有儿子，所以执行计划的入口是 Id=7，整个执行计划中 Id=7 最先执行。

提问：怎么快速找到执行计划的入口？

回答：我们可以利用光标移动大法，先将光标放在 Id=0 这一步，然后一直向下向右移动光标，直到找到没有儿子的 Id，这个 Id 就是执行计划的入口。

提问：怎么判断是哪个表与哪个表进行关联的？

回答：我们先找到表在执行计划中的 Id，然后看这个 Id（或者是这个 Id 的父亲）与谁对齐（利用光标上下移动），它与谁对齐，就与谁进行关联。比如 Id=17 这个表，它本身没有和任何 Id 对齐，但是 Id=17 的父亲是 Id=16，与 Id=14 对齐，Id=14 的儿子是 Id=15，所以 Id=17 这个表是与 Id=15 这个表进行关联的，并且两个表是进行 HASH 连接的。

3.4 运用光标移动大法阅读执行计划

提问：在 SQL 优化实战中，怎么应用光标移动大法优化 SQL？

回答：例如，有如下执行计划。

Id	Operation	Name	Starts	E-Rows	A-Rows	A-Time
0	SELECT STATEMENT		1		1324	00:02:42.23
1	SORT GROUP BY		1	1	1324	00:02:42.23
2	VIEW	VM_NWVW_2	1	1	6808	00:02:42.18
3	HASH UNIQUE		1	1	6808	00:02:42.18
4	NESTED LOOPS		1		5220K	00:02:21.06
5	NESTED LOOPS		1	1	5220K	00:02:00.18
6	NESTED LOOPS		1	1	5220K	00:01:49.74
7	NESTED LOOPS		1	2	5220K	00:01:18.42
8	NESTED LOOPS		1	1	6808	00:00:01.62
9	NESTED LOOPS		1	1	6808	00:00:00.54
10	NESTED LOOPS		1	1	11248	00:00:00.40
*11	HASH JOIN		1	5	11248	00:00:00.07
12	PARTITION LIST SUBQUERY		1	47	25	00:00:00.01
13	INLIST ITERATOR		1		25	00:00:00.01
14	TABLE ACCESS BY LOCAL INDEX ROWID	OPT_ACCT_FDIM	25	47	25	00:00:00.01
*15	INDEX RANGE SCAN	OPT_ACCT_FDIM_NX2	25	47	25	00:00:00.01
16	NESTED LOOPS		1	10482	12788	00:00:00.03
17	NESTED LOOPS		1	1	1	00:00:00.01
*18	INDEX RANGE SCAN	OPT_BUS_UNIT_FDIM_UX2	1	1	1	00:00:00.01
*19	INDEX RANGE SCAN	OPT_BUS_UNIT_FDIM_UX2	1	1	1	00:00:00.01
20	PARTITION LIST ITERATOR		1	10482	12788	00:00:00.03
*21	TABLE ACCESS FULL	OPT_ACTVY_FCT	1	10482	12788	00:00:00.03
*22	TABLE ACCESS BY GLOBAL INDEX ROWID	OPT_PRMTN_FDIM	11248	1	11248	00:00:00.31
*23	INDEX UNIQUE SCAN	OPT_PRMTN_FDIM_PK	11248	1	11248	00:00:00.12
24	TABLE ACCESS BY INDEX ROWID	OPT_CAL_MASTR_DIM	11248	1	6808	00:00:00.14
*25	INDEX UNIQUE SCAN	OPT_CAL_MASTR_DIM_PK	11248	1	11248	00:00:00.05
26	PARTITION LIST ALL		6808	1	6808	00:00:01.08
*27	TABLE ACCESS BY LOCAL INDEX ROWID	OPT_PRMTN_FDIM	115K	1	6808	00:00:01.05
*28	INDEX RANGE SCAN	OPT_PRMTN_FDIM_NX3	115K	4	6808	00:00:00.78
29	TABLE ACCESS BY GLOBAL INDEX ROWID	OPT_PRMTN_PROD_FLTR_LKP	6808	39	5220K	00:01:19.79
*30	INDEX RANGE SCAN	OPT_PRMTN_PROD_FLTR_LKP_NX1	6808	3	5220K	00:00:43.96
*31	TABLE ACCESS BY GLOBAL INDEX ROWID	OPT_ACCT_FDIM	5220K	1	5220K	00:00:23.79
*32	INDEX UNIQUE SCAN	OPT_ACCT_FDIM_PK	5220K	1	5220K	00:00:08.38
*33	INDEX UNIQUE SCAN	OPT_CAL_MASTR_DIM_PK	5220K	1	5220K	00:00:07.58
*34	TABLE ACCESS BY INDEX ROWID	OPT_CAL_MASTR_DIM	5220K	1	5220K	00:00:17.28

如果是 SQL 优化初学者（高手可以一眼看出执行计划哪里有性能问题），可以先利用光标移动大法找到执行计划入口，检查入口 Rows 返回的真实行数与 CBO 估算的行数是否存在较大差异。比如，这里执行计划入口为 Id=15，优化器估算返回 47 行（E-Rows=47），实际上返回了 25 行（A-Rows=25），E-Rows 与 A-Rows 差别不大。找到执行计划入口之后，我们应该从执行计划入口往上检查，Id=15 上面的是 Id=14，Id=14 上面的是 Id=13，这样一直检查到 Id=11。Id=11 估算返回 5 行（E-Rows=5），但是实际上返回了 11 248 行（A-Rows=11 248），所以执行计划 Id=11 这步有问题，由于 Id=11 Rows 估算错误，它会导致后面整个执行计划出错，应该想办法让 CBO 估算出较为准确的 Rows。

我们还可以利用光标移动大法找出哪个表与哪个表进行关联的，例如下面执行计划。

Id=29 的表它与 Id=8 对齐，这表示 Id=29 的表是与一个结果集进行关联的，关联方式为嵌套循环（Id=7，NESTED LOOPS）。从执行计划中我们可以看到 Id=29 是嵌套循环的被驱动表，但是没走索引，走的是全表扫描。如果 Id=29 的表是一个大表，会出现严重的性能问题，因为它会被扫描多次，而且每次扫描的时候都是全表扫描，所以，我们需要在 Id=29 的表中创建一个索引（连接列上创建索引）。

第 3 章 执行计划

Id	Operation	Name	Rows	Bytes	Cost	(%CPU)	Time	Pstart	Pstop
0	SELECT STATEMENT		1	352	1551	(17)	00:00:07		
1	SORT GROUP BY		1	352	1551	(17)	00:00:07		
2	VIEW	VM_NWVW_2	1	352	1550	(17)	00:00:07		
3	HASH UNIQUE		1	652	1550	(17)	00:00:07		
4	NESTED LOOPS								
5	NESTED LOOPS		1	652	1549	(17)	00:00:07		
6	NESTED LOOPS		1	639	1548	(17)	00:00:07		
7	NESTED LOOPS		2	1180	1546	(17)	00:00:07		
8	NESTED LOOPS		1	568	130	(5)	00:00:01		
9	NESTED LOOPS		1	509	109	(6)	00:00:01		
10	NESTED LOOPS		1	484	108	(6)	00:00:01		
* 11	HASH JOIN		5	830	103	(6)	00:00:01		
12	PARTITION LIST SUBQUERY		47	4089	82	(3)	00:00:01	KEY(SQ)	KEY(SQ)
13	INLIST ITERATOR								
14	TABLE ACCESS BY LOCAL INDEX ROWID	OPT_ACCT_FDIM	47	4089	82	(3)	00:00:01	KEY(SQ)	KEY(SQ)
* 15	INDEX RANGE SCAN	OPT_ACCT_FDIM_NX2	47	43	5	(0)	00:00:01	KEY(SQ)	KEY(SQ)
16	NESTED LOOPS		10482	808K	20	(15)	00:00:01		
17	NESTED LOOPS		1	40	2	(0)	00:00:01		
* 18	INDEX RANGE SCAN	OPT_BUS_UNIT_FDIM_UX2	1	26	1	(0)	00:00:01		
* 19	INDEX RANGE SCAN	OPT_BUS_UNIT_FDIM_UX2	1	14	1	(0)	00:00:01		
20	PARTITION LIST ITERATOR		10482	1699K	18	(17)	00:00:01	KEY	KEY
* 21	TABLE ACCESS FULL	OPT_ACTVY_FCT	10482	1699K	18	(17)	00:00:01	KEY	KEY
* 22	TABLE ACCESS BY GLOBAL INDEX ROWID	OPT_PRMTN_FDIM	1	318	1	(0)	00:00:01	ROWID	ROWID
* 23	INDEX UNIQUE SCAN	OPT_PRMTN_FDIM_PK	1		0	(0)	00:00:01		
24	TABLE ACCESS BY INDEX ROWID	OPT_CAL_MASTR_DIM	1	25	1	(0)	00:00:01		
* 25	INDEX UNIQUE SCAN	OPT_CAL_MASTR_DIM_PK	1		0	(0)	00:00:01		
26	PARTITION LIST ALL		1	59	21	(0)	00:00:01	1	17
* 27	TABLE ACCESS BY LOCAL INDEX ROWID	OPT_PRMTN_FDIM	1	59	21	(0)	00:00:01	1	17
* 28	INDEX RANGE SCAN	OPT_PRMTN_FDIM_NX3	4		17	(0)	00:00:01	1	17
29	PARTITION LIST ITERATOR		39	858	1416	(18)	00:00:07	KEY	KEY
* 30	TABLE ACCESS FULL	OPT_PRMTN_PROD_FLTR_LKP	39	858	1416	(18)	00:00:07	KEY	KEY
* 31	TABLE ACCESS BY GLOBAL INDEX ROWID	OPT_ACCT_FDIM	1	49	1	(0)	00:00:01	ROWID	ROWID
* 32	INDEX UNIQUE SCAN	OPT_ACCT_FDIM_PK	1		0	(0)	00:00:01		
* 33	INDEX UNIQUE SCAN	OPT_CAL_MASTR_DIM_PK	1		0	(0)	00:00:01		
* 34	TABLE ACCESS BY INDEX ROWID	OPT_CAL_MASTR_DIM	1	13	1	(0)	00:00:01		

第 4 章 访问路径（ACCESS PATH）

访问路径指的就是通过哪种扫描方式获取数据，比如全表扫描、索引扫描或者直接通过 ROWID 获取数据。想要成为 SQL 优化高手，我们就必须深入理解各种访问路径。本章将会详细介绍常见的访问路径。

4.1 常见访问路径

4.1.1 TABLE ACCESS FULL

TABLE ACCESS FULL 表示全表扫描，一般情况下是多块读，HINT: FULL(表名/别名)。等待事件为 db file scattered read。如果是并行全表扫描，等待事件为 direct path read。在 Oracle11g 中有个新特征，在对一个大表进行全表扫描的时候，会将表直接读入 PGA，绕过 buffer cache，这个时候全表扫描的等待事件也是 direct path read。一般情况下，我们都会禁用该新特征。等待事件 direct path read 在开启了异步 I/O(disk_asynch_io)的情况下统计是不准确的。关于等待事件，本书不做讨论，那毕竟超出了本书范围。

因为 direct path read 统计不准，所以我们在编写本书的时候禁用了 direct path read。

```
SQL> alter system set "_serial_direct_read"=false;
System altered.
```

全表扫描究竟是怎么扫描数据的呢？回忆一下 Oracle 的逻辑存储结构，Oracle 最小的存储单位是块（block），**物理上连续**的块组成了区（extent），区又组成了段（segment）。对于非分区表，如果表中没有 clob/blob 字段，那么一个表就是一个段。全表扫描，其实就是扫描表中所有格式化过的区。**因为区里面的数据块在物理上是连续的，所以全表扫描可以多块读**。全表扫描不能跨区扫描，因为区与区之间的块物理上不一定是连续的。对于分区表，如果表中没有 clob/blob 字段，一个分区就是一个段，分区表扫描方式与非分区表扫描方式是一样的。

对一个非分区表进行并行扫描，其实就是同时扫描表中多个不同区，因为区与区之间的块物理上不连续，所以我们不需要担心扫描到相同数据块。

对一个分区表进行并行扫描，有两种方式。如果需要扫描多个分区，那么是以分区为粒度进行并行扫描的，这时如果分区数据不均衡，会严重影响并行扫描速度；如果只需要扫描单个分区，这时是以区为粒度进行并行扫描的。

如果表中有 clob 字段，clob 会单独存放在一个段中，当全表扫描需要访问 clob 字段时，这时性能会严重下降，因此尽量避免在 Oracle 中使用 clob。我们可以考虑将 clob 字段拆分为多个 varchar2（4000）字段，或者将 clob 存放在 nosql 数据库中，例如 mongodb。

一般的操作系统，一次 I/O 最多只支持读取或者写入 1MB 数据。数据块为 8KB 的时候，一次 I/O 最多能读取 128 个块。数据块为 16KB 的时候，一次 I/O 最多能读取 64 个块，数据块为 32KB 的时候，一次 I/O 最多能读取 32 个块。

如果表中有部分块已经缓存在 buffer cache 中，在进行全表扫描的时候，扫描到已经被缓存的块所在区时，就会引起 I/O 中断。如果一个表不同的区有大量块缓存在 buffer cache 中，这个时候，全表扫描性能会严重下降，因为有大量的 I/O 中断，导致每次 I/O 不能扫描 1MB 数据。

我们以测试表 test 为例，先查看测试表 test 有多少个区。

```
SQL> select extent_id,blocks, block_id
  2    from dba_extents
  3   where segment_name = 'TEST'
  4     and owner = 'SCOTT';

 EXTENT_ID     BLOCKS   BLOCK_ID
---------- ---------- ----------
         0          8        528
         1          8        536
         2          8        544
         3          8        552
         4          8        560
         5          8        568
         6          8        576
         7          8        584
         8          8        592
         9          8        600
        10          8        608
        11          8        616
        12          8        624
        13          8        632
        14          8        640
        15          8        648
        16        128        768
        17        128        896
        18        128       1024
        19        128       1152
        20        128       1280
        21        128       1408
        22        128       1536
        23        128       1664

24 rows selected.
```

测试表 test 一共有 24 个区，而且每个区都没有超过 128 个块。正常情况下，对测试表 test 进行全表扫描需要进行 24 次多块读。现在我们清空 buffer cache 缓存，对 test 表进行全表扫描，同时使用 10046 事件监控等待事件。

```
SQL> show parameter db_file_multiblock

NAME                                 TYPE        VALUE
------------------------------------ ----------- -----
db_file_multiblock_read_count        integer     128

SQL> alter system flush buffer_cache;
```

```
System altered.

SQL> alter session set events '10046 trace name context forever, level 8';

Session altered.

SQL> select count(*) from test;

  COUNT(*)
----------
     72462

SQL> alter session set events '10046 trace name context off';

Session altered.
```

下面是经过 tkprof 格式化后的 10046 trace 文件的部分数据。

```
Rows     Row Source Operation
-------  ---------------------------------------------------
      1  SORT AGGREGATE (cr=1037 pr=1033 pw=0 time=0 us)
  72462    TABLE ACCESS FULL TEST (cr=1037 pr=1033 pw=0 time=7795 us cost=289 size=0 card=72462)


Elapsed times include waiting on following events:
  Event waited on                             Times   Max. Wait  Total Waited
  ------------------------------------------  Waited  ---------  ------------
  SQL*Net message to client                        2       0.00          0.00
  Disk file operations I/O                         1       0.00          0.00
  db file sequential read                          1       0.00          0.00
  db file scattered read                          24       0.00          0.01
  SQL*Net message from client                      2      11.10         11.10
```

正如我们猜想的那样，全表扫描多块读（db file scattered read）耗费了 24 次。

现在我们利用下面 SQL，查找一些介于第 17 个区和第 24 个区之间的 rowid。

```
select rowid,
       dbms_rowid.rowid_relative_fno(rowid) file#,
       dbms_rowid.rowid_block_number(rowid) block#
  from test;
```

我们可以根据 block_id 为边界来判断 rowid 在哪个区。

现在我们清空 buffer cache，选取 4 个不同区的 rowid 访问表中数据，这样就将 4 个不同区的块缓存在 buffer cache 中了，然后对 test 表进行全表扫描，同时使用 10046 事件监控等待事件。

```
SQL> alter system flush buffer_cache;

System altered.

SQL> select count(*)
  2    from test
  3   where rowid in ('AAASNJAAEAAAAMPAAk', 'AAASNJAAEAAAAQRAAn',
  4                  'AAASNJAAEAAAAQ2AAR', 'AAASNJAAEAAAAUhAAM');

  COUNT(*)
----------
         4

SQL> alter session set events '10046 trace name context forever, level 8';
```

```
Session altered.

SQL> select count(*) from test;

  COUNT(*)
----------
     72462

SQL> alter session set events '10046 trace name context off';

Session altered.
```

下面是经过 tkprof 格式化后的 10046 trace 文件的部分数据。

```
Rows     Row Source Operation
-------  ---------------------------------------------------
      1   SORT AGGREGATE (cr=1037 pr=1029 pw=0 time=0 us)
  72462    TABLE ACCESS FULL TEST (cr=1037 pr=1029 pw=0 time=10479 us cost=289 size=0 card=72462)


Elapsed times include waiting on following events:
  Event waited on                             Times   Max. Wait  Total Waited
  ----------------------------------------   Waited  ----------  ------------
  SQL*Net message to client                       2        0.00          0.00
  db file sequential read                         1        0.00          0.00
  db file scattered read                         28        0.00          0.02
  SQL*Net message from client                     2        3.85          3.85
```

因为缓存了 4 个不同区的块在 buffer cache 中，全表扫描的时候需要中断 4 次 I/O，所以全表扫描多块读一共耗费了 28 次。

如果表正在发生大事务，在进行全表扫描的时候，还会从 undo 读取部分数据。从 undo 读取数据是单块读，这种情况下全表扫描效率非常低下。因此，我们建议使用批量游标的方式处理大事务。使用批量游标处理大事务还可以减少对 undo 的使用，防止事务失败回滚太慢。

以示例表 test 为例，我们先在一个会话中更新表中所有数据，模拟一个大事务。

```
SQL> update test set owner='SCOTT';

72462 rows updated.
```

我们开启另一个会话，清空 buffer cache 缓存并且设置 10046 事件，然后运行查询。

```
SQL> alter system flush buffer_cache;

System altered.

SQL> alter session set events '10046 trace name context forever, level 8';

Session altered.

SQL> select count(*) from test;

  COUNT(*)
----------
     72462

SQL> alter session set events '10046 trace name context off';
```

```
Session altered.
```

下面是经过 tkprof 格式化后的 10046 trace 文件的部分数据。

```
Rows     Row Source Operation
-------  ---------------------------------------------------
      1  SORT AGGREGATE (cr=74531 pr=3380 pw=0 time=0 us)
  72462    TABLE ACCESS FULL TEST (cr=74531 pr=3380 pw=0 time=962057 us cost=289 size=
0 card=72462)


Elapsed times include waiting on following events:
  Event waited on                             Times   Max. Wait  Total Waited
  ----------------------------------------    Waited  ---------  ------------
  SQL*Net message to client                      2      0.00          0.00
  Disk file operations I/O                       1      0.00          0.00
  db file sequential read                     2348      0.00          0.41
  db file scattered read                        24      0.00          0.02
  SQL*Net message from client                    2     11.43         11.43
```

db file sequential read 表示单块读，一共读取了 2 348 次，这里的单块读就是大事务产生的 undo 所引起的。

Oracle 行存储数据库在进行全表扫描时会扫描表中所有的列。关于行存储与列存储本书将在后面章节介绍。

4.1.2 TABLE ACCESS BY USER ROWID

TABLE ACCESS BY USER ROWID 表示直接用 ROWID 获取数据，单块读。

该访问路径在 Oracle 所有的访问路径中性能是最好的。

我们以测试表 test 为例，运行下面 SQL 并且查看执行计划。

```
SQL> select * from test where rowid='AAASNJAAEAAAAJqAA3';

Execution Plan
----------------------------------------------------------
Plan hash value: 1358188196

--------------------------------------------------------------------------------------
| Id  | Operation                    | Name | Rows  | Bytes | Cost (%CPU)| Time     |
--------------------------------------------------------------------------------------
|   0 | SELECT STATEMENT             |      |     1 |    97 |     1   (0)| 00:00:01 |
|   1 |  TABLE ACCESS BY USER ROWID  | TEST |     1 |    97 |     1   (0)| 00:00:01 |
--------------------------------------------------------------------------------------
```

在 where 条件中直接使用 rowid 获取数据就会使用该访问路径。

4.1.3 TABLE ACCESS BY ROWID RANGE

TABLE ACCESS BY ROWID RANGE 表示 ROWID 范围扫描，多块读。因为同一个块里面的 ROWID 是连续的，同一个 EXTENT 里面的 ROWID 也是连续的，所以可以多块读。

我们以测试表 test 为例，运行下面 SQL 并且查看执行计划。

```
SQL> select * from test where rowid>='AAASs5AAEAAB+SLAAA';
```

```
72462 rows selected.
Execution Plan
----------------------------------------------------------
Plan hash value: 3472873366

--------------------------------------------------------------------------------
| Id  | Operation                   | Name | Rows  | Bytes | Cost (%CPU)| Time     |
--------------------------------------------------------------------------------
|   0 | SELECT STATEMENT            |      |  3651 |  345K |   186   (1)| 00:00:03 |
|*  1 |  TABLE ACCESS BY ROWID RANGE| TEST |  3651 |  345K |   186   (1)| 00:00:03 |
--------------------------------------------------------------------------------

Predicate Information (identified by operation id):
---------------------------------------------------

   1 - access(ROWID>='AAASs5AAEAAB+SLAAA')
```

where 条件中直接使用 rowid 进行范围扫描就会使用该执行计划。

4.1.4 TABLE ACCESS BY INDEX ROWID

TABLE ACCESS BY INDEX ROWID 表示回表，单块读。

我们在第 1 章中提到过回表，在此不再赘述。

4.1.5 INDEX UNIQUE SCAN

INDEX UNIQUE SCAN 表示索引唯一扫描，单块读。

对唯一索引或者对主键列进行等值查找，就会走 INDEX UNIQUE SCAN。因为对唯一索引或者对主键列进行等值查找，CBO 能确保最多只返回 1 行数据，所以这时可以走索引唯一扫描。

我们以 scott 账户中 emp 表为例，运行下面 SQL 并且查看执行计划。

```
SQL> select * from emp where empno=7369;
Execution Plan
----------------------------------------------------------
Plan hash value: 2949544139

--------------------------------------------------------------------------------
| Id  | Operation                   | Name   | Rows | Bytes | Cost(%CPU)| Time     |
--------------------------------------------------------------------------------
|   0 | SELECT STATEMENT            |        |    1 |    38 |     1  (0)| 00:00:01 |
|   1 |  TABLE ACCESS BY INDEX ROWID| EMP    |    1 |    38 |     1  (0)| 00:00:01 |
|*  2 |   INDEX UNIQUE SCAN         | PK_EMP |    1 |       |     0  (0)| 00:00:01 |
--------------------------------------------------------------------------------

Predicate Information (identified by operation id):
---------------------------------------------------

   2 - access("EMPNO"=7369)
```

因为 empno 是主键列，对 empno 进行等值访问，就走了 INDEX UNIQUE SCAN。INDEX UNIQUE SCAN 最多只返回一行数据，只会扫描"索引高度"个索引块，在所有的 Oracle 访问路径中，其性能仅次于 TABLE ACCESS BY USER ROWID。

4.1.6 INDEX RANGE SCAN

INDEX RANGE SCAN 表示索引范围扫描，单块读，返回的数据是有序的（默认升序）。HINT: INDEX（表名/别名 索引名）。对唯一索引或者主键进行范围查找，对非唯一索引进行等值查找，范围查找，就会发生 INDEX RANGE SCAN。等待事件为 db file sequential read。

我们以测试表 test 为例，运行下面 SQL 并且查看执行计划。

```
SQL> select * from test where object_id=100;

Execution Plan
----------------------------------------------------------
Plan hash value: 3946039639

--------------------------------------------------------------------------------
| Id  | Operation                   | Name   | Rows | Bytes | Cost (%CPU)| Time     |
--------------------------------------------------------------------------------
|   0 | SELECT STATEMENT            |        |    1 |    97 |     2   (0)| 00:00:01 |
|   1 |  TABLE ACCESS BY INDEX ROWID| TEST   |    1 |    97 |     2   (0)| 00:00:01 |
|*  2 |   INDEX RANGE SCAN          | IDX_ID |    1 |       |     1   (0)| 00:00:01 |
--------------------------------------------------------------------------------

Predicate Information (identified by operation id):
---------------------------------------------------

   2 - access("OBJECT_ID"=100)
```

因为索引 IDX_ID 是非唯一索引，对非唯一索引进行等值查找并不能确保只返回一行数据，有可能返回多行数据，所以执行计划会进行索引范围扫描。

索引范围扫描默认是从索引中最左边的叶子块开始，然后往右边的叶子块扫描（从小到大），当检查到不匹配数据的时候，就停止扫描。 现在我们将过滤条件改为小于，并且对过滤列进行降序排序，查看执行计划。

```
SQL> select * from test where object_id<100 order by object_id desc;

98 rows selected.

Execution Plan
----------------------------------------------------------
Plan hash value: 1069979465

---------------------------------------------------------------------------------------
| Id  | Operation                      | Name   | Rows | Bytes | Cost(%CPU)| Time     |
---------------------------------------------------------------------------------------
|   0 | SELECT STATEMENT               |        |   96 |  9312 |     4  (0)| 00:00:01 |
|   1 |  TABLE ACCESS BY INDEX ROWID   | TEST   |   96 |  9312 |     4  (0)| 00:00:01 |
|*  2 |   INDEX RANGE SCAN DESCENDING  | IDX_ID |   96 |       |     2  (0)| 00:00:01 |
---------------------------------------------------------------------------------------

Predicate Information (identified by operation id):
---------------------------------------------------

   2 - access("OBJECT_ID"<100)
```

```
            filter("OBJECT_ID"<100)
```

INDEX RANGE SCAN DECENDING 表示索引降序范围扫描，从右往左扫描，返回的数据是降序显示的。

假设一个索引叶子块能存储 100 行数据，通过索引返回 100 行以内的数据，只扫描"索引高度"个索引块，如果通过索引返回 200 行数据，就需要扫描两个叶子块。通过索引返回的行数越多，扫描的索引叶子块也就越多，随着扫描的叶子块个数的增加，索引范围扫描的性能开销也就越大。如果索引范围扫描需要回表，同样假设一个索引叶子块能存储 100 行数据，通过索引返回 1000 行数据，只需要扫描 10 个索引叶子块（单块读），但是回表可能会需要访问几十个到几百个表块（单块读）。在检查执行计划的时候我们要注意索引范围扫描返回多少行数据，如果返回少量数据，不会出现性能问题；如果返回大量数据，在没有回表的情况下也还好；如果返回大量数据同时还有回表，这时我们应该考虑通过创建组合索引消除回表或者使用全表扫描来代替它。

4.1.7　INDEX SKIP SCAN

INDEX SKIP SCAN 表示索引跳跃扫描，单块读。返回的数据是有序的（默认升序）。HINT: INDEX_SS（表名/别名 索引名）。当组合索引的引导列（第一个列）没有在 where 条件中，并且组合索引的引导列/前几个列的基数很低，where 过滤条件对组合索引中非引导列进行过滤的时候就会发生索引跳跃扫描，等待事件为 db file sequential read。

我们在测试表 test 上创建如下索引。

```
SQL> create index idx_ownerid on test(owner,object_id);

Index created.
```

然后我们删除 object_id 列上的索引 IDX_ID。

```
SQL> drop index idx_id;

Index dropped.
```

我们执行如下 SQL 并且查看执行计划。

```
SQL> select * from test where object_id<100;

98 rows selected.

Execution Plan
----------------------------------------------------------
Plan hash value: 847134193

--------------------------------------------------------------------------------------
| Id  | Operation                   | Name        | Rows | Bytes | Cost(%CPU)| Time     |
--------------------------------------------------------------------------------------
|   0 | SELECT STATEMENT            |             |   96 |  9312 |   100   (0)| 00:00:02 |
|   1 |  TABLE ACCESS BY INDEX ROWID| TEST        |   96 |  9312 |   100   (0)| 00:00:02 |
|*  2 |   INDEX SKIP SCAN           | IDX_OWNERID |   96 |       |    97   (0)| 00:00:02 |
--------------------------------------------------------------------------------------
```

```
Predicate Information (identified by operation id):
---------------------------------------------------

   2 - access("OBJECT_ID"<100)
       filter("OBJECT_ID"<100)
```

从执行计划中我们可以看到上面 SQL 走了索引跳跃扫描。最理想的情况应该是直接走 where 条件列 object_id 上的索引，并且走 INDEX RANGE SCAN。但是因为 where 条件列上面没有直接创建索引，而是间接地被包含在组合索引中，为了避免全表扫描，CBO 就选择了索引跳跃扫描。

INDEX SKIP SCAN 中有个 SKIP 关键字，也就是说它是跳着扫描的。那么想要跳跃扫描，必须是组合索引，如果是单列索引怎么跳？另外，组合索引的引导列不能出现在 where 条件中，如果引导列出现在 where 条件中，它为什么还跳跃扫描呢，直接 INDEX RANGE SCAN 不就可以了？再有，要引导列基数很低，如果引导列基数很高，那么它"跳"的次数就多了，性能就差了。

当执行计划中出现了 INDEX SKIP SCAN，我们可以直接在过滤列上面建立索引，使用 INDEX RANGE SCAN 代替 INDEX SKIP SCAN。

4.1.8 INDEX FULL SCAN

INDEX FULL SCAN 表示索引全扫描，单块读，返回的数据是有序的（默认升序）。HINT：INDEX（表名/别名 索引名）。索引全扫描会扫描索引中所有的叶子块（从左往右扫描），如果索引很大，会产生严重性能问题（因为是单块读）。等待事件为 db file sequential read。

它通常发生在下面 3 种情况。

- ➢ 分页语句，分页语句在本书第 8 章中会详细介绍，这里不做赘述。
- ➢ SQL 语句有 order by 选项，order by 的列都包含在索引中，并且 order by 后列顺序必须和索引列顺序一致。order by 的第一个列不能有过滤条件，如果有过滤条件就会走索引范围扫描（INDEX RANGE SCAN）。同时表的数据量不能太大（数据量太大会走 TABLE ACCESS FULL + SORT ORDER BY）。我们有如下 SQL。

```
select * from test order by object_id,owner;
```

我们创建如下索引（索引顺序必须与排序顺序一致，加 0 是为了让索引能存 NULL）。

```
SQL> create index idx_idowner on test(object_id,owner,0);

Index created.
```

我们执行如下 SQL 并且查看执行计划。

```
SQL> select * from test order by object_id,owner;

72462 rows selected.

Execution Plan
----------------------------------------------------------
Plan hash value: 3870803568
```

```
---------------------------------------------------------------------------------
| Id |Operation                    | Name         | Rows  | Bytes |Cost(%CPU)| Time     |
---------------------------------------------------------------------------------
|  0 |SELECT STATEMENT             |              | 73020 | 6916K|1338   (1)| 00:00:17 |
|  1 | TABLE ACCESS BY INDEX ROWID | TEST         | 73020 | 6916K|1338   (1)| 00:00:17 |
|  2 |  INDEX FULL SCAN            | IDX_IDOWNER  | 73020 |      | 242   (1)| 00:00:03 |
---------------------------------------------------------------------------------
```

> 在进行 SORT MERGE JOIN 的时候，如果表数据量比较小，让连接列走 INDEX FULL SCAN 可以避免排序。例子如下。

```
SQL> select /*+ use_merge(e,d) */
  2     *
  3    from emp e, dept d
  4   where e.deptno = d.deptno;

14 rows selected.

Execution Plan
----------------------------------------------------------
Plan hash value: 844388907

---------------------------------------------------------------------------------
| Id |Operation                    | Name    | Rows | Bytes | Cost(%CPU)| Time     |
---------------------------------------------------------------------------------
|  0 |SELECT STATEMENT             |         |   14 |   812 |    6 (17)| 00:00:01 |
|  1 | MERGE JOIN                  |         |   14 |   812 |    6 (17)| 00:00:01 |
|  2 |  TABLE ACCESS BY INDEX ROWID| DEPT    |    4 |    80 |    2  (0)| 00:00:01 |
|  3 |   INDEX FULL SCAN           | PK_DEPT |    4 |       |    1  (0)| 00:00:01 |
|* 4 |  SORT JOIN                  |         |   14 |   532 |    4 (25)| 00:00:01 |
|  5 |   TABLE ACCESS FULL         | EMP     |   14 |   532 |    3  (0)| 00:00:01 |
---------------------------------------------------------------------------------

Predicate Information (identified by operation id):
---------------------------------------------------

   4 - access("E"."DEPTNO"="D"."DEPTNO")
       filter("E"."DEPTNO"="D"."DEPTNO")
```

当看到执行计划中有 INDEX FULL SCAN，我们首先要检查 INDEX FULL SCAN 是否有回表。

如果 INDEX FULL SCAN 没有回表，我们要检查索引段大小，如果索引段太大（GB 级别），应该使用 INDEX FAST FULL SCAN 代替 INDEX FULL SCAN，因为 INDEX FAST FULL SCAN 是多块读，INDEX FULL SCAN 是单块读，即使使用了 INDEX FAST FULL SCAN 会产生额外的排序操作，也要用 INDEX FAST FULL SCAN 代替 INDEX FULL SCAN。

如果 INDEX FULL SCAN 有回表，大多数情况下，这种执行计划是错误的，因为 INDEX FULL SCAN 是单块读，回表也是单块读。这时应该走全表扫描，因为全表扫描是多块读。如果分页语句走了 INDEX FULL SCAN 然后回表，这时应该没有太大问题，具体原因请大家阅读本书 8.3 节。

4.1.9 INDEX FAST FULL SCAN

INDEX FAST FULL SCAN 表示索引快速全扫描，多块读。HINT：INDEX_FFS（表名/别名 索引名）。当需要从表中查询出大量数据但是只需要获取表中部分列的数据的，我们可以利用索引快速全扫描代替全表扫描来提升性能。索引快速全扫描的扫描方式与全表扫描的扫描方式是一样，都是按区扫描，所以它可以多块读，而且可以并行扫描。等待事件为 db file scattered read，如果是并行扫描，等待事件为 direct path read。

现有如下 SQL。

```
select owner,object_name from test;
```

该 SQL 没有过滤条件，默认情况下会走全表扫描。但是因为 Oracle 是行存储数据库，全表扫描的时候会扫描表中所有的列，而上面查询只访问表中两个列，全表扫描会多扫描额外 13 个列，所以我们可以创建一个组合索引，使用索引快速全扫描代替全表扫描。

```
SQL> create index idx_ownername on test(owner,object_name,0);

Index created.
```

我们查看 SQL 执行计划。

```
SQL> select owner,object_name from test;

72462 rows selected.

Execution Plan
----------------------------------------------------------
Plan hash value: 3888663772

--------------------------------------------------------------------------------
| Id  | Operation             | Name          | Rows  | Bytes | Cost(%CPU)| Time     |
--------------------------------------------------------------------------------
|   0 | SELECT STATEMENT      |               | 73020 | 2210K|    79   (2)| 00:00:01 |
|   1 |  INDEX FAST FULL SCAN | IDX_OWNERNAME | 73020 | 2210K|    79   (2)| 00:00:01 |
--------------------------------------------------------------------------------
```

现有如下 SQL。

```
select object_name from test where object_id<100;
```

该 SQL 有过滤条件，根据过滤条件 where object_id<100 过滤数据之后只返回少量数据，一般情况下我们直接在 object_id 列创建索引，让该 SQL 走 object_id 列的索引即可。

```
SQL> create index idx_id on test(object_id);

Index created.

SQL> select object_name from test where object_id<100;

98 rows selected.

Execution Plan
----------------------------------------------------------
Plan hash value: 3946039639
```

第4章 访问路径（ACCESS PATH）

```
--------------------------------------------------------------------------------
| Id  | Operation                    | Name   | Rows | Bytes | Cost(%CPU)| Time     |
--------------------------------------------------------------------------------
|   0 | SELECT STATEMENT             |        |   96 |  2880 |     4  (0)| 00:00:01 |
|   1 |  TABLE ACCESS BY INDEX ROWID | TEST   |   96 |  2880 |     4  (0)| 00:00:01 |
|*  2 |   INDEX RANGE SCAN           | IDX_ID |   96 |       |     2  (0)| 00:00:01 |
--------------------------------------------------------------------------------

Predicate Information (identified by operation id):
---------------------------------------------------

   2 - access("OBJECT_ID"<100)


Statistics
----------------------------------------------------------
          0  recursive calls
          0  db block gets
         18  consistent gets
          0  physical reads
          0  redo size
       2217  bytes sent via SQL*Net to client
        485  bytes received via SQL*Net from client
          8  SQL*Net roundtrips to/from client
          0  sorts (memory)
          0  sorts (disk)
         98  rows processed
```

因为该 SQL 只查询一个字段，所以我们可以将 select 列放到组合索引中，避免回表。

```
SQL> create index idx_idname on test(object_id,object_name);

Index created.
```

我们再次查看 SQL 的执行计划。

```
SQL> select object_name from test where object_id<100;

98 rows selected.


Execution Plan
----------------------------------------------------------
Plan hash value: 3678957952

--------------------------------------------------------------------------------
| Id  | Operation         | Name       | Rows | Bytes | Cost (%CPU)| Time     |
--------------------------------------------------------------------------------
|   0 | SELECT STATEMENT  |            |   96 |  2880 |     2   (0)| 00:00:01 |
|*  1 |  INDEX RANGE SCAN | IDX_IDNAME |   96 |  2880 |     2   (0)| 00:00:01 |
--------------------------------------------------------------------------------

Predicate Information (identified by operation id):
---------------------------------------------------

   1 - access("OBJECT_ID"<100)


Statistics
----------------------------------------------------------
          0  recursive calls
```

```
       0  db block gets
       9  consistent gets
       0  physical reads
       0  redo size
    2217  bytes sent via SQL*Net to client
     485  bytes received via SQL*Net from client
       8  SQL*Net roundtrips to/from client
       0  sorts (memory)
       0  sorts (disk)
      98  rows processed
```

现有如下 SQL。

```
select object_name from test where object_id>100;
```

以上 SQL 过滤条件是 where object_id>100，返回大量数据，应该走全表扫描，但是因为 SQL 只访问一个字段，所以我们可以走索引快速全扫描来代替全表扫描。

```
SQL> select object_name from test where object_id>100;

72363 rows selected.

Execution Plan
----------------------------------------------------------
Plan hash value: 252646278

--------------------------------------------------------------------------------
| Id  | Operation             | Name       | Rows  | Bytes | Cost (%CPU)| Time     |
--------------------------------------------------------------------------------
|   0 | SELECT STATEMENT      |            | 72924 | 2136K |    73   (2)| 00:00:01 |
|*  1 |  INDEX FAST FULL SCAN | IDX_IDNAME | 72924 | 2136K |    73   (2)| 00:00:01 |
--------------------------------------------------------------------------------

Predicate Information (identified by operation id):
---------------------------------------------------

   1 - filter("OBJECT_ID">100)
```

大家可能会有疑问，以上 SQL 能否走 INDEX RANGE SCAN 呢？INDEX RANGE SCAN 是单块读，SQL 会返回表中大量数据，"几乎"会扫描索引中所有的叶子块。INDEX FAST FULL SCAN 是多块读，会扫描索引中所有的块（根块、所有的分支块、所有的叶子块）。虽然 INDEX RANGE SCAN 与 INDEX FAST FULL SCAN 相比扫描的块少（逻辑读少），但是 INDEX RANGE SCAN 是单块读，耗费的 I/O 次数比 INDEX FAST FULL SCAN 的 I/O 次数多，所以 INDEX FAST FULL SCAN 性能更好。

在做 SQL 优化的时候，我们不要只看逻辑读来判断一个 SQL 性能的好坏，物理 I/O 次数比逻辑读更为重要。有时候逻辑读高的执行计划性能反而比逻辑读低的执行计划性能更好，因为逻辑读高的执行计划物理 I/O 次数比逻辑读低的执行计划物理 I/O 次数低。

在 Oracle 数据库中，INDEX FAST FULL SCAN 是用来代替 TABLE ACCESS FULL 的。因为 Oracle 是行存储数据库，TABLE ACCESS FULL 会扫描表中所有的列，而 INDEX FAST FULL SCAN 只需要扫描表中部分列，INDEX FAST FULL SCAN 就是由 Oracle 是行存储这个"缺陷"而产生的。

如果数据库是 Exadata，INDEX FAST FULL SCAN 几乎没有用武之地，因为 Exadata 是行列混合存储，在全表扫描的时候可以只扫描需要的列（Smart Scan），没必要使用 INDEX FAST FULL SCAN 来代替全表扫描。如果我们在 Exadata 中强行使用 INDEX FAST FUL SCAN 来代替全表扫描，反而会降低数据库性能，因为没办法使用 Exadata 中的 Smart Scan。

如果我们启用了 12c 中的新特性 IN MEMORY OPTION，INDEX FAST FULL SCAN 几乎也没有用武之地了，因为表中的数据可以以列的形式存放在内存中，这时直接访问内存中的数据即可。

4.1.10　INDEX FULL SCAN（MIN/MAX）

INDEX FULL SCAN（MIN/MAX）表示索引最小/最大值扫描、单块读，该访问路径发生在 SELECT MAX（COLUMN）FROM TABLE 或者 SELECT MIN（COLUMN）FROM TABLE 等 SQL 语句中。

INDEX FULL SCAN（MIN/MAX）只会访问"索引高度"个索引块，其性能与 INDEX UNIQUE SCAN 一样，仅次于 TABLE ACCESS BY USER ROWID。

现有如下 SQL。

```
select max(object_id) from t;
```

该 SQL 查询 object_id 的最大值，如果 object_id 列有索引，索引默认是升序排序的，这时我们只需要扫描索引中"最右边"的叶子块就能得到 object_id 的最大值。现在我们查看该 SQL 的执行计划。

```
SQL> select max(object_id) from t;

Elapsed: 00:00:00.00

Execution Plan
----------------------------------------------------------
Plan hash value: 2448092560

--------------------------------------------------------------------------------
| Id  | Operation                  | Name    | Rows  | Bytes | Cost(%CPU)| Time     |
--------------------------------------------------------------------------------
|   0 | SELECT STATEMENT           |         |     1 |    13 |   186   (1)| 00:00:03 |
|   1 |  SORT AGGREGATE            |         |     1 |    13 |            |          |
|   2 |   INDEX FULL SCAN (MIN/MAX)| IDX_T_ID| 67907 |  862K |            |          |
--------------------------------------------------------------------------------

Note
-----
   - dynamic sampling used for this statement (level=2)


Statistics
----------------------------------------------------------
          0  recursive calls
          0  db block gets
          2  consistent gets
          0  physical reads
          0  redo size
```

```
    430  bytes sent via SQL*Net to client
    419  bytes received via SQL*Net from client
      2  SQL*Net roundtrips to/from client
      0  sorts (memory)
      0  sorts (disk)
      1  rows processed
```

现有另外一个 SQL。

```
select max(object_id),min(object_id) from t;
```

该 SQL 要同时查看 object_id 的最大值和最小值，如果想直接从 object_id 列的索引获取数据，我们只需要扫描索引中"最左边"和"最右边"的叶子块就可以。在 Btree 索引中，索引叶子块是双向指向的，如果要一次性获取索引中"最左边"和"最右边"的叶子块，我们就需要连带的扫描"最大值"与"最小值"中间的叶子块，而本案例中，中间叶子块的数据并不是我们需要的。如果该 SQL 走索引，会走 INDEX FAST FULL SCAN，而不会走 INDEX FULL SCAN，因为 INDEX FAST FULL SCAN 可以多块读，而 INDEX FULL SCAN 是单块读，两者性能差距巨大（如果索引已经缓存在 buffer cache 中，走 INDEX FULL SCAN 与 INDEX FAST FULL SCAN 效率几乎一样，因为不需要物理 I/O）。需要注意的是，该 SQL 没有排除 object_id 为 NULL，如果直接运行该 SQL，不会走索引。

```
SQL> select max(object_id),min(object_id) from t;

Elapsed: 00:00:00.02

Execution Plan
----------------------------------------------------------
Plan hash value: 2966233522

--------------------------------------------------------------------------
| Id  | Operation          | Name | Rows  | Bytes | Cost (%CPU)| Time     |
--------------------------------------------------------------------------
|   0 | SELECT STATEMENT   |      |     1 |    13 |   186   (1)| 00:00:03 |
|   1 |  SORT AGGREGATE    |      |     1 |    13 |            |          |
|   2 |   TABLE ACCESS FULL| T    | 67907 |   862K|   186   (1)| 00:00:03 |
--------------------------------------------------------------------------
```

我们排除 object_id 为 NULL，查看执行计划。

```
SQL> select max(object_id),min(object_id) from t where object_id is not null;

Elapsed: 00:00:00.01

Execution Plan
----------------------------------------------------------
Plan hash value: 3570898368

-----------------------------------------------------------------------------------
| Id  | Operation             | Name    | Rows  | Bytes | Cost (%CPU)| Time     |
-----------------------------------------------------------------------------------
|   0 | SELECT STATEMENT      |         |     1 |    13 |    33   (4)| 00:00:01 |
|   1 |  SORT AGGREGATE       |         |     1 |    13 |            |          |
|*  2 |   INDEX FAST FULL SCAN| IDX_T_ID| 67907 |   862K|    33   (4)| 00:00:01 |
-----------------------------------------------------------------------------------

Predicate Information (identified by operation id):
```

```
            2 - filter("OBJECT_ID" IS NOT NULL)
Note
-----
   - dynamic sampling used for this statement (level=2)


Statistics
----------------------------------------------------------
          0  recursive calls
          0  db block gets
        169  consistent gets
          0  physical reads
          0  redo size
        501  bytes sent via SQL*Net to client
        419  bytes received via SQL*Net from client
          2  SQL*Net roundtrips to/from client
          0  sorts (memory)
          0  sorts (disk)
          1  rows processed
```

从上面的执行计划中我们可以看到 SQL 走了 INDEX FAST FULL SCAN，INDEX FAST FULL SCAN 会扫描索引段中所有的块，理想的情况是只扫描索引中"最左边"和"最右边"的叶子块。现在我们将该 SQL 改写为如下 SQL。

```
select (select max(object_id) from t),(select min(object_id) from t) from dual;
```

我们查看后的执行计划。

```
SQL> select (select max(object_id) from t),(select min(object_id) from t) from dual;

Elapsed: 00:00:00.01

Execution Plan
----------------------------------------------------------
Plan hash value: 3622839313

-------------------------------------------------------------------------------------
| Id | Operation                   | Name    | Rows  | Bytes | Cost(%CPU)| Time     |
-------------------------------------------------------------------------------------
|  0 | SELECT STATEMENT            |         |     1 |       |     2  (0)| 00:00:01 |
|  1 |  SORT AGGREGATE             |         |     1 |    13 |           |          |
|  2 |   INDEX FULL SCAN (MIN/MAX) | IDX_T_ID| 67907 |  862K |           |          |
|  3 |  SORT AGGREGATE             |         |     1 |    13 |           |          |
|  4 |   INDEX FULL SCAN (MIN/MAX) | IDX_T_ID| 67907 |  862K |           |          |
|  5 |  FAST DUAL                  |         |     1 |       |     2  (0)| 00:00:01 |
-------------------------------------------------------------------------------------

Note
-----
   - dynamic sampling used for this statement (level=2)


Statistics
----------------------------------------------------------
          0  recursive calls
          0  db block gets
          4  consistent gets
```

```
        0  physical reads
        0  redo size
      527  bytes sent via SQL*Net to client
      419  bytes received via SQL*Net from client
        2  SQL*Net roundtrips to/from client
        0  sorts (memory)
        0  sorts (disk)
        1  rows processed
```

原始 SQL 因为需要 1 次性从索引中取得最大值和最小值，所以导致走了 INDEX FAST FULL SCAN。我们将该 SQL 进行等价改写之后，访问了索引两次，一次取最大值，一次取最小值，从而避免扫描不需要的索引叶子块，大大提升了查询性能。

4.1.11　MAT_VIEW REWRITE ACCESS FULL

MAT_VIEW REWRITE ACCESS FULL 表示物化视图全表扫描、多块读。因为物化视图本质上也是一个表，所以其扫描方式与全表扫描方式一样。如果我们开启了查询重写功能，而且 SQL 查询能够直接从物化视图中获得结果，就会走该访问路径。

现在我们创建一个物化视图 TEST_MV。

```
SQL> create materialized view test_mv
  2      build immediate enable query rewrite
  3      as select object_id,object_name from test;

Materialized view created.
```

有如下 SQL 查询。

```
select object_id,object_name from test;
```

因为物化视图 TEST_MV 已经包含查询需要的字段，所以该 SQL 会直接访问物化视图 TEST_MV。

```
SQL> select object_id,object_name from test;

72462 rows selected.

Execution Plan
----------------------------------------------------------
Plan hash value: 1627509066

--------------------------------------------------------------------------------
| Id | Operation                     | Name    | Rows  | Bytes | Cost(%CPU)| Time     |
--------------------------------------------------------------------------------
|  0 | SELECT STATEMENT              |         | 67036 | 5171K |    65   (2)| 00:00:01 |
|  1 |  MAT_VIEW REWRITE ACCESS FULL | TEST_MV | 67036 | 5171K |    65   (2)| 00:00:01 |
--------------------------------------------------------------------------------
```

4.2　单块读与多块读

单块读与多块读这两个概念对于掌握 SQL 优化非常重要，更准确地说是单块读的物理 I/O 次数和多块读的物理 I/O 次数对于掌握 SQL 优化非常重要。

从磁盘 1 次读取 1 个块到 buffer cache 就叫单块读，从磁盘 1 次读取多个块到 buffer cache 就叫多块读。如果数据块都已经缓存在 buffer cache 中，那就不需要物理 I/O 了，没有物理 I/O 也就不存在单块读与多块读。

绝大多数的平台，一次 I/O 最多只能读取或者写入 1MB 数据，Oracle 的块大小默认是 8k，那么一次 I/O 最多只能写入 128 个块到磁盘，最多只能读取 128 个块到 buffer cache。**在判断哪个访问路径性能好的时候，通常是估算每个访问路径的 I/O 次数，谁的 I/O 次数少，谁的性能就好。在估算 I/O 次数的时候，我们只需要算个大概就可以了，没必要很精确。**

4.3 为什么有时候索引扫描比全表扫描更慢

假设一个表有 100 万行数据，表的段大小为 1GB。如果对表进行全表扫描，最理想的情况下，每次 I/O 都读取 1MB 数据（128 个块），将 1GB 的表从磁盘读入 buffer cache 需要 1 024 次 I/O。在实际情况中，表的段前 16 个 extent，每个 extent 都只有 8 个块，每次 I/O 只能读取 8 个块，而不是 128 个块，表中有部分块会被缓存在 buffer cache 中，会引起 I/O 中断，那么将 1GB 的表从磁盘读入 buffer cache 可能需要耗费 1 500 次物理 I/O。

从表中查询 5 万行数据，走索引。假设一个索引叶子块能存储 100 行数据，那么 5 万行数据需要扫描 500 个叶子块（单块读），也就是需要 500 次物理 I/O，然后有 5 万条数据需要回表，假设索引的集群因子很小（接近表的块数），假设每个数据块存储 50 行数据，那么回表需要耗费 1 000 次物理 I/O（单块读），也就是说从表中查询 5 万行数据，如果走索引，一共需要耗费大概 1 500 次物理 I/O。如果索引的集群因子较大（接近表的总行数），那么回表要耗费更多的物理 I/O，可能是 3 000 次，而不是 1 000 次。

根据上述理论我们知道，走索引返回的数据越多，需要耗费的 I/O 次数也就越多，因此，返回大量数据应该走全表扫描或者是 INDEX FAST FULL SCAN，返回少量数据才走索引扫描。根据上述理论，我们一般建议返回表中总行数 5%以内的数据，走索引扫描，超过 5%走全表扫描。请注意，5%只是一个参考值，适用于绝大多数场景，如有特殊情况，具体问题具体分析。

4.4 DML 对于索引维护的影响

本节主要讨论 DML 对于索引维护的影响。

在 OLTP 高并发 INSERT 环境中，递增列（时间，使用序列的主键列）的索引很容易引起索引热点块争用。递增列的索引会不断地往索引"最右边"的叶子块插入最新数据（因为索引默认升序排序），在高并发 INSERT 的时候，一次只能由一个 SESSION 进行 INSERT，其余 SESSION 会处于等待状态，这样就引起了索引热点块争用。对于递增的主键列索引，我们可以对这个索引进行反转（reverse），这样在高并发 INSERT 的时候，就不会同时插入索引"最右边"的叶子块，而是会均衡地插入到各个不同的索引叶子块中，这样就解决了主键列索引的热点块问题。将索引进行反转之后，索引的集群因子会变得很大（基本上接近于表的总行数），此时索引范围扫描回表会有严重的性能问题。但是一般情况下，主键列都是等值访问，索引走的是索引唯一扫描（INDEX UNIQUE SCAN），不受集群因子的影响，所以对主键列索引进行反转没有任何问题。对于递增的时间列索引，我们不能对这个索引进行反转，因为经常会对时间字段进行范围查找，对时间字段的索引反转之后，索引的集群因子会变得很大，严重影响回表性能。遇到这种情况，我们应该考虑对表根据时间进行范围分区，利用分区裁剪来提升查询性能而不是在时间字段建立索引来提升性能。

在 OLTP 高并发 INSERT 环境中，非递增列索引（比如电话号码）一般不会引起索引热点块争用。非递增列的数据都是随机的（电话号码），在高并发 INSERT 的时候，会随机地插入到索引的各个叶子块中，因此非递增列索引不会引起索引热点块问题，但是如果索引太多会严重影响高并发 INSERT 的性能。

当只有 1 个会话进行 INSERT 时，表中会有 1 个块发生变化，有多少个索引，就会有多少个索引叶子块发生变化（不考虑索引分裂的情况），假设有 10 个索引，那么就有 10 个索引叶子块发生变化。如果有 10 个会话同时进行 INSERT，这时表中最多有 10 个块会发生变化，索引中最多有 100 个块会发生变化（10 个 SESSION 与 10 个索引相乘）。在高并发的 INSERT 环境中，表中的索引越多，INSERT 速度越慢。对于高并发 INSERT，我们一般是采用分库分表、读写分离和消息队列等技术来解决。

在 OLAP 环境中，没有高并发 INSERT 的情况，一般是单进程做批量 INSERT。单进程做批量 INSERT，可以在递增列上建立索引。因为是单进程，没有并发，不会有索引热点块争用，数据也是一直插入的索引中"最右边"的叶子块，所以递增列索引对批量 INSERT 影响不会太大。单进程做批量 INSERT，不能在非递增列建立索引。因为批量 INSERT 几乎会更新索引中所有的叶子块，所以非递增列索引对批量 INSERT 影响很大。在 OLAP 环境中，事实（FACT）表没有主键，时间列一般也是分区字段，所以递增列上面一般是没有索引的，而电话号码等非递增列往往需要索引，为了提高批量 INSERT 的效率，我们可以在 INSERT 之前先禁止索引，等 INSERT 完成之后再重建索引。

第 5 章 表连接方式

本章是本书核心章节，希望大家反复阅读本章内容，直到全部掌握为止。

表（结果集）与表（结果集）之间的连接方式非常重要，如果 CBO 选择了错误的连接方式，本来几秒就能出结果的 SQL 可能执行一天都执行不完。如果想要快速定位超大型 SQL 性能问题，我们就必须深入理解表连接方式。**在多表关联的时候，一般情况下只能是两个表先关联，两表关联之后的结果再和其他表/结果集关联**，如果执行计划中出现了 Filter，这时可以一次性关联多个表。

5.1 嵌套循环（NESTED LOOPS）

嵌套循环的算法：**驱动表返回一行数据，通过连接列传值给被驱动表，驱动表返回多少行，被驱动表就要被扫描多少次**。

嵌套循环可以快速返回两表关联的前几条数据，如果 SQL 中添加了 HINT:FIRST_ROWS，在两表关联的时候，优化器更倾向于嵌套循环。

嵌套循环驱动表应该返回少量数据。如果驱动表返回了 100 万行，那么被驱动表就会被扫描 100 万次。这个时候 SQL 会执行很久，被驱动表会被误认为热点表，被驱动表连接列的索引也会被误认为热点索引。

嵌套循环被驱动表必须走索引。如果嵌套循环被驱动表的连接列没包含在索引中，那么被驱动表就只能走全表扫描，而且是反复多次全表扫描。当被驱动表很大的时候，SQL 就执行不出结果。

嵌套循环被驱动表走索引只能走 INDEX UNIQUE SCAN 或者 INDEX RANGE SCAN。

嵌套循环被驱动表不能走 TABLE ACCESS FULL，不能走 INDEX FULL SCAN，不能走 INDEX SKIP SCAN，也不能走 INDEX FAST FULL SCAN。

嵌套循环被驱动表的连接列基数应该很高。如果被驱动表连接列的基数很低，那么被驱动表就不应该走索引，这样一来被驱动表就只能进行全表扫描了，但是被驱动表也不能走全表扫描。

两表关联返回少量数据才能走嵌套循环。前面提到，嵌套循环被驱动表必须走索引，如果两表关联，返回 100 万行数据，那么被驱动表走索引就会产生 100 万次回表。回表一般是单块读，这个时候 SQL 性能极差，所以两表关联返回少量数据才能走嵌套循环。

我们在测试账号 scott 中运行如下 SQL。

```
SQL> select /*+ gather_plan_statistics use_nl(e,d) leading(e) */
  2    e.ename, e.job, d.dname
  3    from emp e, dept d
```

```
    4  where e.deptno = d.deptno;
```

......省略输出结果......

我们运行下面命令获取带有 A-TIME 的执行计划。

```
SQL> select * from table(dbms_xplan.display_cursor(null,null,'ALLSTATS LAST'));

PLAN_TABLE_OUTPUT
-------------------------------------------------------------------------------------
SQL_ID  g374au8y24mw5, child number 0
-------------------------------------
select /*+ gather_plan_statistics use_nl(e,d) leading(e) */  e.ename,
e.job, d.dname   from emp e, dept d  where e.deptno = d.deptno

Plan hash value: 3625962092

-------------------------------------------------------------------------------------
| Id | Operation                      |Name    |Starts|E-Rows|A-Rows|   A-Time   |Buffers|
-------------------------------------------------------------------------------------
|  0 | SELECT STATEMENT               |        |   1|      |   14|00:00:00.01|   26 |
|  1 |  NESTED LOOPS                  |        |   1|      |   14|00:00:00.01|   26 |
|  2 |   NESTED LOOPS                 |        |   1|   15|   14|00:00:00.01|   12 |
|  3 |    TABLE ACCESS FULL           |EMP     |   1|   15|   14|00:00:00.01|    8 |
|* 4 |    INDEX UNIQUE SCAN           |PK_DEPT |  14|    1|   14|00:00:00.01|    4 |
|  5 |   TABLE ACCESS BY INDEX ROWID  |DEPT    |  14|    1|   14|00:00:00.01|   14 |
-------------------------------------------------------------------------------------

Predicate Information (identified by operation id):
---------------------------------------------------

4 - access("E"."DEPTNO"="D"."DEPTNO")
```

在执行计划中，离 NESTED LOOPS 关键字最近的表就是驱动表。这里 EMP 就是驱动表，DEPT 就是被驱动表。

驱动表 EMP 扫描了一次（Id=3，Starts=1），返回了 14 行数据（Id=3，A-Row），传值 14 次给被驱动表（Id=4），被驱动表扫描了 14 次（Id=4，Id=5，Starts=14）。

下面是嵌套循环的 PLSQL 代码实现。

```
declare
  cursor cur_emp is
    select ename, job, deptno from emp;
  v_dname dept.dname%type;
begin
  for x in cur_emp loop
    select dname into v_dname from dept where deptno = x.deptno;
    dbms_output.put_line(x.ename || ' ' || x.job || ' ' || v_dname);
  end loop;
end;
```

游标 cur_emp 就相当于驱动表 EMP，扫描了一次，一共返回了 14 条记录。该游标循环了 14 次，每次循环的时候传值给 dept，dept 被扫描了 14 次。

为什么嵌套循环被驱动表的连接列要创建索引呢？我们注意观察加粗部分的 PLSQL 代码。

```
declare
  cursor cur_emp is
    select ename, job, deptno from emp;
```

```
    v_dname dept.dname%type;
begin
  for x in cur_emp loop
    select dname into v_dname from dept where deptno = x.deptno;
    dbms_output.put_line(x.ename || ' ' || x.job || ' ' || v_dname);
  end loop;
end;
```

因为扫描被驱动表 dept 次数为 14 次，每次需要通过 deptno 列传值，所以嵌套循环被驱动表的连接列需要创建索引。

虽然本书不讲 PLSQL 优化，但是笔者见过太多的 PLSQL 垃圾代码，因此，提醒大家，在编写 PLSQL 的时候，尽量避免游标循环里面套用 SQL，因为那是纯天然的嵌套循环。假如游标返回 100 万行数据，游标里面的 SQL 会被执行 100 万次。同样的道理，游标里面尽量不要再套游标，如果外层游标循环 1 万次，内层游标循环 1 万次，那么最里面的 SQL 将被执行一亿次。

当两表使用外连接进行关联，如果执行计划是走嵌套循环，那么这时无法更改驱动表，驱动表会被固定为主表，例如下面 SQL。

```
SQL> explain plan for select /*+ use_nl(d,e) leading(e)   */
  2   *
  3    from dept d
  4    left join emp e on d.deptno = e.deptno;

Explained.

SQL> select * from table(dbms_xplan.display);

PLAN_TABLE_OUTPUT
---------------------------------------------------------------------------
Plan hash value: 2022884187

---------------------------------------------------------------------------
| Id  | Operation           | Name | Rows  | Bytes | Cost (%CPU)| Time     |
---------------------------------------------------------------------------
|   0 | SELECT STATEMENT    |      |    14 |   812 |     8   (0)| 00:00:01 |
|   1 |  NESTED LOOPS OUTER |      |    14 |   812 |     8   (0)| 00:00:01 |
|   2 |   TABLE ACCESS FULL | DEPT |     4 |    80 |     3   (0)| 00:00:01 |
|*  3 |   TABLE ACCESS FULL | EMP  |     4 |   152 |     1   (0)| 00:00:01 |
---------------------------------------------------------------------------

   3 - filter("D"."DEPTNO"="E"."DEPTNO"(+))

15 rows selected.
```

use_nl(d,e)表示让两表走嵌套循环，在书写 HINT 的时候，如果表有别名，HINT 中一定要使用别名，否则 HINT 不生效；如果表没有别名，HINT 中就直接使用表名。

leading(e)表示让 EMP 表作为驱动表。

从执行计划中我们可以看到，DEPT 与 EMP 是采用嵌套循环进行连接的，这说明 use_nl(d,e) 生效了。执行计划中驱动表为 DEPT，虽然设置了 leading(e)，但是没有生效。

为什么 leading(e) 没有生效呢？因为 DEPT 与 EMP 是外连接，DEPT 是主表，EMP 是从表，外连接走嵌套循环的时候驱动表只能是主表。

5.1 嵌套循环（NESTED LOOPS）

为什么两表关联是外连接的时候，走嵌套循环无法更改驱动表呢？因为嵌套循环需要传值，主表传值给从表之后，如果发现从表没有关联上，直接显示为 NULL 即可；但是如果是从表传值给主表，没关联上的数据不能传值给主表，不可能传 NULL 给主表，所以两表关联是外连接的时候，走嵌套循环驱动表只能固定为主表。

需要注意的是，如果外连接中从表有过滤条件，那么此时外连接会变为内连接，例如下面 SQL。

```
SQL> select /*+ leading(e) use_nl(d,e) */ *
  2    from dept d
  3    left join emp e on d.deptno = e.deptno
  4   where e.sal < 3000;

11 rows selected.

Execution Plan
----------------------------------------------------------
Plan hash value: 351108634

--------------------------------------------------------------------------------
| Id |Operation                      | Name    | Rows | Bytes | Cost(%CPU)| Time     |
--------------------------------------------------------------------------------
|  0 |SELECT STATEMENT               |         |  12  |  696  |  15   (0)| 00:00:01 |
|  1 | NESTED LOOPS                  |         |  12  |  696  |  15   (0)| 00:00:01 |
|* 2 |  TABLE ACCESS FULL            | EMP     |  12  |  456  |   3   (0)| 00:00:01 |
|  3 |  TABLE ACCESS BY INDEX ROWID  | DEPT    |   1  |   20  |   1   (0)| 00:00:01 |
|* 4 |   INDEX UNIQUE SCAN           | PK_DEPT |   1  |       |   0   (0)| 00:00:01 |
--------------------------------------------------------------------------------

Predicate Information (identified by operation id):
---------------------------------------------------

   2 - filter("E"."SAL"<3000)
   4 - access("D"."DEPTNO"="E"."DEPTNO")
```

HINT 指定了让从表 EMP 作为嵌套循环驱动表，从执行计划中我们看到，EMP 确实是作为嵌套循环的驱动表，而且执行计划中没有 OUTER 关键字，这说明 SQL 已经变为内连接。

为什么外连接的从表有过滤条件会变成内连接呢？因为外连接的从表有过滤条件已经排除了从表与主表没有关联上显示为 NULL 的情况。

提问：两表关联走不走 NL 是看两个表关联之后返回的数据量多少？还是看驱动表返回的数据量多少？

回答：如果两个表是 1∶N 关系，驱动表为 1，被驱动表为 N 并且 N 很大，这时即使驱动表返回数据量很少，也不能走嵌套循环，因为两表关联之后返回的数据量会很多。所以判断两表关联是否应该走 NL 应该直接查看两表关联之后返回的数据量，如果两表关联之后返回的数据量少，可以走 NL；返回的数据量多，应该走 HASH 连接。

提问：大表是否可以当嵌套循环（NL）驱动表？

回答：可以，如果大表过滤之后返回的数据量很少就可以当 NL 驱动表。

提问：select * from a,b where a.id=b.id; 如果 a 有 100 条数据，b 有 100 万行数据，a 与 b 是 1∶N 关系，N 很低，应该怎么优化 SQL？

回答：因为 a 与 b 是 1∶N 关系，N 很低，我们可以在 b 的连接列（id）上创建索引，让 a 与 b 走嵌套循环（a nl b），这样 b 表会被扫描 100 次，但是每次扫描表的时候走的是 id 列的索引（范围扫描）。如果让 a 和 b 进行 HASH 连接，b 表会被全表扫描（因为没有过滤条件），需要查询表中 100 万行数据，而如果让 a 和 b 进行嵌套循环，b 表只需要查询出表中最多几百行数据（100*N）。一般情况下，一个小表与一个大表关联，我们可以考虑让小表 NL 大表，大表走连接列索引（如果大表有过滤条件，需要将过滤条件与连接列组合起来创建组合索引），从而避免大表被全表扫描。

最后，为了加深对嵌套循环的理解，大家可以在 SQLPLUS 中依次运行以下脚本，观察 SQL 执行速度，思考 SQL 为什么会执行缓慢：

```
create table a as selcct * from dba_objects;
create table b as select * from dba_objects;
set timi on
set lines 200 pages 100
set autot trace
select /*+ use_nl(a,b) */ * from a,b where a.object_id=b.object_id;
```

5.2 HASH 连接（HASH JOIN）

上文提到，两表关联返回少量数据应该走嵌套循环，两表关联返回大量数据应该走 HASH 连接。

HASH 连接的算法：两表等值关联，返回大量数据，将较小的表选为驱动表，将驱动表的"select 列和 join 列"读入 PGA 中的 work area，然后对驱动表的连接列进行 hash 运算生成 hash table，当驱动表的所有数据完全读入 PGA 中的 work area 之后，再读取被驱动表（被驱动表不需要读入 PGA 中的 work area），对被驱动表的连接列也进行 hash 运算，然后到 PGA 中的 work area 去探测 hash table，找到数据就关联上，没找到数据就没关联上。哈希连接只支持等值连接。

我们在测试账号 scott 中运行如下 SQL。

```
SQL> select /*+ gather_plan_statistics use_hash(e,d)  */
  2    e.ename, e.job, d.dname
  3    from emp e, dept d
  4   where e.deptno = d.deptno;
```

此处省略输出结果。

我们运行如下命令获取执行计划。

```
SQL> select * from table(dbms_xplan.display_cursor(null,null,'ALLSTATS LAST'));
```

5.2 HASH 连接（HASH JOIN）

```
PLAN_TABLE_OUTPUT
-----------------------------------
SQL_ID  2dj5zrbcps5yu, child number 0
-----------------------------------
select /*+ gather_plan_statistics use_hash(e,d)  */  e.ename, e.job,
d.dname    from emp e, dept d   where e.deptno = d.deptno

Plan hash value: 615168685

--------------------------------------------------------------------------------
| Id |Operation           |Name|Starts|E-Rows|A-Rows|  A-Time   |Buffers|OMem|1Mem|Used-Mem|
--------------------------------------------------------------------------------
|  0 |SELECT STATEMENT    |    |   1|      |   14|00:00:00.01|    15|    |    |        |
|* 1 | HASH JOIN          |    |   1|   15|   14|00:00:00.01|    15|888K|888K|714K(0) |
|  2 |  TABLE ACCESS FULL |DEPT|   1|    4|    4|00:00:00.01|     7|    |    |        |
|  3 |  TABLE ACCESS FULL |EMP |   1|   15|   14|00:00:00.01|     8|    |    |        |
--------------------------------------------------------------------------------

Predicate Information (identified by operation id):
---------------------------------------------------

   1 - access("E"."DEPTNO"="D"."DEPTNO")
```

执行计划中离 HASH 连接关键字最近的表就是驱动表。这里 DEPT 就是驱动表，EMP 就是被驱动表。驱动表 DEPT 只扫描了一次（Id=2，Starts=1），被驱动表 EMP 也只扫描了一次（Id=3，Starts=1）。再次强调，嵌套循环被驱动表需要扫描多次，HASH 连接的被驱动表只需要扫描一次。

Used-Mem 表示 HASH 连接消耗了多少 PGA，当驱动表太大、PGA 不能完全容纳驱动表时，驱动表就会溢出到临时表空间，进而产生磁盘 HASH 连接，这时候 HASH 连接性能会严重下降。嵌套循环不需要消耗 PGA。

嵌套循环每循环一次，会将驱动表连接列传值给被驱动表的连接列，也就是说嵌套循环会进行传值。HASH 连接没有传值的过程。在进行 HASH 连接的时候，被驱动表的连接列会生成 HASH 值，到 PGA 中去探测驱动表所生成的 hash table。HASH 连接的驱动表与被驱动表的连接列都不需要创建索引。

OLTP 环境一般是高并发小事物居多，此类 SQL 返回结果很少，SQL 执行计划多以嵌套循环为主，因此 OLTP 环境 SGA 设置较大，PGA 设置较小（因为嵌套循环不消耗 PGA）。而 OLAP 环境多数 SQL 都是大规模的 ETL，此类 SQL 返回结果集很多，SQL 执行计划通常以 HASH 连接为主，往往要消耗大量 PGA，所以 OLAP 系统 PGA 设置较大。

当两表使用外连接进行关联，如果执行计划走的是 HASH 连接，想要更改驱动表，我们需要使用 swap_join_inputs，而不是 leading，例如下面 SQL。

```
SQL> explain plan for select /*+ use_hash(d,e) leading(e) */
  2   *
  3    from dept d
  4    left join emp e on d.deptno = e.deptno;

Explained.

SQL> select * from table(dbms_xplan.display);
```

```
PLAN_TABLE_OUTPUT
--------------------------------------------------------------------------------
Plan hash value: 3713469723

--------------------------------------------------------------------------------
| Id  | Operation            | Name | Rows | Bytes | Cost (%CPU)| Time     |
--------------------------------------------------------------------------------
|   0 | SELECT STATEMENT     |      |   14 |   812 |    7  (15) | 00:00:01 |
|*  1 |  HASH JOIN OUTER     |      |   14 |   812 |    7  (15) | 00:00:01 |
|   2 |   TABLE ACCESS FULL  | DEPT |    4 |    80 |    3   (0) | 00:00:01 |
|   3 |   TABLE ACCESS FULL  | EMP  |   14 |   532 |    3   (0) | 00:00:01 |
--------------------------------------------------------------------------------

Predicate Information (identified by operation id):
---------------------------------------------------

   1 - access("D"."DEPTNO"="E"."DEPTNO"(+))

15 rows selected.
```

从执行计划中我们可以看到，DEPT 与 EMP 是采用 HASH 连接，这说明 use_hash(d,e)生效了。执行计划中，驱动表为 DEPT，虽然设置了 leading(e)，但是没有生效。现在我们使用 swap_join_inputs 来更改外连接中 HASH 连接的驱动表。

```
SQL> explain plan for select /*+ use_hash(d,e) swap_join_inputs(e) */
  2   *
  3   from dept d
  4   left join emp e on d.deptno = e.deptno;

Explained.

SQL> select * from table(dbms_xplan.display);

PLAN_TABLE_OUTPUT
--------------------------------------------------------------------------------
Plan hash value: 3590956717

--------------------------------------------------------------------------------
| Id  | Operation             | Name | Rows | Bytes | Cost (%CPU)| Time     |
--------------------------------------------------------------------------------
|   0 | SELECT STATEMENT      |      |   14 |   812 |    7  (15) | 00:00:01 |
|*  1 |  HASH JOIN RIGHT OUTER|      |   14 |   812 |    7  (15) | 00:00:01 |
|   2 |   TABLE ACCESS FULL   | EMP  |   14 |   532 |    3   (0) | 00:00:01 |
|   3 |   TABLE ACCESS FULL   | DEPT |    4 |    80 |    3   (0) | 00:00:01 |
--------------------------------------------------------------------------------

Predicate Information (identified by operation id):
---------------------------------------------------

   1 - access("D"."DEPTNO"="E"."DEPTNO"(+))

15 rows selected.
```

从执行计划中我们可以看到，使用 swap_join_inputs 更改了外连接中 HASH 连接的驱动表。

思考：怎么优化 HASH 连接？

回答：因为 HASH 连接需要将驱动表的 select 列和 join 列放入 PGA 中，所以，我们应该尽量避免书写 select * from....语句，将需要的列放在 select list 中，这样可以减少驱动表对 PGA

的占用，避免驱动表被溢出到临时表空间，从而提升查询性能。如果无法避免驱动表被溢出到临时表空间，我们可以将临时表空间创建在 SSD 上或者 RAID 0 上，加快临时数据的交换速度。

当 PGA 采用自动管理，单个进程的 work area 被限制在 1G 以内，如果是 PGA 采用手动管理，单个进程的 work area 不能超过 2GB。如果驱动表比较大，比如驱动表有 4GB，可以开启并行查询至少 parallel(4)，将表拆分为至少 4 份，这样每个并行进程中的 work area 能够容纳 1GB 数据，从而避免驱动表被溢出到临时表空间。如果驱动表非常大，比如有几十 GB，这时开启并行 HASH 也无能为力，这时，应该考虑对表进行拆分，在第 8 章中，我们会为大家详细介绍表的拆分方法。

5.3 排序合并连接（SORT MERGE JOIN）

前文提到 HASH 连接主要用于处理两表等值关联返回大量数据。

排序合并连接主要用于处理两表非等值关联，比如>, >=, <, <=, <>，但是不能用于 instr、substr、like、regexp_like 关联，instr、substr、like、regexp_like 关联只能走嵌套循环。

现有如下 SQL。

```
select * from a,b where a.id>=b.id;
```

A 表有 10 万条数据，B 表有 20 万条数据，A 表与 B 表的 ID 列都是从 1 开始每次加 1。

该 SQL 是非等值连接，因此不能进行 HASH 连接。

假如该 SQL 走的是嵌套循环，A 作为驱动表，B 作为被驱动表，那么 B 表会被扫描 10 万次。前文提到，嵌套循环被驱动表连接列要包含在索引中，那么 B 表的 ID 列需要创建一个索引，嵌套循环会进行传值，当 A 表通过 ID 列传值超过 10 000 的时候，B 表通过 ID 列的索引返回数据每次都会超过 10 000 条，这个时候会造成 B 表大量回表。所以该 SQL 不能走嵌套循环，只能走排序合并连接。

排序合并连接的算法：两表关联，先对两个表根据连接列进行排序，将较小的表作为驱动表（Oracle 官方认为排序合并连接没有驱动表，笔者认为是有的），然后从驱动表中取出连接列的值，到已经排好序的被驱动表中匹配数据，如果匹配上数据，就关联成功。驱动表返回多少行，被驱动表就要被匹配多少次，这个匹配的过程类似嵌套循环，但是嵌套循环是从被驱动表的索引中匹配数据，而排序合并连接是在内存中（PGA 中的 work area）匹配数据。

我们在测试账号 scott 中运行如下 SQL。

```
SQL> select /*+ gather_plan_statistics */ e.ename, e.job,
  2  d.dname   from emp e, dept d  where e.deptno >= d.deptno;
......省略输出结果......
```

我们获取执行计划。

```
SQL> select * from table(dbms_xplan.display_cursor(null,null,'ALLSTATS LAST'));

PLAN_TABLE_OUTPUT
---------------------------------------------------------------------------------
```

```
SQL_ID  f673my5x7tkkg, child number 0
-------------------------------------
select /*+ gather_plan_statistics */  e.ename, e.job, d.dname   from
emp e, dept d  where e.deptno >= d.deptno

Plan hash value: 844388907

----------------------------------------------------------------------------------
| Id |Operation                           |Name    |Starts|E-Rows|A-Rows|   A-Time   |Buffers|
----------------------------------------------------------------------------------
|  0 |SELECT STATEMENT                    |        |    1 |      |   31 |00:00:00.01 |    15 |
|  1 | MERGE JOIN                         |        |    1 |    3 |   31 |00:00:00.01 |    15 |
|  2 |  TABLE ACCESS BY INDEX ROWID       |DEPT    |    1 |    4 |    4 |00:00:00.01 |     8 |
|  3 |   INDEX FULL SCAN                  |PK_DEPT |    1 |    4 |    4 |00:00:00.01 |     4 |
|* 4 |  SORT JOIN                         |        |    4 |   14 |   31 |00:00:00.01 |     7 |
|  5 |   TABLE ACCESS FULL                |EMP     |    1 |   14 |   14 |00:00:00.01 |     7 |
----------------------------------------------------------------------------------

Predicate Information (identified by operation id):
---------------------------------------------------

   4 - access("E"."DEPTNO">="D"."DEPTNO")
       filter("E"."DEPTNO">="D"."DEPTNO")
```

执行计划中离 MERGE JOIN 关键字最近的表就是驱动表。这里 DEPT 就是驱动表,EMP 就是被驱动表。驱动表 DEPT 只扫描了一次(Id=2,Starts=1),被驱动表 EMP 也只扫描了一次(Id=5,Starts=1)。

因为 DEPT 走的是 INDEX FULL SCAN,INDEX FULL SCAN 返回的数据是有序的,所以 DEPT 表就不需要排序了。EMP 走的是全表扫描,返回的数据是无序的,所以 EMP 表在 PGA 中进行了排序。在实际工作中,我们一定要注意 INDEX FULL SCAN 返回了多少行数据,如果 INDEX FULL SCAN 返回的行数太多,应该强制走全表扫描,具体原因请参考本书 4.1.8 节。

现在我们强制 DEPT 表走全表扫描,查看执行计划。

```
SQL> select /*+ full(d) */
  2    e.ename, e.job, d.dname
  3    from emp e, dept d
  4    where e.deptno >= d.deptno;

31 rows selected.

Execution Plan
----------------------------------------------------------
Plan hash value: 1407029907

--------------------------------------------------------------------------
| Id  | Operation           | Name | Rows  | Bytes | Cost (%CPU)| Time     |
--------------------------------------------------------------------------
|   0 | SELECT STATEMENT    |      |     3 |    90 |     8  (25)| 00:00:01 |
|   1 |  MERGE JOIN         |      |     3 |    90 |     8  (25)| 00:00:01 |
|   2 |   SORT JOIN         |      |     4 |    52 |     4  (25)| 00:00:01 |
|   3 |    TABLE ACCESS FULL| DEPT |     4 |    52 |     3   (0)| 00:00:01 |
|*  4 |   SORT JOIN         |      |    14 |   238 |     4  (25)| 00:00:01 |
|   5 |    TABLE ACCESS FULL| EMP  |    14 |   238 |     3   (0)| 00:00:01 |
--------------------------------------------------------------------------

Predicate Information (identified by operation id):
```

```
     4 - access("E"."DEPTNO">="D"."DEPTNO")
         filter("E"."DEPTNO">="D"."DEPTNO")
```

从执行计划中我们看到，DEPT 走的是全表扫描，因为全表扫描返回的数据是无序的，所以 DEPT 在 PGA 中进行了排序。

如果两表是等值关联，一般不建议走排序合并连接。因为排序合并连接需要将两个表放入 PGA 中，而 HASH 连接只需要将驱动表放入 PGA 中，排序合并连接与 HASH 连接相比，需要耗费更多的 PGA。即使排序合并连接中有一个表走的是 INDEX FULL SCAN，另外一个表也需要放入 PGA 中，而这个表往往是大表，如果走 HASH 连接，大表会作为被驱动表，是不会被放入 PGA 中的。因此，两表等值关联，要么走 NL（返回数据量少），要么走 HASH（返回数据量多），一般情况下不要走 SMJ。

思考：怎么优化排序合并连接？

回答：如果两表关联是等值关联，走的是排序合并连接，我们可以将表连接方式改为 HASH 连接。如果两表关联是非等值关联，比如>，>=，<，<=，<>，这时我们应该先从业务上入手，尝试将非等值关联改写为等值关联，因为非等值关联返回的结果集"类似"于笛卡儿积，当两个表都比较大的时候，非等值关联返回的数据量相当"恐怖"。如果没有办法将非等值关联改写为等值关联，我们可以考虑增加两表的限制条件，将两个表数据量缩小，最后可以考虑开启并行查询加快 SQL 执行速度。

表 5-1 列举出了 3 种表连接方式的主要区别。

表 5-1 表连接方式

表连接方式	驱动表	PGA	输出结果集	不等值连接	被驱动表扫描次数
嵌套循环	有（靠近关键字）	不消耗	少	支持	等于驱动表返回行数
哈希连接	有（靠近关键字）	消耗	多	不支持	1
排序合并连接	无	消耗	多	支持	1

5.4 笛卡儿连接（CARTESIAN JOIN）

两个表关联没有连接条件的时候会产生笛卡儿积，这种表连接方式就叫笛卡儿连接。

我们在测试账号 scott 中运行如下 SQL。

```
SQL> set autot trace
SQL> select * from emp, dept;

56 rows selected.

Execution Plan
----------------------------------------------------------
Plan hash value: 2034389985

--------------------------------------------------------------------------
| Id  | Operation          | Name | Rows  | Bytes | Cost (%CPU)| Time     |
--------------------------------------------------------------------------
```

```
|   0 | SELECT STATEMENT      |      | 56 | 3248 |  8  (0)| 00:00:01 |
|   1 |  MERGE JOIN CARTESIAN |      | 56 | 3248 |  8  (0)| 00:00:01 |
|   2 |   TABLE ACCESS FULL   | DEPT |  4 |   80 |  3  (0)| 00:00:01 |
|   3 |   BUFFER SORT         |      | 14 |  532 |  5  (0)| 00:00:01 |
|   4 |    TABLE ACCESS FULL  | EMP  | 14 |  532 |  1  (0)| 00:00:01 |
```

执行计划中 MERGE JOIN CARTESIAN 就表示笛卡儿连接。笛卡儿连接会返回两个表的乘积。DEPT 有 4 行数据，EMP 有 14 行数据，两个表进行笛卡儿连接之后会返回 56 行数据。笛卡儿连接会对两表中其中一个表进行排序，执行计划中的 BUFFER SORT 就表示排序。

在多表关联的时候，两个表没有直接关联条件，但是优化器错误地把某个表返回的 Rows 算为 1 行（注意必须是 1 行），这个时候也可能发生笛卡儿连接。例子如下。

```
SQL> select * from table(dbms_xplan.display());

PLAN_TABLE_OUTPUT
-------------------------------------------------------------------------------
Plan hash value: 710264295
-------------------------------------------------------------------------------
| Id   | Operation                         | Name                      | Rows  |
-------------------------------------------------------------------------------
|   0  | SELECT STATEMENT                  |                           |     1 |
|   1  |  WINDOW SORT                      |                           |     1 |
|*  2  |   TABLE ACCESS BY INDEX ROWID     | F_AGT_GUARANTY_INFO_H     |     1 |
|   3  |    NESTED LOOPS                   |                           |     1 |
|   4  |     NESTED LOOPS                  |                           |     1 |
|   5  |      MERGE JOIN CARTESIAN         |                           |     1 |
|   6  |       TABLE ACCESS FULL           | B_M_BUSINESS_CONTRACT     |     1 |
|   7  |       BUFFER SORT                 |                           | 61507 |
|*  8  |        TABLE ACCESS FULL          | F_AGT_GUARANTY_RELATIVE_H | 61507 |
|   9  |      TABLE ACCESS BY INDEX ROWID  | F_CONTRACT_RELATIVE       |     1 |
|* 10  |       INDEX UNIQUE SCAN           | SYS_C0019578              |     1 |
|* 11  |     INDEX RANGE SCAN              | SYS_C005707               |     1 |
```

执行计划中 Id=6 的表和 Id=8 的表就是进行笛卡儿连接的。

在这个执行计划中，为什么优化器会选择笛卡儿积连接呢？

因为 Id=6 这个表返回的 Rows 被优化器错误地估算为 1 行，优化器认为 1 行的表与任意大小的表进行笛卡儿关联，数据也不会翻番，这是安全的。所以这里优化器选择了笛卡儿连接。

Id=6 这步是全表扫描，而且没过滤条件（因为没有*），优化器认为它只返回 1 行。大家请思考，全表扫描返回 1 行并且无过滤条件，这个可能吗？难道表里面真的就只有 1 行数据？这不符合常识。那么显然是 Id=6 的表没有收集统计信息，导致优化器默认地把该表算为 1 行（当时数据库没开启动态采样）。下面是上述执行计划的 SQL 语句。

```
SELECT b.agmt_id,
       b.corp_org,
       b.cur_cd,
       b.businesstype,
       c.object_no,
       c.guaranty_crsum,
       row_number() over(PARTITION BY b.agmt_id, b.corp_org, c.object_no ORDER BY b.a
gmt_id, b.corp_org, c.object_no) row_no
  FROM b_m_business_contract        b, --合同表
       dwf.f_contract_relative      c, --合同关联表
```

5.4 笛卡儿连接（CARTESIAN JOIN）

```
       dwf.f_agt_guaranty_relative_h r, --业务合同、担保合同与担保物关联表
       dwf.f_agt_guaranty_info_h     g --担保物信息表
 WHERE b.corp_org = c.corp_org
   AND b.agmt_id = c.contract_seqno --业务合同号
   AND c.object_type = 'GuarantyContract'
   AND r.start_dt <= DATE '2012-09-17' /*当天日期*/
   AND r.end_dt > DATE '2012-09-17' /*当天日期*/
   AND c.contract_seqno = r.object_no --业务合同号
   AND c.object_no = r.guaranty_no --担保合同编号
   AND c.corp_org = r.corp_org --企业法人编码
   AND r.object_type = 'BusinessContract'
   AND r.agmt_id = g.agmt_id --担保物编号
   AND r.corp_org = g.corp_org --企业法人编码
   AND g.start_dt <= DATE '2012-09-17' /*当天日期*/
   AND g.end_dt > DATE '2012-09-17' /*当天日期*/
   AND g.guarantytype = '020010' --质押存款
```

执行计划中进行笛卡儿关联的表就是 b 和 r，在 SQL 语句中 b 和 r 没有直接关联条件。

如果两个表有直接关联条件，无法控制两个表进行笛卡儿连接。

如果两个表没有直接关联条件，我们在编写 SQL 的时候将两个表依次放在 from 后面并且添加 HINT：ordered，就可以使两个表进行笛卡儿积关联。

```
SQL> select /*+ ordered */
  2    a.ename, a.sal, a.deptno, b.dname, c.grade
  3    from dept b, salgrade c, emp a
  4   where a.deptno = b.deptno
  5     and a.sal between c.losal and c.hisal;

14 rows selected.


Execution Plan
----------------------------------------------------------
Plan hash value: 2197699399

--------------------------------------------------------------------------------
| Id  | Operation             | Name     | Rows  | Bytes | Cost (%CPU)| Time     |
--------------------------------------------------------------------------------
|   0 | SELECT STATEMENT      |          |     1 |    65 |    12   (9)| 00:00:01 |
|*  1 |  HASH JOIN            |          |     1 |    65 |    12   (9)| 00:00:01 |
|   2 |   MERGE JOIN CARTESIAN|          |    20 |  1040 |     8   (0)| 00:00:01 |
|   3 |    TABLE ACCESS FULL  | DEPT     |     4 |    52 |     3   (0)| 00:00:01 |
|   4 |    BUFFER SORT        |          |     5 |   195 |     5   (0)| 00:00:01 |
|   5 |     TABLE ACCESS FULL | SALGRADE |     5 |   195 |     1   (0)| 00:00:01 |
|   6 |   TABLE ACCESS FULL   | EMP      |    14 |   182 |     3   (0)| 00:00:01 |
--------------------------------------------------------------------------------

Predicate Information (identified by operation id):
---------------------------------------------------

   1 - access("A"."DEPTNO"="B"."DEPTNO")
       filter("A"."SAL">="C"."LOSAL" AND "A"."SAL"<="C"."HISAL")
```

在 SQL 语句中，DEPT 与 SALGRADE 没有直接关联条件，HINT：ordered 表示根据 SQL 语句中 from 后面表的顺序依次关联。因为 DEPT 与 SALGRADE 没有直接关联条件，而且 SQL 语句中添加了 HINT：ordered，再有 SQL 语句中两个表是依次放在 from 后面的，所以 DEPT 与 SALGRADE 只能进行笛卡儿连接。

思考：当执行计划中有笛卡儿连接应该怎么优化呢？

首先应该检查表是否有关联条件，如果表没有关联条件，那么应该询问开发与业务人员为何表没有关联条件，是否为满足业务需求而故意不写关联条件。

其次应该检查离笛卡儿连接最近的表是否真的返回 1 行数据，如果返回行数真的只有 1 行，那么走笛卡儿连接是没有问题的，如果返回行数超过 1 行，那就需要检查为什么 Rows 会估算错误，同时要纠正错误的 Rows。纠正错误的 Rows 之后，优化器就不会走笛卡儿连接了。

我们可以使用 HINT /*+ opt_param('_optimizer_mjc_enabled', 'false') */ 禁止笛卡儿连接。

5.5 标量子查询（SCALAR SUBQUERY）

当一个子查询介于 select 与 from 之间，这种子查询就叫标量子查询，例子如下。

```
select e.ename,
       e.sal,
       (select d.dname from dept d where d.deptno = e.deptno) dname
  from emp e;
```

我们在测试账号 scott 中运行如下 SQL。

```
SQL> select /*+ gather_plan_statistics */ e.ename,
  2         e.sal,
  3         (select d.dname from dept d where d.deptno = e.deptno) dname
  4    from emp e;
......省略输出结果......
SQL> select * from table(dbms_xplan.display_cursor(null,null,'ALLSTATS LAST'));

PLAN_TABLE_OUTPUT
---------------------------------------------------------------------------
SQL_ID  ggmw3tv6xypx1, child number 0
-------------------------------------
select /*+ gather_plan_statistics */ e.ename,         e.sal,
(select d.dname from dept d where d.deptno = e.deptno) dname     from emp e

Plan hash value: 2981343222

---------------------------------------------------------------------------
| Id |Operation                    |Name    |Starts|E-Rows|A-Rows| A-Time      |Buffers|
---------------------------------------------------------------------------
|  0 |SELECT STATEMENT             |        |    1 |      |   14 |00:00:00.01 |     8 |
|  1 | TABLE ACCESS BY INDEX ROWID |DEPT    |    3 |    1 |    3 |00:00:00.01 |     5 |
|* 2 |  INDEX UNIQUE SCAN          |PK_DEPT |    3 |    1 |    3 |00:00:00.01 |     2 |
|  3 | TABLE ACCESS FULL           |EMP     |    1 |   14 |   14 |00:00:00.01 |     8 |
---------------------------------------------------------------------------

Predicate Information (identified by operation id):
---------------------------------------------------

   2 - access("D"."DEPTNO"=:B1)
```

标量子查询类似一个天然的嵌套循环，而且驱动表固定为主表。大家是否还记得：嵌套循环被驱动表的连接列必须包含在索引中。同理，标量子查询中子查询的表连接列也必须包含在索引中。主表 EMP 通过连接列（DEPTNO）传值给子查询中的表（DEPT），执行计划中:B1 就表示传值，这个传值过程一共进行了 3 次，因为主表（EMP）的连接列（DEPTNO）基数等于 3。

```
SQL> select count(distinct deptno) from emp;
COUNT(DISTINCTDEPTNO)
---------------------
                    3
```

我们建议在工作中，尽量避免使用标量子查询，假如主表返回大量数据，主表的连接列基数很高，那么子查询中的表会被多次扫描，从而严重影响 SQL 性能。如果主表数据量小，或者主表的连接列基数很低，那么这个时候我们也可以使用标量子查询，但是记得要给子查询中表的连接列建立索引。

当 SQL 里面有标量子查询，我们可以将标量子查询等价改写为外连接，从而使它们可以进行 HASH 连接。为什么要将标量子查询改写为外连接而不是内连接呢？因为标量子查询是一个传值的过程，如果主表传值给子查询，子查询没有查询到数据，这个时候会显示 NULL。如果将标量子查询改写为内连接，会丢失没有关联上的数据。

现有如下标量子查询。

```
SQL> select d.dname,
  2         d.loc,
  3         (select max(e.sal) from emp e where e.deptno = d.deptno) max_sal
  4    from dept d;

DNAME          LOC              MAX_SAL
-------------- -------------- ---------
ACCOUNTING     NEW YORK            5000
RESEARCH       DALLAS              3000
SALES          CHICAGO             2850
OPERATIONS     BOSTON           ---NULL
```

我们可以将其等价改写为外连接：

```
SQL> select d.dname, d.loc, e.max_sal
  2    from dept d
  3    left join (select max(sal) max_sal,
  4                      deptno
  5                 from emp
  6                group by deptno)e
  7      on d.deptno = e.deptno;

DNAME          LOC              MAX_SAL
-------------- -------------- ---------
ACCOUNTING     NEW YORK            5000
RESEARCH       DALLAS              3000
SALES          CHICAGO             2850
OPERATIONS     BOSTON           ---NULL
```

当然了，如果主表的连接列是外键，而子查询的连接列是主键，我们就没必要改写为外连接了，因为外键不可能存 NULL 值，可以直接改写为内连接。例如本书中所用的标量子查询示例就可以改写为内连接，因为 DEPT 与 EMP 有主外键关系。

```sql
select e.ename, e.sal, d.dname
  from emp e
  inner join dept d on e.deptno = d.deptno;
```

在 Oracle12c 中，简单的标量子查询会被优化器等价改写为外连接。

5.6 半连接（SEMI JOIN）

两表关联只返回一个表的数据就叫半连接。半连接一般就是指的 in 和 exists。在 SQL 优化实战中，半连接的优化是最为复杂的。

5.6.1 半连接等价改写

in 和 exists 一般情况下都可以进行等价改写。

半连接 in 的写法如下。

```
SQL> select * from dept where deptno in (select deptno from emp);

    DEPTNO DNAME          LOC
---------- -------------- -------------
        10 ACCOUNTING     NEW YORK
        20 RESEARCH       DALLAS
        30 SALES          CHICAGO
```

半连接 exists 的写法如下。

```
SQL> select * from dept where exists (select null from emp where dept.deptno=emp.dept
no);

    DEPTNO DNAME          LOC
---------- -------------- -------------
        10 ACCOUNTING     NEW YORK
        20 RESEARCH       DALLAS
        30 SALES          CHICAGO
```

in 和 exists 有时候也可以等价地改写为内连接，例如，上面查询语句可以改写为如下写法。

```
SQL> select d.*
  2    from dept d, (select deptno from emp group by deptno) e
  3   where d.deptno = e.deptno;

    DEPTNO DNAME          LOC
---------- -------------- -------------
        10 ACCOUNTING     NEW YORK
        20 RESEARCH       DALLAS
        30 SALES          CHICAGO
```

注意：上面内连接的写法性能没有半连接写法性能高，因为多了 GROUP BY 去重操作。

在将半连接改写为内连接的时候，我们要注意主表与子表（子查询中的表）的关系。这里 DEPT 与 EMP 是 1：n 关系。在半连接的写法中，返回的是 DEPT 表的数据，也就是说返回的数据是属于 1 的关系。然而在使用内连接的写法中，由于 DEPT 与 EMP 是 1：n 关系，两表关联之后会返回 n（有重复数据），所以我们需要加上 GROUP BY 去掉重复数据。

如果半连接中主表属于 1 的关系，子表（子查询中的表）属于 n 的关系，我们在改写为内连接的时候，需要加上 GROUP BY 去重。注意：这个时候半连接性能高于内连接。

如果半连接中主表属于 n 的关系，子表（子查询中的表）属于 1 的关系，我们在改写为内连接的时候，就不需要去重了。注意：这个时候半连接与内连接性能一样。

如果半连接中主表属于 n 的关系，子表（子查询中的表）也属于 n 的关系，这时我们可以先对子查询去重，将子表转换为 1 的关系，然后再关联，千万不能先关联再去重。

作者的个人技术博客上记载了一篇半连接被优化器改写为内连接而导致查询变慢的经典案例，如果大家有兴趣可以阅读参考：http://blog.csdn.net/robinson1988/article/details/51148332。

5.6.2 控制半连接执行计划

我们先来查看示例 SQL 的原始执行计划。

```
SQL> select * from dept where deptno in (select deptno from emp);

Execution Plan
----------------------------------------------------------
Plan hash value: 1090737117

--------------------------------------------------------------------------------
| Id  | Operation                     | Name     | Rows  | Bytes | Cost(%CPU)| Time     |
--------------------------------------------------------------------------------
|   0 | SELECT STATEMENT              |          |     3 |    69 |     6  (17)| 00:00:01 |
|   1 |  MERGE JOIN SEMI              |          |     3 |    69 |     6  (17)| 00:00:01 |
|   2 |   TABLE ACCESS BY INDEX ROWID | DEPT     |     4 |    80 |     2   (0)| 00:00:01 |
|   3 |    INDEX FULL SCAN            | PK_DEPT  |     4 |       |     1   (0)| 00:00:01 |
|*  4 |   SORT UNIQUE                 |          |    14 |    42 |     4  (25)| 00:00:01 |
|   5 |    TABLE ACCESS FULL          | EMP      |    14 |    42 |     3   (0)| 00:00:01 |
--------------------------------------------------------------------------------

Predicate Information (identified by operation id):
---------------------------------------------------

4 - access("DEPTNO"="DEPTNO")
    filter("DEPTNO"="DEPTNO")
```

执行计划中 DEPT 与 EMP 是采用排序合并连接进行关联的。

我们现在让 DEPT 与 EMP 进行嵌套循环连接，同时让 DEPT 当驱动表。

```
SQL> select /*+ use_nl(emp@a,dept) leading(dept) */
  2  *
  3    from dept
  4   where deptno in (select /*+ qb_name(a) */ deptno from emp);

Execution Plan
----------------------------------------------------------
Plan hash value: 2645846736

--------------------------------------------------------------------------------
| Id  | Operation          | Name  | Rows  | Bytes | Cost (%CPU)| Time     |
--------------------------------------------------------------------------------
|   0 | SELECT STATEMENT   |       |     3 |    69 |     8   (0)| 00:00:01 |
```

```
|   1 |  NESTED LOOPS SEMI  |      |    3 |   69 |     8   (0)| 00:00:01 |
|   2 |   TABLE ACCESS FULL | DEPT |    4 |   80 |     3   (0)| 00:00:01 |
|*  3 |   TABLE ACCESS FULL | EMP  |    9 |   27 |     1   (0)| 00:00:01 |
-----------------------------------------------------------------------

Predicate Information (identified by operation id):
---------------------------------------------------

   3 - filter("DEPTNO"="DEPTNO")
```

有读者可能会好奇，为何不写 HINT /*+ use_nl(dept,emp) leading(dept) */？

因为在 Oracle 数据库中，每个子查询都会自动生成一个查询块（query block），子查询里面的表会自动地被优化器取别名。这里 from 后面的表只有 DEPT，而 EMP 在子查询中，HINT 写成 use_nl(dept,emp)会导致 CBO 无法识别 EMP，为了让 CBO 能识别到 EMP，在子查询中添加了 qb_name 这个 HINT，给子查询取别名为 a，再在主查询中使用 use_nl(emp@a,dept)，就能使两表进行嵌套循环关联。

如果不想使用 qb_name 这个 HINT，我们也可以参考如下操作。

```
SQL> explain plan for select * from dept where deptno in (select deptno from emp);

Explained.

SQL> select * from table(dbms_xplan.display(null, null, 'advanced -projection -outlin
e -predicate'));

PLAN_TABLE_OUTPUT
--------------------------------------------------------------------------------
Plan hash value: 1090737117

--------------------------------------------------------------------------------
| Id  | Operation                    | Name    | Rows | Bytes | Cost (%CPU)|Time     |
--------------------------------------------------------------------------------
|   0 | SELECT STATEMENT             |         |    3 |   69 |     6  (17)|00:00:01|
|   1 |  MERGE JOIN SEMI             |         |    3 |   69 |     6  (17)|00:00:01|
|   2 |   TABLE ACCESS BY INDEX ROWID| DEPT    |    4 |   80 |     2   (0)|00:00:01|
|   3 |    INDEX FULL SCAN           | PK_DEPT |    4 |      |     1   (0)|00:00:01|
|   4 |   SORT UNIQUE                |         |   14 |   42 |     4  (25)|00:00:01|
|   5 |    TABLE ACCESS FULL         | EMP     |   14 |   42 |     3   (0)|00:00:01|
--------------------------------------------------------------------------------

Query Block Name / Object Alias (identified by operation id):
-------------------------------------------------------------

   1 - SEL$5DA710D3
   2 - SEL$5DA710D3 / DEPT@SEL$1
   3 - SEL$5DA710D3 / DEPT@SEL$1
   5 - SEL$5DA710D3 / EMP@SEL$2

20 rows selected.

SQL> select /*+ use_nl(dept,emp@sel$2) leading(dept) */
  2    *
  3    from dept
  4   where deptno in (select deptno from emp);

Execution Plan
----------------------------------------------------------
```

5.6 半连接（SEMI JOIN）

```
Plan hash value: 2645846736

---------------------------------------------------------------------------
| Id  | Operation           | Name | Rows  | Bytes | Cost (%CPU)| Time     |
---------------------------------------------------------------------------
|   0 | SELECT STATEMENT    |      |     3 |    69 |     8   (0)| 00:00:01 |
|   1 |  NESTED LOOPS SEMI  |      |     3 |    69 |     8   (0)| 00:00:01 |
|   2 |   TABLE ACCESS FULL | DEPT |     4 |    80 |     3   (0)| 00:00:01 |
|*  3 |   TABLE ACCESS FULL | EMP  |     9 |    27 |     1   (0)| 00:00:01 |
---------------------------------------------------------------------------

Predicate Information (identified by operation id):
---------------------------------------------------

   3 - filter("DEPTNO"="DEPTNO")
```

现在我们让 DEPT 与 EMP 进行 HASH 连接，同时让 EMP 作为驱动表。

```
SQL> select /*+ use_hash(dept,emp@sel$2) leading(emp@sel$2) */
  2  *
  3    from dept
  4   where deptno in (select deptno from emp);

Execution Plan
----------------------------------------------------------
Plan hash value: 300394613

---------------------------------------------------------------------------
| Id  | Operation            | Name | Rows  | Bytes | Cost (%CPU)| Time     |
---------------------------------------------------------------------------
|   0 | SELECT STATEMENT     |      |     3 |    69 |     8  (25)| 00:00:01 |
|*  1 |  HASH JOIN           |      |     3 |    69 |     8  (25)| 00:00:01 |
|   2 |   SORT UNIQUE        |      |    14 |    42 |     3   (0)| 00:00:01 |
|   3 |    TABLE ACCESS FULL | EMP  |    14 |    42 |     3   (0)| 00:00:01 |
|   4 |   TABLE ACCESS FULL  | DEPT |     4 |    80 |     3   (0)| 00:00:01 |
---------------------------------------------------------------------------

Predicate Information (identified by operation id):
---------------------------------------------------

   1 - access("DEPTNO"="DEPTNO")
```

让 EMP 表作为驱动表之后，CBO 先对 EMP 进行了去重（SORT UNIQUE）操作，这里 CBO 其实对该 SQL 进行了等价改写，将半连接等价改写为内连接（因为执行计划中没有 SEMI 关键字），在改写的过程中，因为 EMP 属于 N 的关系，所以对 EMP 进行了去重。

5.6.3 读者思考

现有如下 SQL。

```
select * from a where a.id in (select id from b);
```

假设 a 有 1000 万，b 有 100 行，请问如何优化该 SQL？
假设 a 有 100 行，b 有 1000 万，请问如何优化该 SQL？
假设 a 有 100 万，b 有 1000 万，请问如何优化该 SQL？

5.7 反连接（ANTI JOIN）

两表关联只返回主表的数据，而且只返回主表与子表没关联上的数据，这种连接就叫反连接。反连接一般就是指的 not in 和 not exists。

5.7.1 反连接等价改写

not in 与 not exists 一般情况下也可以进行等价改写。

not in 的写法如下。

```
SQL> select * from dept where deptno not in (select deptno from emp);

    DEPTNO DNAME          LOC
---------- -------------- -------------------------
        40 OPERATIONS     BOSTON
```

not exists 的写法如下。

```
SQL> select *
  2    from dept
  3   where not exists (select null from emp where dept.deptno = emp.deptno);

    DEPTNO DNAME          LOC
---------- -------------- -------------------------
        40 OPERATIONS     BOSTON
```

需要注意的是，not in 里面如果有 null，整个查询会返回空，而 in 里面有 null，查询不受 null 影响，例子如下。

```
SQL> select * from dept where deptno not in (10,null);

no rows selected

SQL> select * from dept where deptno in (10,null);

    DEPTNO DNAME          LOC
---------- -------------- -------------------------
        10 ACCOUNTING     NEW YORK
```

所以在将 not exists 等价改写为 not in 的时候，要注意 null。一般情况下，如果反连接采用 not in 写法，我们需要在 where 条件中剔除 null。

```
select *
  from dept
 where deptno not in (select deptno from emp where deptno is not null);
```

not in 与 not exists 除了可以相互等价改写以外，还可以等价地改写为外连接，例如，上面查询可以等价改写为如下写法。

```
SQL> select d.*
  2    from dept d
  3    left join emp e on d.deptno = e.deptno
```

5.7 反连接(ANTI JOIN)

```
  4    where e.deptno is null;

    DEPTNO DNAME              LOC
---------- -------------- -------------
        40 OPERATIONS         BOSTON
```

为什么反连接可以改写为"外连接+子表连接条件 is null"?我们再来回顾一下反连接定义:两表关联只返回主表的数据,而且只返回主表与子表没有关联上的数据。根据反连接定义,翻译为标准 SQL 写法就是"外连接+子表连接条件 is null"。与半连接改写为内连接不同的是,反连接改写为外连接不需要考虑两表之间的关系。

5.7.2 控制反连接执行计划

我们先来查看示例 SQL 的原始执行计划。

```
SQL> select * from dept where deptno not in (select deptno from emp);

Execution Plan
----------------------------------------------------------
Plan hash value: 2230682264

-------------------------------------------------------------------------------------
| Id  | Operation                     | Name    | Rows  | Bytes | Cost(%CPU)| Time     |
-------------------------------------------------------------------------------------
|   0 | SELECT STATEMENT              |         |     1 |    23 |     6  (17)| 00:00:01|
|   1 |  MERGE JOIN ANTI NA           |         |     1 |    23 |     6  (17)| 00:00:01|
|   2 |   SORT JOIN                   |         |     4 |    80 |     2   (0)| 00:00:01|
|   3 |    TABLE ACCESS BY INDEX ROWID| DEPT    |     4 |    80 |     2   (0)| 00:00:01|
|   4 |     INDEX FULL SCAN           | PK_DEPT |     4 |       |     1   (0)| 00:00:01|
|*  5 |   SORT UNIQUE                 |         |    14 |    42 |     4  (25)| 00:00:01|
|   6 |    TABLE ACCESS FULL          | EMP     |    14 |    42 |     3   (0)| 00:00:01|
-------------------------------------------------------------------------------------

Predicate Information (identified by operation id):
---------------------------------------------------

   5 - access("DEPTNO"="DEPTNO")
       filter("DEPTNO"="DEPTNO")
```

原始执行计划中 DEPT 与 EMP 是采用排序合并连接进行关联的。
我们现在让 DEPT 与 EMP 使用嵌套循环进行关联,不指定驱动表。

```
SQL> select /*+ use_nl(dept,emp@a) */ *
  2    from dept
  3   where deptno not in (select /*+ qb_name(a) */
  4                         deptno
  5                          from emp);

Execution Plan
----------------------------------------------------------
Plan hash value: 1831344308

-------------------------------------------------------------------------------------
| Id  | Operation          | Name | Rows  | Bytes | Cost (%CPU)| Time     |
-------------------------------------------------------------------------------------
|   0 | SELECT STATEMENT   |      |     1 |    23 |    11   (0)| 00:00:01 |
```

第 5 章　表连接方式

```
|*  1 |  FILTER                    |       |     |     |         |          |
|   2 |   NESTED LOOPS ANTI SNA|       |   1 |  23 |  11 (28)| 00:00:01 |
|   3 |    TABLE ACCESS FULL       | DEPT  |   4 |  80 |   3 (0) | 00:00:01 |
|*  4 |    TABLE ACCESS FULL       | EMP   |   9 |  27 |   1 (0) | 00:00:01 |
|*  5 |   TABLE ACCESS FULL        | EMP   |   1 |   3 |   3 (0) | 00:00:01 |
-----------------------------------------------------------------------------

Predicate Information (identified by operation id):
---------------------------------------------------

   1 - filter( NOT EXISTS (SELECT /*+ QB_NAME ("A") */ 0 FROM "EMP"
              "EMP" WHERE "DEPTNO" IS NULL))
   4 - filter("DEPTNO"="DEPTNO")
   5 - filter("DEPTNO" IS NULL)
```

执行计划居然变成了 FILTER，我们指定的 HINT 被 CBO 忽略了。这究竟是什么原因呢？注意观察 FILTER 对应的谓词部分我们就能发现原因。因为子表 EMP 的连接列 DEPTNO 没有排除存在 null 的情况，所以 CBO 选择了 FILTER。现在我们给子查询加上语句 where deptno is not null 再看一下执行计划。

```
SQL> select /*+ use_nl(dept,emp@a) */ *
  2    from dept
  3   where deptno not in (select /*+ qb_name(a) */
  4                          deptno
  5                          from emp where deptno is not null);

Execution Plan
----------------------------------------------------------
Plan hash value: 1522491139

-----------------------------------------------------------------------------
| Id  | Operation            | Name | Rows | Bytes | Cost (%CPU)| Time     |
-----------------------------------------------------------------------------
|   0 | SELECT STATEMENT     |      |    1 |   23  |    8  (0) | 00:00:01 |
|   1 |  NESTED LOOPS ANTI   |      |    1 |   23  |    8  (0) | 00:00:01 |
|   2 |   TABLE ACCESS FULL  | DEPT |    4 |   80  |    3  (0) | 00:00:01 |
|*  3 |   TABLE ACCESS FULL  | EMP  |    9 |   27  |    1  (0) | 00:00:01 |
-----------------------------------------------------------------------------

Predicate Information (identified by operation id):
---------------------------------------------------

   3 - filter("DEPTNO" IS NOT NULL AND "DEPTNO"="DEPTNO")
```

现在我们将 not in 改写为 not exists，加上 HINT，再查看执行计划。

```
SQL> select /*+ use_nl(dept,emp@a) */ *
  2    from dept
  3   where not exists
  4       (select /*+ qb_name(a) */ null from emp where emp.deptno = dept.deptno);

Execution Plan
----------------------------------------------------------
Plan hash value: 1522491139

-----------------------------------------------------------------------------
| Id  | Operation            | Name | Rows | Bytes | Cost (%CPU)| Time     |
-----------------------------------------------------------------------------
```

5.7 反连接（ANTI JOIN）

```
|   0 | SELECT STATEMENT   |      |    1 |    23 |     8   (0)| 00:00:01 |
|   1 |  NESTED LOOPS ANTI |      |    1 |    23 |     8   (0)| 00:00:01 |
|   2 |   TABLE ACCESS FULL| DEPT |    4 |    80 |     3   (0)| 00:00:01 |
|*  3 |   TABLE ACCESS FULL| EMP  |    9 |    27 |     1   (0)| 00:00:01 |
---------------------------------------------------------------------------

Predicate Information (identified by operation id):
---------------------------------------------------

   3 - filter("EMP"."DEPTNO"="DEPT"."DEPTNO")
```

在执行计划中，DEPT 是嵌套循环的驱动表，EMP 是嵌套循环的被驱动表。现在我们让 DEPT 与 EMP 还进行嵌套循环连接，但是让 EMP 作为驱动表。

```
SQL> select /*+ use_nl(dept,emp@a) leading(emp@a) */ *
  2   from dept
  3   where not exists
  4   (select /*+ qb_name(a) */ null from emp where emp.deptno = dept.deptno);

Execution Plan
----------------------------------------------------------
Plan hash value: 1522491139

---------------------------------------------------------------------------
| Id  | Operation          | Name | Rows  | Bytes | Cost (%CPU)| Time     |
---------------------------------------------------------------------------
|   0 | SELECT STATEMENT   |      |    1 |    23 |     8   (0)| 00:00:01 |
|   1 |  NESTED LOOPS ANTI |      |    1 |    23 |     8   (0)| 00:00:01 |
|   2 |   TABLE ACCESS FULL| DEPT |    4 |    80 |     3   (0)| 00:00:01 |
|*  3 |   TABLE ACCESS FULL| EMP  |    9 |    27 |     1   (0)| 00:00:01 |
---------------------------------------------------------------------------

Predicate Information (identified by operation id):
---------------------------------------------------

   3 - filter("EMP"."DEPTNO"="DEPT"."DEPTNO")
```

注意观察执行计划，虽然我们使用了 leading(emp@a) 强制让 EMP 作为驱动表，但是执行计划中驱动表还是 DEPT。这是为什么呢？因为反连接等价于"外连接+子表连接条件 is null"，大家是否还记得：当两表关联是外连接，使用嵌套循环进行关联的时候无法更改驱动表，驱动表会被固定为主表。

现在我们让 DEPT 与 EMP 进行 HASH 连接，而且让 EMP 作为驱动表。

```
SQL> select /*+ use_hash(dept,emp@a) leading(emp@a) */ *
  2   from dept
  3   where not exists
  4   (select /*+ qb_name(a) */ null from emp where emp.deptno = dept.deptno);

Execution Plan
----------------------------------------------------------
Plan hash value: 474461924

---------------------------------------------------------------------------
| Id  | Operation          | Name | Rows  | Bytes | Cost (%CPU)| Time     |
---------------------------------------------------------------------------
|   0 | SELECT STATEMENT   |      |    1 |    23 |     7  (15)| 00:00:01 |
```

107

```
|*  1 |  HASH JOIN ANTI   |       |   1 |   23 |    7  (15)| 00:00:01 |
|   2 |   TABLE ACCESS FULL| DEPT |   4 |   80 |    3   (0)| 00:00:01 |
|   3 |   TABLE ACCESS FULL| EMP  |  14 |   42 |    3   (0)| 00:00:01 |
---------------------------------------------------------------------

Predicate Information (identified by operation id):
---------------------------------------------------

   1 - access("EMP"."DEPTNO"="DEPT"."DEPTNO")
```

虽然 DEPT 与 EMP 采用的是 HASH 连接，但是驱动表还是 DEPT。为什么 leading(emp@a) 失效了呢？因为两表关联如果是外连接，要改变 HASH 连接的驱动表必须使用 swap_join_inputs。现在我们使用 swap_join_inputs 来更改 HASH 连接的驱动表。

```
SQL> select /*+ use_hash(dept,emp@a) swap_join_inputs(emp@a) */ *
  2    from dept
  3   where not exists
  4   (select /*+ qb_name(a) */ null from emp where emp.deptno = dept.deptno);

Execution Plan
----------------------------------------------------------
Plan hash value: 152508289

---------------------------------------------------------------------
| Id  | Operation           | Name | Rows | Bytes | Cost (%CPU)| Time     |
---------------------------------------------------------------------
|   0 | SELECT STATEMENT    |      |   1 |   23 |    7  (15)| 00:00:01 |
|*  1 |  HASH JOIN RIGHT ANTI|     |   1 |   23 |    7  (15)| 00:00:01 |
|   2 |   TABLE ACCESS FULL | EMP  |  14 |   42 |    3   (0)| 00:00:01 |
|   3 |   TABLE ACCESS FULL | DEPT |   4 |   80 |    3   (0)| 00:00:01 |
---------------------------------------------------------------------

Predicate Information (identified by operation id):
---------------------------------------------------

   1 - access("EMP"."DEPTNO"="DEPT"."DEPTNO")
```

5.7.3 读者思考

现有如下 SQL。

```
select * from a where a.id not in (select id from b where id is not null);
```

假设 a 有 1 000 万条，b 有 1 000 条，请问如何优化该 SQL？
假设 a 有 1 000 条，b 有 1 000 万条，请问如何优化该 SQL？
假设 a 有 100 万条，b 有 1 000 万条，请问如何优化该 SQL？

5.8 FILTER

如果子查询（in/exists/not in/not exists）没能展开（unnest），在执行计划中就会产生 FILTER，FILTER 类似嵌套循环，FILTER 的算法与标量子查询一模一样。

现有如下 SQL 以及其执行计划。

5.8　FILTER

```
SQL> select ename, deptno
  2    from emp
  3   where exists (select deptno
  4                   from dept
  5                  where emp.deptno = dept.deptno
  6                    and dname = 'RESEARCH'
  7                    and rownum = 1);

Execution Plan
----------------------------------------------------------
Plan hash value: 3414630506

--------------------------------------------------------------------------------
| Id  | Operation                     | Name    | Rows | Bytes | Cost(%CPU)|Time     |
--------------------------------------------------------------------------------
|   0 | SELECT STATEMENT              |         |    5 |    45 |     6  (0)|00:00:01|
|*  1 |  FILTER                       |         |      |       |           |         |
|   2 |   TABLE ACCESS FULL           | EMP     |   14 |   126 |     3  (0)|00:00:01|
|*  3 |   COUNT STOPKEY               |         |      |       |           |         |
|*  4 |    TABLE ACCESS BY INDEX ROWID| DEPT    |    1 |    13 |     1  (0)|00:00:01|
|*  5 |     INDEX UNIQUE SCAN         | PK_DEPT |    1 |       |     0  (0)|00:00:01|
--------------------------------------------------------------------------------

Predicate Information (identified by operation id):
---------------------------------------------------

   1 - filter( EXISTS (SELECT 0 FROM "DEPT" "DEPT" WHERE ROWNUM=1 AND
              "DEPT"."DEPTNO"=:B1 AND "DNAME"='RESEARCH'))
   3 - filter(ROWNUM=1)
   4 - filter("DNAME"='RESEARCH')
   5 - access("DEPT"."DEPTNO"=:B1)
```

执行计划中，Id=1 就是 FILTER。注意观察 FILTER 所对应的谓词信息，FILTER 对应的谓词中包含有 **EXISTS**（子查询:B1）。运用光标移动大法我们可以知道 FILTER 下面有两个儿子（Id=2，Id=3）。

现在我们来看一下上面 SQL 带有 A-Time 的执行计划。

```
SQL> alter session set statistics_level=all;

Session altered.

SQL> select ename, deptno
  2    from emp
  3   where exists (select deptno
  4                   from dept
  5                  where emp.deptno = dept.deptno
  6                    and dname = 'RESEARCH'
  7                    and rownum = 1);

ENAME                              DEPTNO
------------------------------ ----------
SMITH                                  20
JONES                                  20
SCOTT                                  20
ADAMS                                  20
FORD                                   20

SQL> select * from table(dbms_xplan.display_cursor(null,null,'ALLSTATS LAST'));
```

```
PLAN_TABLE_OUTPUT
--------------------------------------------------------------------------------
SQL_ID  6mq67by27udgm, child number 1
-------------------------------------
select ename, deptno    from emp  where exists (select deptno
from dept          where emp.deptno = dept.deptno           and dname
= 'RESEARCH'              and rownum = 1)

Plan hash value: 3414630506

--------------------------------------------------------------------------------
| Id  | Operation                      | Name    | Starts | E-Rows | A-Rows |
--------------------------------------------------------------------------------
|   0 | SELECT STATEMENT               |         |      1 |        |      5 |
|*  1 |  FILTER                        |         |      1 |        |      5 |
|   2 |   TABLE ACCESS FULL            | EMP     |      1 |     14 |     14 |
|*  3 |   COUNT STOPKEY                |         |      3 |        |      1 |
|*  4 |    TABLE ACCESS BY INDEX ROWID | DEPT    |      3 |      1 |      1 |
|*  5 |     INDEX UNIQUE SCAN          | PK_DEPT |      3 |      1 |      3 |
--------------------------------------------------------------------------------

Predicate Information (identified by operation id):
---------------------------------------------------

   1 - filter( IS NOT NULL)
   3 - filter(ROWNUM=1)
   4 - filter("DNAME"='RESEARCH')
   5 - access("DEPT"."DEPTNO"=:B1)
```

为了方便排版，执行计划中省略了部分内容。Id=2 以及 Id=3 都是 FILTER 的儿子。Id=2 靠近 FILTER，我们可以把 Id=2 理解为 FILTER 的驱动表；Id=3 离 FILTER 比较远，可以把 Id=3 理解为 FILTER 的被驱动表。驱动表 EMP 只扫描了一次（Id=2，Starts=1），被驱动表被扫描了 3 次（Id=3，Starts=3）。

FILTER 的算法与标量子查询一模一样，驱动表都是固定的（固定为主表），不可更改。

从执行计划中我们可以看到，主表（EMP）通过连接列（DEPTNO）传值给子表（DEPT），:B1 就表示传值，主表（EMP）的连接列（DEPTNO）基数为 3，所以被驱动表（DEPT）被扫描了 3 次。FILTER 一般在整个 SQL 的快要执行完毕的时候执行（Filter 的 Id 一般小于等于 3）。

请注意，执行计划中还有一种 FILTER，这类 FILTER 只起过滤作用，这类 FILTER 下面只有一个儿子，谓词中没有 exists，也没有绑定变量:B1，例子如下。

```
PLAN_TABLE_OUTPUT
--------------------------------------------------------------------------------

--------------------------------------------------------------------------------
| Id  | Operation                      | Name               | Rows | Bytes |Cost |
--------------------------------------------------------------------------------
|   0 | SELECT STATEMENT               |                    |    1 |    81 | 1618|
|   1 |  SORT AGGREGATE                |                    |    1 |    81 |     |
|*  2 |   FILTER                       |                    |      |       |     |
|*  3 |    HASH JOIN OUTER             |                    |      |       |     |
|   4 |     NESTED LOOPS OUTER         |                    |  642 | 38520 |  838|
|*  5 |      INDEX FAST FULL SCAN      | PK_T_SEND_VEHICLE  |  413 |  8260 |   12|
|   6 |      TABLE ACCESS BY INDEX ROWID| T_TASK_HEAD       |    2 |    80 |    2|
|*  7 |       INDEX RANGE SCAN         | IDX_TASK_VEHICLE_NO|    2 |       |    1|
```

```
|  8 |     TABLE ACCESS FULL         | T_TASK_DETAIL      |   162K|  3337K|  777|
-------------------------------------------------------------------------------------

Predicate Information (identified by operation id):
---------------------------------------------------

   2 - filter("TTASKDETAI2_"."IS_REAL"='N' OR "TTASKDETAI2_"."IS_REAL" IS NULL)
   3 - access("TRANSTASKH0_"."TRANS_TASK_NO"="TTASKDETAI2_"."TRANS_TASK_NO"(+))
   5 - filter(TRIM("SENDVEHICL1_"."SEND_VEHICLE_NO")='010370412122280054')
   7 - access("TRANSTASKH0_"."SEND_VEHICLE_NO"(+)="SENDVEHICL1_"."SEND_VEHICLE_NO")
```

我们在做 SQL 优化的时候，一般只需要关注 FILTER 下面有两个或者两个以上儿子这种 FILTER。关于如何避免执行计划中产生 FILTER 以及执行计划中产生了 FILTER 怎么优化，请参阅本书 7.1 节。

5.9 IN 与 EXISTS 谁快谁慢

我相信很多人都受到过 in 与 exists 谁快谁慢的困扰。如果执行计划中没有产生 FILTER，那么我们可以参考以下思路：in 与 exists 是半连接，半连接也属于表连接，那么既然是表连接，我们需要关心两表的大小以及两表之间究竟走什么连接方式，还要控制两表的连接方式，才能随心所欲优化 SQL，而不是去记什么时候 in 跑得快，什么时候 exists 跑得快。如果执行计划中产生了 FILTER，大家还需阅读 7.1 节才能彻底知道答案。

5.10 SQL 语句的本质

前文提到，标量子查询可以改写为外连接（需要注意表与表之间关系，去重），半连接可以改写为内连接（需要注意表与表之间关系，去重），反连接可以改写为外连接（不需要注意表与表之间关系，也不需要去重）。SQL 语句中几乎所有的子查询都能改写为表连接的方式，所以我们提出这个观点：SQL 语句其本质就是表连接（内连接与外连接），以及表与表之间是几比几关系再加上 GROPU BY。

第 6 章 成本计算

6.1 优化 SQL 需要看 COST 吗

很多人在做 SQL 优化的时候都会去看 Cost。很多人经常问：为什么 Cost 很小，但是 SQL 就是跑很久不出结果呢？在这里告诉大家，做 SQL 优化的时候根本不需要去看 Cost，因为 Cost 是根据统计信息、根据一些数学公式计算出来的。正是因为 Cost 是基于统计信息、基于数学公式计算出来的，那么一旦统计信息有误差，数学公式有缺陷，Cost 就算错了。而一旦 Cost 计算错误，执行计划也就错了。当 SQL 需要优化的时候，Cost 往往是错误的，既然是错误的 Cost，我们干什么还要去看 Cost 呢？

本章带领大家手动计算全表扫描以及索引扫描成本，同时由此引出 SQL 优化核心思想。

6.2 全表扫描成本计算

本实验基于 Oracle11.2.0.1 Scott 账户。

```
SQL> select * from v$version where rownum=1;

BANNER
--------------------------------------------------------------
Oracle Database 11g Enterprise Edition Release 11.2.0.1.0 - Production
```

我们先创建一个表，名为 t_fullscan_cost（注意，只需要表结构，不要数据）。

```
SQL> create table t_fullscan_cost as select * from dba_objects where 1=0;
Table created.
```

我们设置表的 pctfree 为 99%，让表的一个块（8k）只能存储 82byte 数据。

```
SQL> alter table t_fullscan_cost pctfree 99 pctused 1;
Table altered.
```

这里只插入一行数据。

```
SQL> insert into t_fullscan_cost select * from dba_objects where rownum<2;
1 row created.
```

我们确保表中一个块只存一行数据。

```
SQL> alter table t_fullscan_cost minimize records_per_block;
Table altered.
```

6.2 全表扫描成本计算

我们再插入 999 行数据。

```
SQL> insert into t_fullscan_cost select * from dba_objects where rownum<1000;
999 rows created.
```

接下来提交数据。

```
SQL> commit;
Commit complete.
```

我们收集表的统计信息。

```
SQL> BEGIN
  2    DBMS_STATS.GATHER_TABLE_STATS(ownname          => 'SCOTT',
  3                                  tabname          => 'T_FULLSCAN_COST',
  4                                  estimate_percent => 100,
  5                                  method_opt       => 'for all columns size 1',
  6                                  degree           => 1,
  7                                  cascade          => TRUE);
  8  END;
  9  /

PL/SQL procedure successfully completed.
```

我们查看表的块数。

```
SQL> select owner, blocks
  2    from dba_tables
  3   where owner = 'SCOTT'
  4     and table_name = 'T_FULLSCAN_COST';

OWNER                BLOCKS
--------------- ----------
SCOTT                 1000
```

这里设置多块读参数为 16。

```
SQL> alter session set db_file_multiblock_read_count=16;

Session altered.
```

我们查看下面 SQL 语句执行计划。

```
SQL> set autot trace
SQL> select count(*) from t_fullscan_cost;

Execution Plan
----------------------------------------------------------
Plan hash value: 387824861

--------------------------------------------------------------------------------
| Id  | Operation           | Name            | Rows  | Cost (%CPU)| Time     |
--------------------------------------------------------------------------------
|   0 | SELECT STATEMENT    |                 |     1 |   220   (0)| 00:00:03 |
|   1 |  SORT AGGREGATE     |                 |     1 |            |          |
|   2 |   TABLE ACCESS FULL | T_FULLSCAN_COST |  1000 |   220   (0)| 00:00:03 |
--------------------------------------------------------------------------------
```

执行计划中 T_FULLSCAN_COST 走的是全表扫描，Cost 为 220。那么这 220 是怎么算出来的呢？我们先来看一下全表扫描成本计算公式。

全表扫描成本的计算方式如下。

```
Cost = (
        #SRds * sreadtim +
        #MRds * mreadtim +
        CPUCycles / cpuspeed
        ) / sreadtime

#SRds - number of single block reads    表示单块读次数
#MRds - number of multi block reads     表示多块读次数
#CPUCyles - number of CPU cycles        CPU 时钟周期数
sreadtim - single block read time       一次单块读耗时，单位毫秒
mreadtim - multi block read time        一次多块读耗时，单位毫秒
cpuspeed - CPU cycles per second        每秒 CPU 时钟周期数
```

注意：如果没有收集过系统统计信息（系统的 CPU 速度，磁盘 I/O 速度等），那么 Oracle 采用非工作量方式来计算成本。如果收集了系统统计信息，那么 Oracle 采用工作量统计方式来计算成本。一般我们是不会收集系统的统计信息的。所以默认情况下都是采用非工作量（noworkload）方式来计算成本。

现在我们来看一下系统的 CPU 和 I/O 情况。

```
SQL> select pname, pval1 from sys.aux_stats$ where sname='SYSSTATS_MAIN';

PNAME                PVAL1
---------------   ----------
CPUSPEED
CPUSPEEDNW       1683.65129    ---cpuspeed
IOSEEKTIM                10    ---I/O 寻道寻址耗时
IOTFRSPEED             4096    ---I/O 传输速度
MAXTHR
MBRC
MREADTIM
SLAVETHR
SREADTIM
```

因为 MBRC 为 NULL，所以 CBO 采用了非工作量来计算成本。

在全表扫描成本计算公式中，#SRds=0，因为是全表扫描一般都是多块读，#MRds=表的块数/多块读参数=1000/16，sreadtim=ioseektim+db_block_size/iotfrspeed，单块读耗时=I/O 寻道寻址耗时+块大小/I/O 传输速度，所以单块读耗时为 12 毫秒。

```
SQL> select (select pval1 from sys.aux_stats$ where pname = 'IOSEEKTIM') +
  2         (select value from v$parameter where name = 'db_block_size') /
  3         (select pval1 from sys.aux_stats$ where pname = 'IOTFRSPEED') "sreadtim"
  4    from dual;

  sreadtim
----------
        12
```

我们根据单块读耗时算法，查询到单块读耗时需要 12 毫秒。

mreadtim=ioseektim+db_file_multiblock_count*db_block_size/iotftspeed

多块读耗时= I/O 寻道寻址耗时+多块读参数*块大小/I/O 传输速度

6.2 全表扫描成本计算

```
SQL> select (select pval1 from sys.aux_stats$ where pname = 'IOSEEKTIM') +
  2         (select value
  3           from v$parameter
  4          where name = 'db_file_multiblock_read_count') *
  5         (select value from v$parameter where name = 'db_block_size') /
  6         (select pval1 from sys.aux_stats$ where pname = 'IOTFRSPEED') "mreadtim"
  7    from dual;

  mreadtim
----------
        42
```

我们根据多块读耗时算法，查询到多块读耗时需要 42 毫秒。

CPUCycles 等于 PLAN_TABL/V$SQL_PLAN 里面的 CPU_COST。

```
SQL> explain plan for select count(*) from t_fullscan_cost;

Explained.

SQL> select cpu_cost from plan_table where rownum<=1;

  CPU_COST
----------
   7271440
```

根据以上信息，我们现在来计算全表扫描成本。

```
SQL> select (0 * 12 + 1000 / 16 * 42 / 12 + 7271440 / (1683.65129 * 1000) / 12) cost
  2    from dual;

      COST
----------
219.109904
```

手动计算出来的 COST 值为 219，和我们看到的 220 相差 1。这是由隐含参数 _tablescan_cost_plus_one 造成的（请用 sys 运行下面的 SQL）。

```
SQL> SELECT x.ksppinm NAME, y.ksppstvl VALUE, x.ksppdesc describ
  2    FROM x$ksppi x, x$ksppcv y
  3   WHERE x.inst_id = USERENV('Instance')
  4     AND y.inst_id = USERENV('Instance')
  5     AND x.indx = y.indx
  6     AND x.ksppinm LIKE '%_table_scan_cost_plus_one%';

NAME                        VALUE            DESCRIB
--------------------------- ---------------- -----------------------------
_table_scan_cost_plus_one   TRUE             bump estimated full table scan
                                             and index ffs cost by one
```

该参数表示在 TABLE FULL SCAN 或者在 INDEX FAST FULL SCAN 的时候将 Cost 加 1。到此，我们终于人工计算出全表扫描成本。

全表扫描成本计算公式究竟是什么含义呢？我们再来看一下全表扫描成本计算公式。

```
Cost = (
        #SRds * sreadtim +
        #MRds * mreadtim +
        CPUCycles / cpuspeed
       ) / sreadtime
```

因为全表扫描没有单块读,所以#SRds=0,CPU 耗费的成本基本上可以忽略不计,所以我们将全表扫描公式变换如下。

```
Cost = (
       #MRds * mreadtim
       ) / sreadtime
```

#MRds 表示多块读 I/O 次数,那么现在我们得到一个结论:全表扫描成本公式的本质含义就是多块读的物理 I/O 次数乘以多块读耗时与单块读耗时的比值。

全表扫描成本计算公式是在 Oracle9i(2000 年左右)开始引入的,当时的 I/O 设备性能远远落后于现在的 I/O 设备(磁盘阵列),随着 SSD 的出现,寻道寻址时间已经可以忽略不计,磁盘阵列的性能已经有较大提升,因此认为在现代的 I/O 设备(磁盘阵列)中,单块读与多块读耗时几乎可以认为是一样的,全表扫描成本计算公式本质含义就是多块读物理 I/O 次数。

6.3 索引范围扫描成本计算

本实验基于 Oracle11.2.0.1 Scott 账户。

```
SQL> select * from v$version where rownum=1;

BANNER
----------------------------------------------------------------
Oracle Database 11g Enterprise Edition Release 11.2.0.1.0 - Production
```

我们先创建一个表名为 t_indexscan_cost。

```
SQL> create table t_indexscan_cost as select * from dba_objects;

Table created.
```

我们在 object_id 列上建立索引如下。

```
SQL> create index idx_cost on t_indexscan_cost(object_id);

Index created.
```

收集表统计信息如下。

```
SQL> BEGIN
  2    DBMS_STATS.GATHER_TABLE_STATS(ownname          => 'SCOTT',
  3                                  tabname          => 'T_INDEXSCAN_COST',
  4                                  estimate_percent => 100,
  5                                  method_opt       => 'for all columns size 1',
  6                                  degree           => 1,
  7                                  cascade          => TRUE);
  8  END;
  9  /
PL/SQL procedure successfully completed.
```

我们查看表总行数、object_id 最大值、object_id 最小值以及 null 值个数。

6.3 索引范围扫描成本计算

```
SQL> select b.num_rows,
  2         a.num_distinct,
  3         a.num_nulls,
  4         utl_raw.cast_to_number(high_value) high_value,
  5         utl_raw.cast_to_number(low_value) low_value,
  6         utl_raw.cast_to_number(high_value) -
  7         utl_raw.cast_to_number(low_value) "HIGH_VALUE-LOW_VALUE"
  8    from dba_tab_col_statistics a, dba_tables b
  9   where a.owner = b.owner
 10     and a.table_name = b.table_name
 11     and a.owner = 'SCOTT'
 12     and a.table_name = ('T_INDEXSCAN_COST')
 13     and a.column_name = 'OBJECT_ID';

  NUM_ROWS NUM_DISTINCT  NUM_NULLS HIGH_VALUE  LOW_VALUE HIGH_VALUE-LOW_VALUE
---------- ------------ ---------- ---------- ---------- --------------------
     72645        72645          0      76239          2                76237
```

我们查看下面 SQL 语句执行计划。

```
SQL> select owner from t_indexscan_cost where object_id<1000;

942 rows selected.

Execution Plan
----------------------------------------------------------
Plan hash value: 1756649757

--------------------------------------------------------------------------------
| Id | Operation                    | Name             | Rows | Bytes | Cost (%CPU)|
--------------------------------------------------------------------------------
|  0 | SELECT STATEMENT             |                  |  951 | 10461 |   19   (0)|
|  1 |  TABLE ACCESS BY INDEX ROWID | T_INDEXSCAN_COST |  951 | 10461 |   19   (0)|
|* 2 |   INDEX RANGE SCAN           | IDX_COST         |  951 |       |    4   (0)|
--------------------------------------------------------------------------------

Predicate Information (identified by operation id):
---------------------------------------------------

   2 - access("OBJECT_ID"<1000)
```

执行计划中，T_INDEXSCAN_COST 表走的是索引范围扫描。Cost 为 19。那么这 Cost 是怎么算出来的呢？我们先来看一下索引范围扫描的成本计算公式。

```
cost =
 blevel +
 celiling(leaf_blocks *effective index selectivity) +
 celiling(clustering_factor * effective table selectivity)
```

索引扫描成本计算公式中，blevel、leaf_blocks、clustering_factor 都可以通过下面查询得到。

```
SQL> select leaf_blocks, blevel, clustering_factor
  2    from dba_indexes
  3   where owner = 'SCOTT'
  4     and index_name = 'IDX_COST';

LEAF_BLOCKS     BLEVEL CLUSTERING_FACTOR
----------- ---------- -----------------
        161          1              1113
```

blevel 表示索引的二元高度，blevel 等于索引高度-1，leaf_blocks 表示索引的叶子块个数，

clustering_factor 表示索引的集群因子，effective index selectivity 表示索引有效选择性，effective table selectivity 表示表的有效选择性。

<的有效选择性为：

(limit-low_value)/(high_value-low_value)

(where 限制条件−最低值)/(最高值−最低值)

那么这里有效选择性=(1000−2)/(76239−2)。

执行计划中，CBO 估算返回的 Rows 为 951，这 951 是怎么算出来的呢？

CBO 预估的基数=有效选择性*(总行数−NULL 数)。

```
SQL> select ceil((1000-2)/(76239-2)*(72645-0)) from dual;

CEIL((1000-2)/(76239-2)*(72645-0))
----------------------------------
                               951
```

现在大家应该理解为什么我们曾在 1.3 节中提出执行计划中的 Rows 都是假的这个观点了。如果 where 条件较多，那么 CBO 在估算 Rows 的时候就会出现较大偏差，而且通常将 Rows 算小。因为当 where 条件变多的时候，CBO 估算返回的 Rows=某列选择性*某列选择性*某列选择性*...*表总行数。选择性一般来说都是小于 1 的分数，当 where 条件变多变复杂之后，CBO 估算的 Rows=小于 1 的分数*小于 1 的分数*小于 1 的分数*...*表的总行数，这种情况下 Rows 当然会越算越小（很多时候 Rows 经常被估算为 1）。

根据上述信息，现在我们来计算索引扫描的成本。

```
SQL> select 1+ceil(161*998/76237)+ceil(1113*998/76237) from dual;

1+CEIL(161*998/76237)+CEIL(1113*998/76237)
------------------------------------------
                                        19
```

手动计算出来的成本为 19，正好与执行计划中的 Cost 吻合。

在 1.4 节中我们曾经提到，如果回表次数太多，就不应该索引扫描，而应该走全表扫描。我们也可以从索引扫描的成本公式中验证该理论。clustering_factor * effective table selectivity 表示回表的 Cost，在示例中，回表的 Cost 为 15，回表的 Cost 占据整个索引扫描 Cost 的 79%。这就是回表次数太多不能走索引扫描的原因。

索引范围扫描成本计算公式的本质含义是什么呢？我们再来看一下索引范围扫描的成本计算公式。

```
cost =
blevel +
celiling(leaf_blocks *effective index selectivity) +
celiling(clustering_factor * effective table selectivity)
```

在 Oracle 数据库中，Btree 索引是树形结构，索引范围扫描需要从根扫描到分支，再扫描到叶子。叶子与叶子之间是双向指向的。blevel 等于索引高度−1，正好是索引根块到分支块的距离。leaf_blocks *effective index selectivity 表示可能需要扫描多少叶子块。clustering_factor * effective table selectivity 表示回表可能需要耗费多少 I/O。

索引范围扫描是单块读，回表也是单块读，因此，我们得到如下结论：索引扫描成本计算公式其本质就是单块读物理 I/O 次数。

为什么全表扫描成本计算公式要除以单块读耗时呢？上文提到，全表扫描 COST=多块读物理 I/O 次数*多块读耗时/单块读耗时，索引范围扫描 COST=单块读物理 I/O 次数。现在我们对全表扫描 COST 以及索引范围扫描 COST 都乘以单块读耗时：

全表扫描 COST*单块读耗时=多块读物理 I/O 次数*多块读耗时=全表扫描总耗时

索引范围扫描 COST*单块读耗时=单块读物理 I/O 次数*单块读耗时=索引扫描总耗时

到此，大家应该明白优化器何时选择全表扫描，何时选择索引扫描，就是比较走全表扫描的总耗时与走索引扫描的总耗时，哪个快就选哪个。

6.4 SQL 优化核心思想

现在的 IT 系统中，CPU 的发展日新月异，内存技术的更新也越来越频繁，只有磁盘技术发展最为迟缓，磁盘（I/O）已经成为整个 IT 系统的瓶颈。在 6.2 节中，我们提到全表扫描的成本其本质含义就是多块读的物理 I/O 次数，在 6.3 节中，我们提到索引范围扫描的成本其本质含义就是单块读的物理 I/O 次数。我们在判断究竟应该走全表扫描还是索引扫描的时候，往往会根据两种不同的扫描方式所耗费的物理 I/O 次数来做出选择，哪种扫描方式耗费的物理 I/O 次数少，就选择哪种扫描方式。在进行 SQL 优化的时候，我们也是根据哪种执行计划所耗费的物理 I/O 次数最少而选择哪种执行计划。

基于上述理论，我们给出整本书的核心观点：SQL 优化的核心思想就是想方设法减少 SQL 的物理 I/O 次数（不管是单块读次数还是多块读次数）。

第 7 章　必须掌握的查询变换

7.1　子查询非嵌套

子查询非嵌套（Subquery Unnesting）：当 where 子查询中有 in、not in、exists、not exists 等，CBO 会尝试将子查询展开（unnest），从而消除 FILTER，这个过程就叫作子查询非嵌套。**子查询非嵌套的目的就是消除 FILTER**。

现有如下 SQL 及其执行计划（Oracle11.2.0.1）。

```
SQL> select ename, deptno
  2    from emp
  3   where exists (select deptno
  4                   from dept
  5                  where dname = 'CHICAGO'
  6                    and emp.deptno = dept.deptno
  7                  union
  8                 select deptno
  9                   from dept
 10                  where loc = 'CHICAGO'
 11                    and dept.deptno = emp.deptno);

6 rows selected.

Execution Plan
----------------------------------------------------------
Plan hash value: 2705207488

--------------------------------------------------------------------------------------
| Id  | Operation                     | Name     | Rows  | Bytes | Cost(%CPU)| Time     |
--------------------------------------------------------------------------------------
|   0 | SELECT STATEMENT              |          |     5 |    45 |    15  (40)| 00:00:01|
|*  1 |  FILTER                       |          |       |       |            |         |
|   2 |   TABLE ACCESS FULL           | EMP      |    14 |   126 |     3   (0)| 00:00:01|
|   3 |   SORT UNIQUE                 |          |     2 |    24 |     4  (75)| 00:00:01|
|   4 |    UNION-ALL                  |          |       |       |            |         |
|*  5 |     TABLE ACCESS BY INDEX ROWID| DEPT    |     1 |    13 |     1   (0)| 00:00:01|
|*  6 |      INDEX UNIQUE SCAN        | PK_DEPT  |     1 |       |     0   (0)| 00:00:01|
|*  7 |     TABLE ACCESS BY INDEX ROWID| DEPT    |     1 |    11 |     1   (0)| 00:00:01|
|*  8 |      INDEX UNIQUE SCAN        | PK_DEPT  |     1 |       |     0   (0)| 00:00:01|
--------------------------------------------------------------------------------------

Predicate Information (identified by operation id):
---------------------------------------------------

   1 - filter( EXISTS ( (SELECT "DEPTNO" FROM "DEPT" "DEPT" WHERE
              "DEPT"."DEPTNO"=:B1 AND "DNAME"='CHICAGO')UNION (SELECT "DEPTNO" FROM "
DEPT"
              "DEPT" WHERE "DEPT"."DEPTNO"=:B2 AND "LOC"='CHICAGO')))
   5 - filter("DNAME"='CHICAGO')
   6 - access("DEPT"."DEPTNO"=:B1)
   7 - filter("LOC"='CHICAGO')
```

```
    8 - access("DEPT"."DEPTNO"=:B1)
```

执行计划中出现了 FILTER，驱动表因此被固定为 EMP。假设 EMP 有几百万甚至几千万行数据，那么该 SQL 效率就非常差。

现在将上述 SQL 改写如下。

```
SQL> select ename, deptno
  2    from emp
  3   where exists (select 1
  4                   from (select deptno
  5                           from dept
  6                          where dname = 'CHICAGO'
  7                          union
  8                         select deptno from dept where loc = 'CHICAGO') a
  9                  where a.deptno = emp.deptno);

6 rows selected.

Execution Plan
----------------------------------------------------------
Plan hash value: 4243948922

---------------------------------------------------------------------------
| Id  | Operation           | Name | Rows  | Bytes | Cost (%CPU)| Time     |
---------------------------------------------------------------------------
|   0 | SELECT STATEMENT    |      |     5 |   110 |    12  (25)| 00:00:01 |
|*  1 |  HASH JOIN SEMI     |      |     5 |   110 |    12  (25)| 00:00:01 |
|   2 |   TABLE ACCESS FULL | EMP  |    14 |   126 |     3   (0)| 00:00:01 |
|   3 |   VIEW              |      |     2 |    26 |     8  (25)| 00:00:01 |
|   4 |    SORT UNIQUE      |      |     1 |    24 |     8  (63)| 00:00:01 |
|   5 |     UNION-ALL       |      |       |       |            |          |
|*  6 |      TABLE ACCESS FULL| DEPT |   1 |    13 |     3   (0)| 00:00:01 |
|*  7 |      TABLE ACCESS FULL| DEPT |   1 |    11 |     3   (0)| 00:00:01 |
---------------------------------------------------------------------------

Predicate Information (identified by operation id):
---------------------------------------------------

   1 - access("A"."DEPTNO"="EMP"."DEPTNO")
   6 - filter("DNAME"='CHICAGO')
   7 - filter("LOC"='CHICAGO')
```

对 SQL 进行等价改写之后，消除了 FILTER。为什么要消除 FILTER 呢？因为 FILTER 的驱动表是固定的，一旦驱动表被固定，那么执行计划也就被固定了。对于 DBA 来说这并不是好事，因为一旦固定的执行计划本身是错误的（低效的），就会引起性能问题，想要提升性能必须改写 SQL 语句，但是这时 SQL 已经上线，无法更改，所以，一定要消除 FILTER。

很多公司都有开发 DBA，开发 DBA 很大一部分的工作职责就是：必须保证 SQL 上线之后，每个 SQL 语句的执行计划都是可控的，这样才能尽可能避免系统中 SQL 越跑越慢。

下面我们继续对上述 SQL 进行等价改写。

```
SQL> select ename, deptno
  2    from emp
  3   where deptno in (select deptno
  4                      from dept
  5                     where dname = 'CHICAGO'
  6                     union
```

```
         7                  select deptno from dept where loc = 'CHICAGO');

6 rows selected.

Execution Plan
----------------------------------------------------------
Plan hash value: 2842951954

---------------------------------------------------------------------------
| Id  | Operation            | Name     | Rows | Bytes | Cost (%CPU)| Time     |
---------------------------------------------------------------------------
|   0 | SELECT STATEMENT     |          |    9 |   198 |   12  (25)| 00:00:01 |
|*  1 |  HASH JOIN           |          |    9 |   198 |   12  (25)| 00:00:01 |
|   2 |   VIEW               | VW_NSO_1 |    2 |    26 |    8  (25)| 00:00:01 |
|   3 |    SORT UNIQUE       |          |    2 |    24 |    8  (63)| 00:00:01 |
|   4 |     UNION-ALL        |          |      |       |           |          |
|*  5 |      TABLE ACCESS FULL| DEPT    |    1 |    13 |    3   (0)| 00:00:01 |
|*  6 |      TABLE ACCESS FULL| DEPT    |    1 |    11 |    3   (0)| 00:00:01 |
|   7 |   TABLE ACCESS FULL  | EMP      |   14 |   126 |    3   (0)| 00:00:01 |
---------------------------------------------------------------------------

Predicate Information (identified by operation id):
---------------------------------------------------

   1 - access("DEPTNO"="DEPTNO")
   5 - filter("DNAME"='CHICAGO')
   6 - filter("LOC"='CHICAGO')
```

将 SQL 改写为 in 之后，也消除了 FILTER。

如何才能产生 FILTER 呢？我们只需要在子查询中添加 /*+ no_unnest */。

```
SQL> select ename, deptno
  2    from emp
  3   where deptno in (select /*+ no_unnest */ deptno
  4                      from dept
  5                     where dname = 'CHICAGO'
  6                     union
  7                    select deptno from dept where loc = 'CHICAGO');

6 rows selected.

Execution Plan
----------------------------------------------------------
Plan hash value: 2705207488

---------------------------------------------------------------------------
| Id  |Operation                          | Name     | Rows | Bytes | Cost(%CPU)|Time     |
---------------------------------------------------------------------------
|   0 |SELECT STATEMENT                   |          |    5 |    45 |   15  (40)|00:00:01|
|*  1 | FILTER                            |          |      |       |           |         |
|   2 |  TABLE ACCESS FULL                | EMP      |   14 |   126 |    3   (0)|00:00:01|
|   3 |  SORT UNIQUE                      |          |    2 |    24 |    4  (75)|00:00:01|
|   4 |   UNION-ALL                       |          |      |       |           |         |
|*  5 |    TABLE ACCESS BY INDEX ROWID    | DEPT     |    1 |    13 |    1   (0)|00:00:01|
|*  6 |     INDEX UNIQUE SCAN             | PK_DEPT  |    1 |       |    0   (0)|00:00:01|
|*  7 |    TABLE ACCESS BY INDEX ROWID    | DEPT     |    1 |    11 |    1   (0)|00:00:01|
|*  8 |     INDEX UNIQUE SCAN             | PK_DEPT  |    1 |       |    0   (0)|00:00:01|
---------------------------------------------------------------------------

Predicate Information (identified by operation id):
```

```
--------------------------------------------------
   1 - filter( EXISTS ( (SELECT /*+ NO_UNNEST */ "DEPTNO" FROM "DEPT" "DEPT"
              WHERE "DEPTNO"=:B1 AND "DNAME"='CHICAGO')UNION (SELECT "DEPTNO" FROM "D
EPT"
              "DEPT" WHERE "DEPTNO"=:B2 AND "LOC"='CHICAGO')))
   5 - filter("DNAME"='CHICAGO')
   6 - access("DEPTNO"=:B1)
   7 - filter("LOC"='CHICAGO')
   8 - access("DEPTNO"=:B1)
```

大家可能会问，既然能通过 HINT(NO_UNNEST)让执行计划产生 FILTER，那么执行计划中如果产生了 FILTER，能否通过 HINT(UNNEST)消除 FILTER 呢？执行计划中的 FILTER 很少能够通过 HINT 消除，一般需要通过 SQL 等价改写来消除。

现在我们对产生 FILTER 的 SQL 添加 HINT(UNNEST)来尝试消除 FILTER。

```
SQL> select ename, deptno
  2    from emp
  3   where exists (select /*+ unnest */ deptno
  4                   from dept
  5                  where dname = 'CHICAGO'
  6                    and emp.deptno = dept.deptno
  7                  union
  8                 select deptno
  9                   from dept
 10                  where loc = 'CHICAGO'
 11                    and dept.deptno = emp.deptno);

6 rows selected.

Execution Plan
----------------------------------------------------------
Plan hash value: 2705207488

--------------------------------------------------------------------------------
| Id |Operation                        | Name    | Rows | Bytes | Cost(%CPU)|Time     |
--------------------------------------------------------------------------------
|  0 |SELECT STATEMENT                 |         |    5 |   45  |   15  (40)|00:00:01|
|* 1 | FILTER                          |         |      |       |           |        |
|  2 |  TABLE ACCESS FULL              | EMP     |   14 |  126  |    3   (0)|00:00:01|
|  3 |  SORT UNIQUE                    |         |    2 |   24  |    4  (75)|00:00:01|
|  4 |   UNION-ALL                     |         |      |       |           |        |
|* 5 |    TABLE ACCESS BY INDEX ROWID  | DEPT    |    1 |   13  |    1   (0)|00:00:01|
|* 6 |     INDEX UNIQUE SCAN           | PK_DEPT |    1 |       |    0   (0)|00:00:01|
|* 7 |    TABLE ACCESS BY INDEX ROWID  | DEPT    |    1 |   11  |    1   (0)|00:00:01|
|* 8 |     INDEX UNIQUE SCAN           | PK_DEPT |    1 |       |    0   (0)|00:00:01|
--------------------------------------------------------------------------------

Predicate Information (identified by operation id):
--------------------------------------------------

   1 - filter( EXISTS ( (SELECT /*+ UNNEST */ "DEPTNO" FROM "DEPT" "DEPT" WHERE
              "DEPT"."DEPTNO"=:B1 AND "DNAME"='CHICAGO')UNION (SELECT "DEPTNO" FROM "
DEPT"
              "DEPT" WHERE "DEPT"."DEPTNO"=:B2 AND "LOC"='CHICAGO')))
   5 - filter("DNAME"='CHICAGO')
   6 - access("DEPT"."DEPTNO"=:B1)
   7 - filter("LOC"='CHICAGO')
   8 - access("DEPT"."DEPTNO"=:B1)
```

执行计划中还是有 FILTER。再次强调：执行计划中如果产生了 FILTER，一般是无法通过 HINT 消除的，一定要注意执行计划中的 FILTER。

请注意，虽然我们一直强调要消除执行计划中的 FILTER，本意是要保证执行计划是可控的，并不意味着执行计划产生了 FILTER 就一定性能差，相反有时候我们还可以用 FILTER 来优化 SQL。

哪些 SQL 写法容易产生 FILTER 呢？当子查询语句含有 exists 或者 not exists 时，子查询中有固化子查询关键词（union/union all/start with connect by/rownum/cube/rollup），那么执行计划中就容易产生 FILTER，例如，exists 中有 rownum 产生 FILTER。

```
SQL> select ename, deptno
  2    from emp
  3   where exists (select deptno
  4                   from dept
  5                  where loc = 'CHICAGO'
  6                    and dept.deptno = emp.deptno
  7                    and rownum <= 1);

6 rows selected.

Execution Plan
----------------------------------------------------------
Plan hash value: 3414630506

--------------------------------------------------------------------------
| Id  | Operation                    | Name    | Rows | Bytes | Cost (%CPU)| Time     |
--------------------------------------------------------------------------
|   0 | SELECT STATEMENT             |         |    5 |    45 |     6   (0)| 00:00:01 |
|*  1 |  FILTER                      |         |      |       |            |          |
|   2 |   TABLE ACCESS FULL          | EMP     |   14 |   126 |     3   (0)| 00:00:01 |
|*  3 |   COUNT STOPKEY              |         |      |       |            |          |
|*  4 |    TABLE ACCESS BY INDEX ROWID| DEPT   |    1 |    11 |     1   (0)| 00:00:01 |
|*  5 |     INDEX UNIQUE SCAN        | PK_DEPT |    1 |       |     0   (0)| 00:00:01 |
--------------------------------------------------------------------------

Predicate Information (identified by operation id):
---------------------------------------------------

   1 - filter( EXISTS (SELECT 0 FROM "DEPT" "DEPT" WHERE ROWNUM<=1 AND
              "DEPT"."DEPTNO"=:B1 AND "LOC"='CHICAGO'))
   3 - filter(ROWNUM<=1)
   4 - filter("LOC"='CHICAGO')
   5 - access("DEPT"."DEPTNO"=:B1)
```

exists 中有树形查询产生 FILTER。

```
SQL> select *
  2    from dept
  3   where exists (select null
  4                   from emp
  5                  where dept.deptno = emp.deptno
  6                  start with empno = 7698
  7                 connect by prior empno = mgr);

Execution Plan
----------------------------------------------------------
```

```
Plan hash value: 4210865686

-------------------------------------------------------------------------------------
| Id |Operation                                    | Name | Rows  | Bytes |Cost (%CPU)|
-------------------------------------------------------------------------------------
|  0 |SELECT STATEMENT                             |      |    1  |   20  |   9   (0)|
|* 1 |  FILTER                                     |      |       |       |           |
|  2 |    TABLE ACCESS FULL                        | DEPT |    4  |   80  |   3   (0)|
|* 3 |    FILTER                                   |      |       |       |           |
|* 4 |     CONNECT BY NO FILTERING WITH SW (UNIQUE)|      |       |       |           |
|  5 |      TABLE ACCESS FULL                      | EMP  |   14  |  154  |   3   (0)|
-------------------------------------------------------------------------------------

Predicate Information (identified by operation id):
---------------------------------------------------

   1 - filter( EXISTS (SELECT 0 FROM "EMP" "EMP" WHERE "EMP"."DEPTNO"=:B1 START WITH
               "EMPNO"=7698 CONNECT BY "MGR"=PRIOR "EMPNO"))
   3 - filter("EMP"."DEPTNO"=:B1)
   4 - access("MGR"=PRIOR "EMPNO")
       filter("EMPNO"=7698)
```

为什么 exists/not exists 容易产生 FILTER，而 in 很少会产生 FILTER 呢？当子查询中有固化关键字（union/union all/start with connect by/rownum/cube/rollup），子查询会被固化为一个整体，采用 exists/not exists 这种写法，这时子查询中有主表连接列，只能是主表通过连接列传值给子表，所以 CBO 只能选择 FILTER。而我们如果将 SQL 改写为 in/not in 这种写法，子查询虽然被固化为整体，但是子查询中没有主表连接列字段，这个时候 CBO 就不会选择 FILTER。

7.2 视图合并

视图合并（View Merge）：当 SQL 语句中有内联视图（in-line view，from 后面的子查询），或者 SQL 语句中有用 create view 创建的视图，CBO 会尝试将内联视图/视图拆开，进行等价的改写，这个过程就叫作视图合并。如果没有发生视图合并，在执行计划中，我们可以看到 VIEW 关键字，而且视图/子查询会作为一个整体。如果发生了视图合并，那么视图/子查询就会被拆开，而且执行计划中视图/子查询部分就没有 VIEW 关键字。

现有如下 SQL 及其执行计划（Oracle11.2.0.1）。

```
SQL> select a.*, c.grade
  2    from (select ename, sal, a.deptno, b.dname
  3            from emp a, dept b
  4           where a.deptno = b.deptno) a,
  5         salgrade c
  6   where a.sal between c.losal and c.hisal;

14 rows selected.

Execution Plan
----------------------------------------------------------
Plan hash value: 3095952880

-------------------------------------------------------------------------------
| Id  | Operation                   | Name    | Rows  | Bytes | Cost (%CPU)|
```

```
---------------------------------------------------------------------------
|   0 | SELECT STATEMENT            |          |    1 |    65 |    9  (23)|
|   1 |  NESTED LOOPS               |          |      |       |           |
|   2 |   NESTED LOOPS              |          |    1 |    65 |    9  (23)|
|   3 |    MERGE JOIN               |          |    1 |    52 |    8  (25)|
|   4 |     SORT JOIN               |          |    5 |   195 |    4  (25)|
|   5 |      TABLE ACCESS FULL      | SALGRADE |    5 |   195 |    3   (0)|
|*  6 |     FILTER                  |          |      |       |           |
|*  7 |      SORT JOIN              |          |   14 |   182 |    4  (25)|
|   8 |       TABLE ACCESS FULL     | EMP      |   14 |   182 |    3   (0)|
|*  9 |    INDEX UNIQUE SCAN        | PK_DEPT  |    1 |       |    0   (0)|
|  10 |   TABLE ACCESS BY INDEX ROWID| DEPT    |    1 |    13 |    1   (0)|
---------------------------------------------------------------------------

Predicate Information (identified by operation id):
---------------------------------------------------

   6 - filter("SAL"<="C"."HISAL")
   7 - access("SAL">="C"."LOSAL")
       filter("SAL">="C"."LOSAL")
   9 - access("A"."DEPTNO"="B"."DEPTNO")
```

SQL 语句中有内联视图，但是执行计划中没有 VIEW 关键字，说明发生了视图合并。内联视图中 EMP 表是与 DEPT 表关联的，但是执行计划中，EMP 表是与 SALGRADE 先关联的，EMP 表与 SALGRADE 关联之后得到一个结果集，再与 DEPT 表进行的关联，**这说明发生了视图合并之后，有可能会打乱视图/子查询中表的原本连接顺序。**

现在我们添加 HINT:no_merge（子查询别名/视图别名）禁止视图合并，再看执行计划。

```
SQL> select /*+ no_merge(a) */
  2    a.*, c.grade
  3    from (select ename, sal, a.deptno, b.dname
  4            from emp a, dept b
  5           where a.deptno = b.deptno) a,
  6         salgrade c
  7   where a.sal between c.losal and c.hisal;

14 rows selected.


Execution Plan
----------------------------------------------------------
Plan hash value: 4110645763

------------------------------------------------------------------------------------
| Id  | Operation                      | Name     | Rows  | Bytes | Cost (%CPU)|
------------------------------------------------------------------------------------
|   0 | SELECT STATEMENT               |          |     1 |    81 |    11  (28)|
|   1 |  MERGE JOIN                    |          |     1 |    81 |    11  (28)|
|   2 |   SORT JOIN                    |          |     5 |   195 |     4  (25)|
|   3 |    TABLE ACCESS FULL           | SALGRADE |     5 |   195 |     3   (0)|
|*  4 |   FILTER                       |          |       |       |            |
|*  5 |    SORT JOIN                   |          |    14 |   588 |     7  (29)|
|   6 |     VIEW                       |          |    14 |   588 |     6  (17)|
|   7 |      MERGE JOIN                |          |    14 |   364 |     6  (17)|
|   8 |       TABLE ACCESS BY INDEX ROWID| DEPT   |     4 |    52 |     2   (0)|
|   9 |        INDEX FULL SCAN         | PK_DEPT  |     4 |       |     1   (0)|
|* 10 |       SORT JOIN                |          |    14 |   182 |     4  (25)|
|  11 |        TABLE ACCESS FULL       | EMP      |    14 |   182 |     3   (0)|
------------------------------------------------------------------------------------
```

```
Predicate Information (identified by operation id):
---------------------------------------------------

   4 - filter("A"."SAL"<="C"."HISAL")
   5 - access("A"."SAL">="C"."LOSAL")
       filter("A"."SAL">="C"."LOSAL")
  10 - access("A"."DEPTNO"="B"."DEPTNO")
       filter("A"."DEPTNO"="B"."DEPTNO")
```

执行计划中有 VIEW 关键字，而且 EMP 是与 DEPT 进行关联的，这说明执行计划中没有发生视图合并。

我们也可以直接在子查询里面添加 HINT:no_merge 禁止视图合并。

```
SQL> select a.*, c.grade
  2    from (select /*+ no_merge */
  3                 ename, sal, a.deptno, b.dname
  4            from emp a, dept b
  5           where a.deptno = b.deptno) a,
  6         salgrade c
  7   where a.sal between c.losal and c.hisal;

14 rows selected.


Execution Plan
----------------------------------------------------------
Plan hash value: 4110645763

---------------------------------------------------------------------------------------
| Id  | Operation                       | Name     | Rows  | Bytes | Cost (%CPU)|
---------------------------------------------------------------------------------------
|   0 | SELECT STATEMENT                |          |     1 |    81 |    11  (28)|
|   1 |  MERGE JOIN                     |          |     1 |    81 |    11  (28)|
|   2 |   SORT JOIN                     |          |     5 |   195 |     4  (25)|
|   3 |    TABLE ACCESS FULL            | SALGRADE |     5 |   195 |     3   (0)|
|*  4 |   FILTER                        |          |       |       |            |
|*  5 |    SORT JOIN                    |          |    14 |   588 |     7  (29)|
|   6 |     VIEW                        |          |    14 |   588 |     6  (17)|
|   7 |      MERGE JOIN                 |          |    14 |   364 |     6  (17)|
|   8 |       TABLE ACCESS BY INDEX ROWID| DEPT    |     4 |    52 |     2   (0)|
|   9 |        INDEX FULL SCAN          | PK_DEPT  |     4 |       |     1   (0)|
|* 10 |       SORT JOIN                 |          |    14 |   182 |     4  (25)|
|  11 |        TABLE ACCESS FULL        | EMP      |    14 |   182 |     3   (0)|
---------------------------------------------------------------------------------------

Predicate Information (identified by operation id):
---------------------------------------------------

   4 - filter("A"."SAL"<="C"."HISAL")
   5 - access("A"."SAL">="C"."LOSAL")
       filter("A"."SAL">="C"."LOSAL")
  10 - access("A"."DEPTNO"="B"."DEPTNO")
       filter("A"."DEPTNO"="B"."DEPTNO")
```

当视图/子查询中有多个表关联，发生视图合并之后一般会将视图/子查询内部表关联顺序打乱。

大家可能遇到过类似案例，例如下面 SQL 所示。

```
select ... from () a,() b where a.id=b.id;
```

单独执行子查询 a，速度非常快，单独执行子查询 b，速度也非常快，但是把上面两个子查询组合在一起，速度反而很慢，这就是典型的视图合并引起的性能问题。遇到类似问题，我们可以添加 HINT:no_merge 禁止视图合并，也可以让子查询 a 与子查询 b 进行 HASH 连接，当子查询 a 与子查询 b 进行 HASH 连接之后，就不会发生视图合并了。

```
select /*+ use_hash(a,b) */ ... from () a,() b where a.id=b.id;
```

为什么让子查询 a 与子查询 b 进行 HASH 连接能使 SQL 变快呢？大家再回忆一下 HASH 连接的算法，嵌套循环会传值（驱动表传值给被驱动表，通过连接列），HASH 连接不会传值。因为 HASH 连接不传值，所以当子查询 a 与子查询 b 进行 HASH 连接之后，会自动地把子查询 a 与子查询 b 作为一个整体。

与子查询非嵌套一样，当视图中有固化子查询关键字的时候，就不能发生视图合并。

固化子查询的关键字包括 union、union all、start with connect by、rownum、cube、rollup。

现在我们对示例 SQL 添加 union all，查看 SQL 执行计划。

```
SQL> select a.*, c.grade
  2   from (select ename, sal, a.deptno, b.dname
  3           from emp a, dept b
  4          where a.deptno = b.deptno
  5          union all
  6         select 'SMITH', 1600, 10, 'ACCOUNTING' from dual) a,
  7        salgrade c
  8   where a.sal between c.losal and c.hisal;

15 rows selected.

Execution Plan
----------------------------------------------------------
Plan hash value: 1428389312

--------------------------------------------------------------------------------
| Id  | Operation                       | Name     | Rows  | Bytes | Cost (%CPU)|
--------------------------------------------------------------------------------
|   0 | SELECT STATEMENT                |          |    1  |   81  |  13  (24)|
|   1 |  MERGE JOIN                     |          |    1  |   81  |  13  (24)|
|   2 |   SORT JOIN                     |          |    5  |  195  |   4  (25)|
|   3 |    TABLE ACCESS FULL            | SALGRADE |    5  |  195  |   3   (0)|
|*  4 |   FILTER                        |          |       |       |          |
|*  5 |    SORT JOIN                    |          |   15  |  630  |   9  (23)|
|   6 |     VIEW                        |          |   15  |  630  |   8  (13)|
|   7 |      UNION-ALL                  |          |       |       |          |
|   8 |       MERGE JOIN                |          |   14  |  364  |   6  (17)|
|   9 |        TABLE ACCESS BY INDEX ROWID| DEPT   |    4  |   52  |   2   (0)|
|  10 |         INDEX FULL SCAN         | PK_DEPT  |    4  |       |   1   (0)|
|* 11 |        SORT JOIN                |          |   14  |  182  |   4  (25)|
|  12 |         TABLE ACCESS FULL       | EMP      |   14  |  182  |   3   (0)|
|  13 |       FAST DUAL                 |          |    1  |       |   2   (0)|
--------------------------------------------------------------------------------

Predicate Information (identified by operation id):
---------------------------------------------------

   4 - filter("A"."SAL"<="C"."HISAL")
   5 - access("A"."SAL">="C"."LOSAL")
```

```
           filter("A"."SAL">="C"."LOSAL")
  11 - access("A"."DEPTNO"="B"."DEPTNO")
       filter("A"."DEPTNO"="B"."DEPTNO")
```

从执行计划中我们可以看到，添加了 union all 之后，子查询被固化，没有发生视图合并。现在我们对 SQL 添加 rownum，查看 SQL 执行计划。

```
SQL> select a.*, c.grade
  2    from (select ename, sal, a.deptno, b.dname
  3            from emp a, dept b
  4           where a.deptno = b.deptno
  5             and rownum >= 1) a,
  6         salgrade c
  7   where a.sal between c.losal and c.hisal;

14 rows selected.

Execution Plan
----------------------------------------------------------
Plan hash value: 819637296

--------------------------------------------------------------------------------
| Id  | Operation                      | Name     | Rows | Bytes | Cost (%CPU)|
--------------------------------------------------------------------------------
|   0 | SELECT STATEMENT               |          |    1 |    72 |   11  (28)|
|   1 |  MERGE JOIN                    |          |    1 |    72 |   11  (28)|
|   2 |   SORT JOIN                    |          |    5 |   195 |    4  (25)|
|   3 |    TABLE ACCESS FULL           | SALGRADE |    5 |   195 |    3   (0)|
|*  4 |   FILTER                       |          |      |       |           |
|*  5 |    SORT JOIN                   |          |   14 |   462 |    7  (29)|
|   6 |     VIEW                       |          |   14 |   462 |    6  (17)|
|   7 |      COUNT                     |          |      |       |           |
|*  8 |       FILTER                   |          |      |       |           |
|   9 |        MERGE JOIN              |          |   14 |   364 |    6  (17)|
|  10 |         TABLE ACCESS BY INDEX ROWID| DEPT |    4 |    52 |    2   (0)|
|  11 |          INDEX FULL SCAN       | PK_DEPT  |    4 |       |    1   (0)|
|* 12 |         SORT JOIN              |          |   14 |   182 |    4  (25)|
|  13 |          TABLE ACCESS FULL     | EMP      |   14 |   182 |    3   (0)|
--------------------------------------------------------------------------------

Predicate Information (identified by operation id):
---------------------------------------------------

   4 - filter("A"."SAL"<="C"."HISAL")
   5 - access("A"."SAL">="C"."LOSAL")
       filter("A"."SAL">="C"."LOSAL")
   8 - filter(ROWNUM>=1)
  12 - access("A"."DEPTNO"="B"."DEPTNO")
       filter("A"."DEPTNO"="B"."DEPTNO")
```

从执行计划中我们可以看到，添加了 rownum 之后，子查询同样被固化，没有发生视图合并。

7.3 谓词推入

谓词推入（Pushing Predicate）：当 SQL 语句中包含不能合并的视图，同时视图有谓词过滤（也就是 where 过滤条件），CBO 会将谓词过滤条件推入视图中，这个过程就叫作谓词推入。

谓词推入的主要目的就是让 Oracle 尽可能早地过滤掉无用的数据，从而提升查询性能。

为什么谓词推入必须要有不能被合并的视图呢？因为一旦视图被合并了，执行计划中根本找不到视图，这个时候谓词往哪里推呢？所以谓词推入的必要前提是 SQL 中要有不能合并的视图。

我们先创建一个不能被合并的视图（视图中有 union all）。

```
SQL> create or replace view v_pushpredicate as
  2    select * from test
  3    union all
  4    select * from test where rownum>=1;

View created.
```

然后我们运行下面的 SQL，同时查看执行计划。

```
SQL> select * from v_pushpredicate where object_id<10;

16 rows selected.

Execution Plan
----------------------------------------------------------
Plan hash value: 669161224

--------------------------------------------------------------------------------
| Id  | Operation                     | Name            | Rows  | Bytes | Cost (%CPU)|
--------------------------------------------------------------------------------
|   0 | SELECT STATEMENT              |                 | 72470 |   14M |   238   (1)|
|*  1 |  VIEW                         | V_PUSHPREDICATE | 72470 |   14M |   238   (1)|
|   2 |   UNION-ALL                   |                 |       |       |            |
|   3 |    TABLE ACCESS BY INDEX ROWID| TEST            |     8 |   776 |     3   (0)|
|*  4 |     INDEX RANGE SCAN          | IDX_ID          |     8 |       |     2   (0)|
|   5 |    COUNT                      |                 |       |       |            |
|*  6 |     FILTER                    |                 |       |       |            |
|   7 |      TABLE ACCESS FULL        | TEST            | 72462 | 6864K |   235   (1)|
--------------------------------------------------------------------------------

Predicate Information (identified by operation id):
---------------------------------------------------

   1 - filter("OBJECT_ID"<10)
   4 - access("OBJECT_ID"<10)
   6 - filter(ROWNUM>=1)
```

SQL 语句中，where 过滤条件是针对视图过滤的，但是从执行计划中（Id=4）我们可以看到，where 过滤条件跑到视图中的表中进行过滤了，这就是谓词推入。因为视图中第二个表有rownum，rownum 会阻止谓词推入，所以第二个表走的是全表扫描，需要到视图上进行过滤（Id=1）。

我们在看执行计划的时候，如果 VIEW 前面有"*"号，这就说明有谓词没有推入到视图中。

一般情况下，常量的谓词推入对性能的提升都是有益的。那么什么是常量的谓词推入呢？常量的谓词推入就是谓词是正常的过滤条件，而非连接列。

在 2011 年我们曾帮网友做过一次常量谓词推入优化，因为实在是太简单，所以没有将其纳入书中。有兴趣的读者可以参考博客：http://blog.csdn.net/robinson1988/article/details/6613851。

还有一种谓词推入，是把连接列当作谓词推入到视图中，这种谓词推入我们一般叫作连接列谓词推入，此类谓词推入最容易产生性能问题。

现在我们将上面视图中的 rownum 去掉（为了使连接列能推入视图）。

```
SQL> create or replace view v_pushpredicate as
  2    select * from test
  3    union all
  4    select * from test;

View created.
```

我们添加 HINT:push_pred 提示将连接列推入到视图中。

```
SQL> select /*+ push_pred(b) */ *
  2    from test a, v_pushpredicate b
  3   where a.object_id = b.object_id
  4     and a.owner = 'SCOTT';

14 rows selected.

Execution Plan
----------------------------------------------------------
Plan hash value: 2131469559

--------------------------------------------------------------------------------
| Id  | Operation                     | Name           | Rows  | Bytes | Cost(%CPU)|
--------------------------------------------------------------------------------
|   0 | SELECT STATEMENT              |                |  4997 | 1444K | 10073  (1)|
|   1 |  NESTED LOOPS                 |                |  4997 | 1444K | 10073  (1)|
|   2 |   TABLE ACCESS BY INDEX ROWID | TEST           |  2499 |  236K |    73  (0)|
|*  3 |    INDEX RANGE SCAN           | IDX_OWNER      |  2499 |       |     6  (0)|
|   4 |   VIEW                        | V_PUSHPREDICATE|     1 |   199 |     4  (0)|
|   5 |    UNION ALL PUSHED PREDICATE |                |       |       |           |
|   6 |     TABLE ACCESS BY INDEX ROWID| TEST          |     1 |    97 |     2  (0)|
|*  7 |      INDEX RANGE SCAN         | IDX_ID         |     1 |       |     1  (0)|
|   8 |     TABLE ACCESS BY INDEX ROWID| TEST          |     1 |    97 |     2  (0)|
|*  9 |      INDEX RANGE SCAN         | IDX_ID         |     1 |       |     1  (0)|
--------------------------------------------------------------------------------

Predicate Information (identified by operation id):
---------------------------------------------------

   3 - access("A"."OWNER"='SCOTT')
   7 - access("OBJECT_ID"="A"."OBJECT_ID")
   9 - access("OBJECT_ID"="A"."OBJECT_ID")
```

将连接列推入到视图中这种谓词推入，一般在执行计划中都能看到 PUSHED PREDICATE 或者 VIEW PUSHED PREDICATE，而且视图一般作为嵌套循环的被驱动表，同时视图中谓词被推入列有索引。这种谓词推入对性能有好有坏。为什么连接列谓词推入，被推入的视图一般都作为嵌套循环的被驱动表呢？这是因为连接列谓词推入需要传值（传值到视图里面），而有传值操作的表连接方法只有嵌套循环或者 FILTER。FILTER 是专门针对半连接或者反连接的（where 后面的子查询），谓词推入是专门针对 from 后面的子查询，所以连接列谓词推入，被推入的视图一般都作为嵌套循环的被驱动表。

在本书示例中，连接列谓词推入的执行计划是最优执行计划。驱动表 test 过滤之后（owner='SCOTT'）只返回 7 行数据，然后通过连接列传值 7 次，传入视图中，视图里面的表走的是索引扫描，因为驱动表 7 次传值，所以被驱动表（视图）一共被扫描了 7 次，但是每次扫描都是索引扫描。

现在我们去掉 HINT:push_pred。

```
SQL> select *
  2    from test a, v_pushpredicate b
  3    where a.object_id = b.object_id
  4    and a.owner = 'SCOTT';

14 rows selected.

Execution Plan
----------------------------------------------------------
Plan hash value: 1745523384

--------------------------------------------------------------------------------
| Id  | Operation                    | Name           | Rows  | Bytes | Cost (%CPU)|
--------------------------------------------------------------------------------
|   0 | SELECT STATEMENT             |                |  4997 | 1483K|   544   (1)|
|*  1 |  HASH JOIN                   |                |  4997 | 1483K|   544   (1)|
|   2 |   TABLE ACCESS BY INDEX ROWID| TEST           |  2499 |  236K|    73   (0)|
|*  3 |    INDEX RANGE SCAN          | IDX_OWNER      |  2499 |       |     6   (0)|
|   4 |   VIEW                       | V_PUSHPREDICATE|  144K |   28M|   470   (1)|
|   5 |    UNION-ALL                 |                |       |       |            |
|   6 |     TABLE ACCESS FULL        | TEST           | 72462 | 6864K|   235   (1)|
|   7 |     TABLE ACCESS FULL        | TEST           | 72462 | 6864K|   235   (1)|
--------------------------------------------------------------------------------

Predicate Information (identified by operation id):
---------------------------------------------------

   1 - access("A"."OBJECT_ID"="B"."OBJECT_ID")
   3 - access("A"."OWNER"='SCOTT')
```

在本书示例中，我们如果不将连接列推入到视图中，视图里面的表就只能全表扫描，这时性能远不如索引扫描，所以本书示例最佳执行计划就是连接列谓词推入的执行计划。

笔者经常遇到连接列谓词推入引起 SQL 性能问题。大家在工作中，如果遇到执行计划中 VIEW PUSHED PREDICATE 一定要注意，如果 SQL 执行很快，不用理会；如果 SQL 执行很慢，可以先关闭连接列谓词推入（alter session set "_push_join_predicate" = false）功能，再逐步分析为什么连接列谓词推入之后，SQL 性能很差。连接列谓词推入性能变差一般是 CBO 将驱动表 Rows 计算错误（算少），导致视图作为嵌套循环被驱动表，然后一直反复被扫描；也有可能是视图太过复杂，视图本身存在性能问题，这时需要单独优化视图。例如视图单独执行耗时 1 秒，在进行谓词推入之后，视图会被扫描多次，假设扫描 1 000 次，每次执行时间从 1 秒提升到了 0.5 秒，但是视图被执行了 1 000 次，总的耗时反而多了，这时谓词推入反而降低性能。

一定要注意，当视图中有 rownum 会导致无法谓词推入，所以一般情况下，我们不建议在视图中使用 rownum。为什么 rownum 会导致无法谓词推入呢？这是因为当谓词推入之后，rownum 的值已经发生改变，已经改变了 SQL 结果集，任何查询变换必须是在不改变 SQL 结果集的前提下才能进行。

第 8 章 调优技巧

8.1 查看真实的基数（Rows）

在 1.3 节中提到，执行计划中的 Rows 是假的，是 CBO 根据统计信息和数学公式估算出来的，所以在看执行计划的时候，一定要注意嵌套循环驱动表的 Rows 是否估算准确，同时也要注意执行计划的入口 Rows 是否算错。因为一旦嵌套循环驱动表的 Rows 估算错误，执行计划就错了。如果执行计划的入口 Rows 估算错误，那执行计划也就不用看了，后面全错。

现有如下执行计划。

```
SQL> select * from table(dbms_xplan.display);

PLAN_TABLE_OUTPUT
--------------------------------------------------------------------------------

Plan hash value: 3215660883

--------------------------------------------------------------------------------
| Id  | Operation                    | Name                 | Rows  | Bytes | Cost(%CPU)|
--------------------------------------------------------------------------------
|   0 | SELECT STATEMENT             |                      |    78 |  4212 | 15507   (1)|
|   1 |  HASH GROUP BY               |                      |    78 |  4212 | 15507   (1)|
|   2 |   NESTED LOOPS               |                      |       |       |            |
|   3 |    NESTED LOOPS              |                      |  3034 |  159K | 15506   (1)|
|*  4 |     TABLE ACCESS FULL        |OPT_REF_UOM_TEMP_SDIM |  2967 |  101K |   650  (14)|
|*  5 |     INDEX RANGE SCAN         |PROD_DIM_PK           |     3 |       |     2   (0)|
|*  6 |    TABLE ACCESS BY INDEX ROWID|PROD_DIM             |     1 |    19 |     5   (0)|
--------------------------------------------------------------------------------

Predicate Information (identified by operation id):
---------------------------------------------------

   4 - filter("UOM"."RELTV_CURR_QTY"=1)
   5 - access("PROD"."PROD_SKID"="UOM"."PROD_SKID")
   6 - filter("PROD"."BUOM_CURR_SKID" IS NOT NULL AND "PROD"."PROD_END_DATE"=TO_DATE('
              9999-12-31 00:00:00', 'syyyy-mm-dd hh24:mi:ss') AND "PROD"."CURR_IND"='
Y' AND
              "PROD"."BUOM_CURR_SKID"="UOM"."UOM_SKID")

22 rows selected.
```

执行计划中 Id=4 是嵌套循环的驱动表，同时也是执行计划的入口，CBO 估算它只返回 2 967 行数据。Id=4 前面有"*"号，表示有谓词过滤 4 - filter("UOM"."RELTV_CURR_QTY"=1)。

根据执行计划中 Id=4 的谓词信息，手动计算 Id=4 应该返回真正的 Rows 如下。

```
SQL> select count(*) from OPT_REF_UOM_TEMP_SDIM where "RELTV_CURR_QTY"=1;
```

```
  COUNT(*)
----------
    946432
```

手动计算出的 Rows 返回了 946 432 行数据，与执行计划中的 2 967 行相差巨大，所以本示例中，执行计划是错误的。

8.2 使用 UNION 代替 OR

当 SQL 语句中同时有 or 和子查询，这种情况下子查询无法展开（unnest），只能走 FILTER。遇到这种情况我们可以将 SQL 改写为 union，从而消除 FILTER。

带有 or 子查询的写法与执行计划如下。

```
SQL> select *
  2    from t1
  3   where owner = 'SCOTT'
  4      or object_id in (select object_id from t2);

72571 rows selected.

Execution Plan
----------------------------------------------------------
Plan hash value: 895956251

--------------------------------------------------------------------------
| Id  | Operation          | Name | Rows  | Bytes | Cost (%CPU)| Time     |
--------------------------------------------------------------------------
|   0 | SELECT STATEMENT   |      |  3378 |  682K |   235   (1)| 00:00:03 |
|*  1 |  FILTER            |      |       |       |            |          |
|   2 |   TABLE ACCESS FULL| T1   | 56766 |   11M |   235   (1)| 00:00:03 |
|*  3 |   TABLE ACCESS FULL| T2   |   734 |  9542 |     2   (0)| 00:00:01 |
--------------------------------------------------------------------------

Predicate Information (identified by operation id):
---------------------------------------------------

   1 - filter("OWNER"='SCOTT' OR  EXISTS (SELECT 0 FROM "T2" "T2" WHERE
              "OBJECT_ID"=:B1))
   3 - filter("OBJECT_ID"=:B1)
```

改写为 union 的写法如下。

```
SQL> select * from t1 where owner='SCOTT'
  2  union
  3  select * from t1 where object_id in(select object_id from t2);

72571 rows selected.

Execution Plan
----------------------------------------------------------
Plan hash value: 696035008

--------------------------------------------------------------------------
| Id  | Operation        | Name | Rows  | Bytes |TempSpc| Cost (%CPU)|
--------------------------------------------------------------------------
|   0 | SELECT STATEMENT |      | 56778 |   11M |       | 4088  (95)|
|   1 |  SORT UNIQUE     |      | 56778 |   11M |   12M | 4088  (95)|
```

```
|   2 |   UNION-ALL              |    |       |      |      |      |     |
|*  3 |    TABLE ACCESS FULL     | T1 |    12 |  2484|      |   234| (1) |
|*  4 |    HASH JOIN             |    | 56766 |   11M| 1800K|  1146| (1) |
|   5 |     TABLE ACCESS FULL    | T2 | 73407 |  931K|      |   234| (1) |
|   6 |     TABLE ACCESS FULL    | T1 | 56766 |   11M|      |   235| (1) |
--------------------------------------------------------------------------

Predicate Information (identified by operation id):
---------------------------------------------------

   3 - filter("OWNER"='SCOTT')
   4 - access("OBJECT_ID"="OBJECT_ID")
```

改写为 union 之后，消除了 FILTER。如果无法改写 SQL，那么 SQL 就只能走 FILTER，这时我们需要在子查询表的连接列（t2.object_id）建立索引。

8.3 分页语句优化思路

分页语句最能考察一个人究竟会不会 SQL 优化，因为分页语句优化几乎囊括了 SQL 优化必须具备的知识。

8.3.1 单表分页优化思路

我们先创建一个测试表 T_PAGE。

```
SQL> create table t_page as select * from dba_objects;

Table created.
```

现有如下 SQL（没有过滤条件，只有排序），要将查询结果分页显示，每页显示 10 条。

```
select * from t_page order by object_id;
```

大家可能会采用以下这种分页框架（错误的分页框架）。

```
select *
  from (select t.*, rownum rn from (需要分页的 SQL) t)
 where rn >= 1
   and rn <= 10;
```

采用这种分页框架会产生严重的性能问题。现在将 SQL 语句代入错误的分页框架中。

```
SQL> select *
  2    from (select t.*, rownum rn
  3            from (select * from t_page order by object_id) t)
  4   where rn >= 1
  5     and rn <= 10;

10 rows selected.

Execution Plan
----------------------------------------------------------
Plan hash value: 3603170480

--------------------------------------------------------------------------------
| Id  | Operation                   | Name | Rows  | Bytes |TempSpc| Cost (%CPU)|
```

```
----------------------------------------------------------------------
|   0 | SELECT STATEMENT    |        | 61800 |  12M|      | 3020  (1)|
|*  1 |  VIEW               |        | 61800 |  12M|      | 3020  (1)|
|   2 |   COUNT             |        |       |     |      |          |
|   3 |    VIEW             |        | 61800 |  12M|      | 3020  (1)|
|   4 |     SORT ORDER BY   |        | 61800 |  12M|  14M | 3020  (1)|
|   5 |      TABLE ACCESS FULL| T_PAGE | 61800 |  12M|      |  236  (1)|
----------------------------------------------------------------------

Predicate Information (identified by operation id):
---------------------------------------------------

   1 - filter("RN"<=10 AND "RN">=1)
```

从执行计划中我们可以看到该 SQL 走了全表扫描，假如 T_PAGE 有上亿条数据，先要将该表（上亿条的表）进行排序（SORT ORDER BY），再取出其中 10 行数据，这时该 SQL 会产生严重的性能问题。所以该 SQL 不能走全表扫描，必须走索引扫描。

该 SQL 没有过滤条件，只有排序，我们可以利用索引已经排序这个特性来优化分页语句，也就是说要将分页语句中的 SORT ORDER BY 消除。一般分页语句中都有排序。

现在我们对排序列 object_id 建立索引，在索引中添加一个常量 0，注意 0 不能放前面。

```
SQL> create index idx_page on t_page(object_id,0);

Index created.
```

为什么要在索引中添加一个常量 0 呢？这是因为 object_id 列允许为 null，如果不添加常量（不一定是 0，可以是 1、2、3，也可以是英文字母），索引中就不能存储 null 值，然而 SQL 并没有写成以下写法。

```
select * from t_page where object_id is not null order by object_id;
```

因为 SQL 中并没有剔除 null 值，所以我们必须要添加一个常量，让索引存储 null 值，这样才能使 SQL 走索引。现在我们来看一下强制走索引的 A-Rows 执行计划（因为涉及到排版和美观，执行计划中删掉了 A-Time 等数据）。

```
SQL> select * from table(dbms_xplan.display_cursor(null,null,'ALLSTATS LAST'));

PLAN_TABLE_OUTPUT
--------------------------------------------------------------------------------
SQL_ID  fw6ym4n8njxqf, child number 0
-------------------------------------
select *   from (select t.*, rownum rn         from (select
    /*+ index(t_page idx_page) */  *
 from t_page              order by object_id) t)  where rn >= 1
and rn <= 10

Plan hash value: 3119682446

--------------------------------------------------------------------------------
| Id |Operation        | Name | Starts | E-Rows | A-Rows | Buffers |
--------------------------------------------------------------------------------
|  0 |SELECT STATEMENT |      |    1   |        |    10  |  1287   |
|* 1 | VIEW            |      |    1   |  61800 |    10  |  1287   |
|  2 |  COUNT          |      |    1   |        | 72608  |  1287   |
|  3 |   VIEW          |      |    1   |  61800 | 72608  |  1287   |
```

```
|   4 |   TABLE ACCESS BY INDEX ROWID| T_PAGE   |      1 |  61800 |  72608 |   1287 |
|   5 |    INDEX FULL SCAN           | IDX_PAGE |      1 |  61800 |  72608 |    183 |
-----------------------------------------------------------------------------------

Predicate Information (identified by operation id):
---------------------------------------------------

   1 - filter(("RN"<=10 AND "RN">=1))
```

因为 SQL 语句中没有 where 过滤条件，强制走索引只能走 INDEX FULL SCAN，无法走索引范围扫描（INDEX RANGE SCAN）。我们注意看执行计划中 A-Rows 这列，INDEX FULL SCAN 扫描了索引中所有叶子块，因为 INDEX FULL SCAN 返回了 72 608 行数据（表的总行数），一共耗费了 1 287 个逻辑读（Buffers=1287）。理想的执行计划是：INDEX FULL SCAN 只扫描 1 个（最多几个）索引叶子块，扫描 10 行数据（A-Rows=10）就停止了。为什么没有走最理想的执行计划呢？这是因为分页框架错了！

下面才是正确的分页框架。

```
select *
  from (select *
          from (select a.*, rownum rn
                  from (需要分页的 SQL) a)
         where rownum <= 10)
 where rn >= 1;
```

现在将 SQL 代入正确的分页框架中，强制走索引，查看 A-Rows 的执行计划（因为涉及到排版和美观，执行计划中删掉了 A-Time 等数据）。

```
SQL> select * from table(dbms_xplan.display_cursor(null,null,'ALLSTATS LAST'));

PLAN_TABLE_OUTPUT
--------------------------------------------------------------------------------
SQL_ID  4vyrpd0h4w30z, child number 0
--------------------------------------
select *    from (select *            from (select a.*, rownum rn
          from (select /*+ index(t_page idx_page) */
     *                            from t_page
order by object_id) a)        where rownum <= 10)   where rn >= 1

Plan hash value: 1201925926

-----------------------------------------------------------------------------------
| Id |Operation                      | Name     | Starts | E-Rows | A-Rows |Buffers|
-----------------------------------------------------------------------------------
|  0 |SELECT STATEMENT               |          |      1 |        |     10 |     5|
|* 1 | VIEW                          |          |      1 |     10 |     10 |     5|
|* 2 |  COUNT STOPKEY                |          |      1 |        |     10 |     5|
|  3 |   VIEW                        |          |      1 |  61800 |     10 |     5|
|  4 |    COUNT                      |          |      1 |        |     10 |     5|
|  5 |     VIEW                      |          |      1 |  61800 |     10 |     5|
|  6 |      TABLE ACCESS BY INDEX ROWID| T_PAGE |      1 |  61800 |     10 |     5|
|  7 |       INDEX FULL SCAN         | IDX_PAGE |      1 |  61800 |     10 |     3|
-----------------------------------------------------------------------------------

Predicate Information (identified by operation id):
---------------------------------------------------
```

```
 1 - filter("RN">=1)
 2 - filter(ROWNUM<=10)
```

从执行计划中我们可以看到，SQL 走了 INDEX FULL SCAN，只扫描了 10 条数据（Id=7 A-Rows=10）就停止了（Id=2 COUNT STOPKEY），一共只耗费了 5 个逻辑读（Buffers=5）。该执行计划利用索引已经排序特性（执行计划中没有 SORT ORDER BY），扫描索引获取了 10 条数据；然后再利用了 COUNT STOPKEY 特性，获取到分页语句需要的数据，SQL 立即停止运行，这才是最佳执行计划。

为什么错误的分页框架会导致性能很差呢？因为错误的分页框架这种写法没有 COUNT STOPKEY(where rownum<=...)功能，COUNT STOPKEY 就是当扫描到指定行数的数据之后，SQL 就停止运行。

现在我们得到分页语句的优化思路：如果分页语句中有排序（order by），要利用索引已经排序特性，将 order by 的列包含在索引中，同时也要利用 rownum 的 COUNT STOPKEY 特性来优化分页 SQL。如果分页中没有排序，可以直接利用 rownum 的 COUNT STOPKEY 特性来优化分页 SQL。

现有如下 SQL（注意，过滤条件是等值过滤，当然也有 order by），现在要将查询结果分页显示，每页显示 10 条。

```
select * from t_page where owner = 'SCOTT' order by object_id;
select * from t_page where owner = 'SYS' order by object_id;
```

第一条 SQL 语句的过滤条件是 where owner='SCOTT'，该过滤条件能过滤掉表中绝大部分数据。第二条 SQL 语句的过滤条件是 where owner='SYS'，该过滤条件能过滤表中一半数据。

我们将上述 SQL 代入正确的分页框架中强制走索引（object_id 列的索引，因为到目前为止 t_page 只有该列建立了索引），查看 A-Rows 的执行计划（因为涉及到排版和美观，执行计划中删掉了 A-Time 等数据）。

```
SQL> select * from table(dbms_xplan.display_cursor(null,null,'ALLSTATS LAST'));

PLAN_TABLE_OUTPUT
---------------------------------------------------------------------------------
SQL_ID  7s4mhq8sz19da, child number 0
---------------------------------------
select *    from (select *         from (select a.*, rownum rn
         from (select /*+ index(t_page idx_page) */
      *                         from t_page
where owner = 'SCOTT'                      order by object_id) a)
        where rownum <= 10)    where rn >= 1

Plan hash value: 1201925926

---------------------------------------------------------------------------------
| Id  | Operation                |Name   | Starts | E-Rows | A-Rows |Buffers|
---------------------------------------------------------------------------------
|   0 | SELECT STATEMENT         |       |      1 |        |     10 |   1273|
|*  1 |  VIEW                    |       |      1 |     10 |     10 |   1273|
|*  2 |   COUNT STOPKEY          |       |      1 |        |     10 |   1273|
```

```
|   3 |      VIEW                     |        |     1 |       57 |     10 |   1273|
|   4 |       COUNT                   |        |     1 |          |     10 |   1273|
|   5 |        VIEW                   |        |     1 |       57 |     10 |   1273|
|*  6 |         TABLE ACCESS BY INDEX ROWID|T_PAGE |     1 |       57 |     10 |   1273|
|   7 |          INDEX FULL SCAN      |IDX_PAGE|     1 |    61800 |  72427 |    183|
-------------------------------------------------------------------------------

Predicate Information (identified by operation id):
---------------------------------------------------

   1 - filter("RN">=1)
   2 - filter(ROWNUM<=10)
   6 - filter("OWNER"='SCOTT')

SQL> select * from table(dbms_xplan.display_cursor(null,null,'ALLSTATS LAST'));

PLAN_TABLE_OUTPUT
-------------------------------------------------------------------------------
SQL_ID  bn5k602hpdcq1, child number 0
-------------------------------------
select *    from (select *           from (select a.*, rownum rn
         from (select /*+ index(t_page idx_page) */
       *                        from t_page
where owner = 'SYS'                         order by object_id) a)
      where rownum <= 10)   where rn >= 1

Plan hash value: 1201925926

-------------------------------------------------------------------------------
| Id |Operation                    | Name   | Starts | E-Rows | A-Rows | Buffers|
-------------------------------------------------------------------------------
|  0 |SELECT STATEMENT             |        |     1 |          |     10 |      5|
|* 1 | VIEW                        |        |     1 |       10 |     10 |      5|
|* 2 |  COUNT STOPKEY              |        |     1 |          |     10 |      5|
|  3 |   VIEW                      |        |     1 |    28199 |     10 |      5|
|  4 |    COUNT                    |        |     1 |          |     10 |      5|
|  5 |     VIEW                    |        |     1 |    28199 |     10 |      5|
|* 6 |      TABLE ACCESS BY INDEX ROWID|T_PAGE |     1 |    28199 |     10 |      5|
|  7 |       INDEX FULL SCAN       |IDX_PAGE|     1 |    61800 |     10 |      3|
-------------------------------------------------------------------------------

Predicate Information (identified by operation id):
---------------------------------------------------

   1 - filter("RN">=1)
   2 - filter(ROWNUM<=10)
   6 - filter("OWNER"='SYS')
```

从执行计划中我们可以看到，两条 SQL 都走了 index full scan，第一条 SQL 从索引中扫描了 72 427 条数据（Id=7 A-Rows=72427），在回表的时候对数据进行了大量过滤（Id=6），最后得到 10 条数据，耗费了 1 273 个逻辑读（Buffers=1273）。第二条 SQL 从索引中扫描了 10 条数据，耗费了 5 个逻辑读（Buffers=5）。显而易见，第二条 SQL 的执行计划是正确的，而第一条 SQL 的执行计划是错误的，应该尽量在索引扫描的时候就取得 10 行数据。

为什么仅仅是过滤条件不一样，两条 SQL 在效率上有这么大区别呢？这是因为第一条 SQL 过滤条件是 owner='SCOTT'，owner='SCOTT'在表中只有很少数据，通过扫描 object_id 列的索引，然后回表再去匹配 owner='SCOTT'，因为 owner='SCOTT'数据量少，要搜索大量数据才能匹配上。而第二条 SQL 的过滤条件是 owner='SYS'，因为 owner='SYS' 数据量多，只需要搜索少量数据就能匹配上。

想要优化第一条 SQL，就需要让其在索引扫描的时候读取少量数据块就取得 10 行数据，这就需要将过滤列（owner）包含在索引中，排序列是 object_id，那么现在我们创建组合索引。

```
SQL> create index idx_page_ownerid on t_page(owner,object_id);

Index created.
```

我们查看强制走索引（idx_page_ownerid）带有 A-Rows 的执行计划（省略了部分数据）。

```
SQL> select * from table(dbms_xplan.display_cursor(null,null,'ALLSTATS LAST'));

PLAN_TABLE_OUTPUT
-------------------------------------------------------------------------------
SQL_ID  a1g16uafr05qf, child number 0
-------------------------------------
select *   from (select *         from (select a.*, rownum rn
         from (select /*+ index(t_page idx_page_ownerid) */
              *                              from t_page
      where owner = 'SCOTT'                         order by
object_id) a)         where rownum <= 10)   where rn >= 1

Plan hash value: 4175643597

-------------------------------------------------------------------------------
| Id |Operation                     |Name             |Starts|E-Rows|A-Rows|Buffers|
-------------------------------------------------------------------------------
|  0 |SELECT STATEMENT              |                 |    1 |      |   10 |     6 |
|* 1 | VIEW                         |                 |    1 |   10 |   10 |     6 |
|* 2 |  COUNT STOPKEY               |                 |    1 |      |   10 |     6 |
|  3 |   VIEW                       |                 |    1 |   57 |   10 |     6 |
|  4 |    COUNT                     |                 |    1 |      |   10 |     6 |
|  5 |     VIEW                     |                 |    1 |   57 |   10 |     6 |
|  6 |      TABLE ACCESS BY INDEX ROWID|T_PAGE         |    1 |   57 |   10 |     6 |
|* 7 |       INDEX RANGE SCAN       |IDX_PAGE_OWNERID |    1 |   57 |   10 |     3 |
-------------------------------------------------------------------------------

Predicate Information (identified by operation id):
---------------------------------------------------

   1 - filter("RN">=1)
   2 - filter(ROWNUM<=10)
   7 - access("OWNER"='SCOTT')
```

从执行计划中我们可以看到，SQL 走了索引范围扫描，从索引中扫描了 10 条数据，一共耗费了 6 个逻辑读。这说明该执行计划是正确的。大家可能会问：可不可以在创建索引的时候将 object_id 放在前面、owner 放在后面？现在我们来创建另外一个索引，将 object_id 列放在前面，owner 放在后面。

```
SQL> create index idx_page_idowner on t_page(object_id,owner);

Index created.
```

我们查看强制走索引（idx_page_idowner）带有 A-Rows 的执行计划（省略了部分数据）。

```
SQL> select * from table(dbms_xplan.display_cursor(null,null,'ALLSTATS LAST'));
```

8.3 分页语句优化思路

```
PLAN_TABLE_OUTPUT
--------------------------------------------------------------------------------
SQL_ID  djdnfyyznp3tf, child number 0
--------------------------------------------------------------------------------
select * from (select *           from (select a.*, rownum rn
        from (select /*+ index(t_page idx_page_idowner) */ *
              from t_page                       where owner =
'SCOTT'                           order by object_id) a)          where
rownum <= 10)   where rn >= 1

Plan hash value: 2811585238

--------------------------------------------------------------------------------
| Id  |Operation                       |Name             |Starts|E-Rows|A-Rows|Buffers|
--------------------------------------------------------------------------------
|  0  |SELECT STATEMENT                |                 |    1 |      |   10 |   224 |
|* 1  | VIEW                           |                 |    1 |   10 |   10 |   224 |
|* 2  |  COUNT STOPKEY                 |                 |    1 |      |   10 |   224 |
|  3  |   VIEW                         |                 |    1 |   57 |   10 |   224 |
|  4  |    COUNT                       |                 |    1 |      |   10 |   224 |
|  5  |     VIEW                       |                 |    1 |   57 |   10 |   224 |
|  6  |      TABLE ACCESS BY INDEX ROWID|T_PAGE          |    1 |   57 |   10 |   224 |
|* 7  |       INDEX FULL SCAN          |IDX_PAGE_IDOWNER |    1 |  247 |   10 |   221 |
--------------------------------------------------------------------------------

Predicate Information (identified by operation id):
---------------------------------------------------

   1 - filter("RN">=1)
   2 - filter(ROWNUM<=10)
   7 - access("OWNER"='SCOTT')
       filter("OWNER"='SCOTT')
```

从执行计划中我们看到，SQL 走了索引全扫描，从索引中扫描了 10 条数据，但是索引全扫描耗费了 221 个逻辑读，因为要边扫描索引边过滤数据（owner='SCOTT'），SQL 一共耗费了 224 个逻辑读，与走 object_id 列的执行计划（耗费了 1 273 个逻辑读）相比，虽然也提升了性能，但是性能最好的是走 idx_page_ownerid 这个索引的执行计划（逻辑读为 6）。

大家可能还会问，可不可以只在 owner 列创建索引呢？也就是说不将排序列包含在索引中。如果过滤条件能过滤掉大部分数据（owner='SCOTT'），那么这时不将排序列包含在索引中也是可以的，因为这时只需要对少量数据进行排序，少量数据排序几乎对性能没有什么影响。但是如果过滤条件只能过滤掉一部分数据，也就是说返回数据量很多（owner='SYS'），这时我们必须将排序列包含在索引中，如果不将排序列包含在索引中，就需要对大量数据进行排序。在实际生产环境中，过滤条件一般都是绑定变量，我们无法控制传参究竟传入哪个值，这就不能确定返回数据究竟是多还是少，所以为了保险起见，建议最好将排序列包含在索引中！

另外要注意，如果排序列有多个列，创建索引的时候，我们要将所有的排序列包含在索引中，并且要注意排序列先后顺序（语句中是怎么排序的，创建索引的时候就对应排序），而且还要注意列是升序还是降序。如果分页语句中排序列只有一个列，但是是降序显示的，创建索引的时候就没必要降序创建了，我们可以使用 HINT: index_desc 让索引降序扫描就行。

现有如下分页语句。

```
select *
  from (select *
          from (select a.*, rownum rn
                  from (select *
                          from t_page
                         order by object_id, object_name desc) a)
         where rownum <= 10)
 where rn >= 1;
```

创建索引的时候，只能是 object_id 列在前，object_name 列在后面，另外 object_name 是降序显示的，那么在创建索引的时候，我们还要指定 object_name 列降序排序。此外该 SQL 没有过滤条件，在创建索引的时候，我们还要加个常量。现在我们创建如下索引。

```
SQL> create index idx_page_idname on t_page(object_id,object_name desc,0);

Index created.
```

我们查看强制走索引（idx_page_idname）带有 A-Rows 的执行计划（省略了部分数据）。

```
SQL> select * from table(dbms_xplan.display_cursor(null,null,'ALLSTATS LAST'));

PLAN_TABLE_OUTPUT
---------------------------------------------------------------------------
SQL_ID  20yk62bptjrs9, child number 0
-------------------------------------
select *   from (select *           from (select a.*, rownum rn
        from (select   /*+ index(t_page idx_page_idname)*/
           *                              from t_page
                   order by object_id, object_name desc) a)            where
rownum <= 10)   where rn >= 1

Plan hash value: 445348578

---------------------------------------------------------------------------
| Id |Operation                          |Name            |Starts|E-Rows|A-Rows|Buffers|
---------------------------------------------------------------------------
|  0 |SELECT STATEMENT                   |                |    1 |      |   10 |    5 |
|* 1 | VIEW                              |                |    1 |   10 |   10 |    5 |
|* 2 |  COUNT STOPKEY                    |                |    1 |      |   10 |    5 |
|  3 |   VIEW                            |                |    1 |61800 |   10 |    5 |
|  4 |    COUNT                          |                |    1 |      |   10 |    5 |
|  5 |     VIEW                          |                |    1 |61800 |   10 |    5 |
|  6 |      TABLE ACCESS BY INDEX ROWID  |T_PAGE          |    1 |61800 |   10 |    5 |
|  7 |       INDEX FULL SCAN             |IDX_PAGE_IDNAME |    1 |61800 |   10 |    3 |
---------------------------------------------------------------------------

Predicate Information (identified by operation id):
---------------------------------------------------

   1 - filter("RN">=1)
   2 - filter(ROWNUM<=10)
```

如果创建索引的时候将 object_name 放在前面，object_id 放在后面，这个时候，索引中列先后顺序与分页语句中排序列先后顺序不一致，强制走索引的时候，执行计划中会出现 SORT ORDER BY 关键字。因为索引的顺序与排序的顺序不一致，所以需要从索引中获取数据之后再排序，有排序就会出现 SORT ORDER BY。现在我们创建如下索引。

```
SQL> create index idx_page_nameid on t_page(object_name,object_id,0);
```

8.3 分页语句优化思路

Index created.

现在查看强制走索引（idx_page_nameid）带有 A-Rows 的执行计划（省略了部分数据）。

```
SQL> select * from table(dbms_xplan.display_cursor(null,null,'ALLSTATS LAST'));

PLAN_TABLE_OUTPUT
--------------------------------------------------------------------------------
SQL_ID  8b8nwayah0z68, child number 0
--------------------------------------
select *    from (select *        from (select a.*, rownum rn
        from (select /*+ index(t_page idx_page_nameid)*/
             *                              from t_page
    order by object_id, object_name desc) a)          where rownum <=
10)  where rn >= 1

Plan hash value: 2869317785

-------------------------------------------------------------------------------------
| Id |Operation                        |Name            |Starts|E-Rows|A-Rows|Buffers|
-------------------------------------------------------------------------------------
|  0 |SELECT STATEMENT                 |                |    1 |      |   10 | 37397 |
|* 1 | VIEW                            |                |    1 |   10 |   10 | 37397 |
|* 2 |  COUNT STOPKEY                  |                |    1 |      |   10 | 37397 |
|  3 |   VIEW                          |                |    1 | 61800|   10 | 37397 |
|  4 |    COUNT                        |                |    1 |      |   10 | 37397 |
|  5 |     VIEW                        |                |    1 | 61800|   10 | 37397 |
|  6 |      SORT ORDER BY              |                |    1 | 61800|   10 | 37397 |
|  7 |       TABLE ACCESS BY INDEX ROWID|T_PAGE         |    1 | 61800| 72608| 37397 |
|  8 |        INDEX FULL SCAN          |IDX_PAGE_NAMEID |    1 | 61800| 72608|   431 |
-------------------------------------------------------------------------------------

Predicate Information (identified by operation id):
---------------------------------------------------

   1 - filter("RN">=1)
   2 - filter(ROWNUM<=10)
```

如果创建索引的时候没有指定 object_name 列降序排序，那么执行计划中也会出现 SORT ORDER BY。因为索引中排序和分页语句中排序不一致。现在我们创建如下索引。

```
SQL> create index idx_page_idname1 on t_page(object_id,object_name,0);

Index created.
```

我们查看强制走索引（idx_page_idname1）带有 A-Rows 的执行计划（省略了部分数据）。

```
SQL> select * from table(dbms_xplan.display_cursor(null,null,'ALLSTATS LAST'));

PLAN_TABLE_OUTPUT
--------------------------------------------------------------------------------
SQL_ID  2dsmtc9b65a7v, child number 0
--------------------------------------
select *    from (select *        from (select a.*, rownum rn
        from (select /*+ index(t_page idx_page_idname1)*/
             *                              from t_page
    order by object_id, object_name desc) a)          where rownum <=
10)  where rn >= 1

Plan hash value: 170538223
```

```
-------------------------------------------------------------------------------
| Id |Operation                       | Name             |Starts|E-Rows|A-Rows|Buffers|
-------------------------------------------------------------------------------
|  0 |SELECT STATEMENT                |                  |    1|      |    10|  1533|
|* 1 | VIEW                           |                  |    1|    10|    10|  1533|
|* 2 |  COUNT STOPKEY                 |                  |    1|      |    10|  1533|
|  3 |   VIEW                         |                  |    1| 61800|    10|  1533|
|  4 |    COUNT                       |                  |    1|      |    10|  1533|
|  5 |     VIEW                       |                  |    1| 61800|    10|  1533|
|  6 |      SORT ORDER BY             |                  |    1| 61800|    10|  1533|
|  7 |       TABLE ACCESS BY INDEX ROWID|T_PAGE          |    1| 61800| 72608|  1533|
|  8 |        INDEX FULL SCAN         |IDX_PAGE_IDNAME1  |    1| 61800| 72608|   430|
-------------------------------------------------------------------------------

Predicate Information (identified by operation id):
---------------------------------------------------

   1 - filter("RN">=1)
   2 - filter(ROWNUM<=10)
```

分页语句中如果出现了 SORT ORDER BY，这就意味着分页语句没有利用到索引已经排序的特性，执行计划一般是错误的，这时需要创建正确的索引。

现有如下 SQL（注意，过滤条件有等值条件，也有非等值条件，当然也有 order by），现在要将查询结果分页显示，每页显示 10 条。

```
select * from t_page where owner = 'SYS' and object_id > 1000 order by object_name;
```

大家请思考，应该怎么创建索引，从而优化上面的分页语句呢？上文提到，如果分页语句中有排序列，创建索引的时候，要将排序列包含在索引中。所以现在我们只需要将过滤列 owner、object_id 以及排序列 object_name 组合起来创建索引中即可。

因为 owner 是等值过滤，object_id 是非等值过滤，创建索引的时候，我们要优先将等值过滤列和排序列组合在一起，然后再将非等值过滤列放到后面。

```
SQL> create index idx_ownernameid on t_page(owner,object_name,object_id);

Index created.
```

让我们查看强制走索引（idx_ownernameid）带有 A-Rows 的执行计划（省略了部分数据）。

```
SQL> select * from table(dbms_xplan.display_cursor(null,null,'ALLSTATS LAST'));

PLAN_TABLE_OUTPUT
--------------------------------------------------------------------------------
SQL_ID  07z0dkm4a9qdz, child number 0
--------------------------------------
select *      from (select *           from (select a.*, rownum rn
         from (select /*+ index(t_page idx_ownernameid) */
               *                       from t_page
         where owner = 'SYS'                       and object_id >
1000                    order by object_name) a)         where
rownum <= 10)    where rn >= 1

Plan hash value: 2090516350

--------------------------------------------------------------------------------
```

8.3 分页语句优化思路

```
| Id  | Operation                      | Name           |Starts|E-Rows| A-Rows |Buffers|
---------------------------------------------------------------------------------------
|  0  | SELECT STATEMENT               |                |   1|      |   10 |   14|
|* 1  |  VIEW                          |                |   1|   10 |   10 |   14|
|* 2  |   COUNT STOPKEY                |                |   1|      |   10 |   14|
|  3  |    VIEW                        |                |   1| 26937|   10 |   14|
|  4  |     COUNT                      |                |   1|      |   10 |   14|
|  5  |      VIEW                      |                |   1| 26937|   10 |   14|
|  6  |       TABLE ACCESS BY INDEX ROWID|T_PAGE        |   1| 26937|   10 |   14|
|* 7  |        INDEX RANGE SCAN        |IDX_OWNERNAMEID |   1|  254 |   10 |    4|
---------------------------------------------------------------------------------------

Predicate Information (identified by operation id):
---------------------------------------------------

   1 - filter("RN">=1)
   2 - filter(ROWNUM<=10)
   7 - access("OWNER"='SYS' AND "OBJECT_ID">1000)
       filter("OBJECT_ID">1000)
```

执行计划中没有 SORT ORDER BY，逻辑读也才 14 个，说明执行计划非常理想。也许大家会问，为何不创建如下这样索引呢？

```
SQL> create index idx_owneridname on t_page(owner,object_id,object_name);

Index created.
```

我们查看强制走索引（idx_owneridname）带有 A-Rows 的执行计划（省略了部分数据）。

```
SQL> select * from table(dbms_xplan.display_cursor(null,null,'ALLSTATS LAST'));

PLAN_TABLE_OUTPUT
-----------------------------------------------------------------------------------
SQL_ID  7bm9sf2u94uxa, child number 0
-----------------------------------
select *   from (select *         from (select a.*, rownum rn
       from (select /*+ index(t_page idx_owneridname) */
             *                          from t_page
     where owner = 'SYS'                         and object_id >
1000                    order by object_name) a)              where
rownum <= 10)   where rn >= 1

Plan hash value: 2498002320

-----------------------------------------------------------------------------------
| Id  | Operation                      | Name           |Starts|E-Rows| A-Rows |Buffers|
-----------------------------------------------------------------------------------
|  0  | SELECT STATEMENT               |                |   1|      |   10 | 1002|
|* 1  |  VIEW                          |                |   1|   10 |   10 | 1002|
|* 2  |   COUNT STOPKEY                |                |   1|      |   10 | 1002|
|  3  |    VIEW                        |                |   1| 26937|   10 | 1002|
|  4  |     COUNT                      |                |   1|      |   10 | 1002|
|  5  |      VIEW                      |                |   1| 26937|   10 | 1002|
|  6  |       SORT ORDER BY            |                |   1| 26937|   10 | 1002|
|  7  |        TABLE ACCESS BY INDEX ROWID|T_PAGE       |   1| 26937| 29919| 1002|
|* 8  |         INDEX RANGE SCAN       |IDX_OWNERIDNAME |   1| 26937| 29919|  189|
-----------------------------------------------------------------------------------

Predicate Information (identified by operation id):
---------------------------------------------------
```

```
1 - filter("RN">=1)
2 - filter(ROWNUM<=10)
8 - access("OWNER"='SYS' AND "OBJECT_ID">1000 AND "OBJECT_ID" IS NOT NULL)
```

该执行计划中有 SORT ORDER BY，说明没有用到索引已经排序特性，而且逻辑读为 1 002 个，这说明该执行计划是错误的。为什么该执行计划是错误的呢？这是因为该分页语句是根据 object_name 进行排序的，但是创建索引的时候是按照 owner、object_id、object_name 顺序创建索引的，索引中前 5 条数据如下。

```
SQL> select *
  2    from (select rownum rn, owner, object_id, object_name
  3            from t_page
  4           where owner = 'SYS'
  5             and object_id > 1000
  6           order by owner, object_id, object_name)
  7   where rownum <= 5;

        RN OWNER  OBJECT_ID OBJECT_NAME
---------- ----- ---------- ------------
         1 SYS        1001 NOEXP$
         2 SYS        1002 EXPPKGOBJ$
         3 SYS        1003 I_OBJTYPE
         4 SYS        1004 EXPPKGACT$
         5 SYS        1005 I_ACTPACKAGE
```

在这前 5 条数据中，我们按照分页语句排序条件 object_name 进行排序，应该是第 4 行数据显示为第一行数据，但是它在索引中排到了第 4 行，所以索引中数据的顺序并不能满足分页语句中的排序要求，这就产生了 SORT ORDER BY，进而导致执行计划错误。为什么按照 owner、object_name、object_id 顺序创建索引，执行计划是对的呢？现在我们取索引中前 5 条数据。

```
SQL> select *
  2    from (select rownum rn, owner, object_id, object_name
  3            from t_page
  4           where owner = 'SYS'
  5             and object_id > 1000
  6           order by owner,object_name,object_id)
  7   where rownum <= 5;

        RN OWNER  OBJECT_ID OBJECT_NAME
---------- ----- ---------- ----------------------------------
         1 SYS       34042 /1000323d_DelegateInvocationHa
         2 SYS       44844 /1000e8d1_LinkedHashMapValueIt
         3 SYS       23397 /1005bd30_LnkdConstant
         4 SYS       19737 /10076b23_OraCustomDatumClosur
         5 SYS       45460 /100c1606_StandardMidiFileRead
```

索引中的数据顺序完全符合分页语句中的排序要求，这就不需要我们进行 SORT ORDER BY 了，所以该执行计划是对的。

现在我们继续完善分页语句的优化思路：如果分页语句中有排序（order by），要利用索引已经排序特性，将 order by 的列按照排序的先后顺序包含在索引中，同时要注意排序是升序还是降序。如果分页语句中有过滤条件，我们要注意过滤条件是否有等值过滤条件，如果有等值过滤条件，要将等值过滤条件优先组合在一起，然后将排序列放在等值过滤条件后面，最后将非等值过滤列放排序列后面。如果分页语句中没有等值过滤条件，我们应该先将排序列放在索引前面,将非等值过滤列放后面,最后利用 rownum 的 COUNT STOPKEY 特性来优化分页 SQL。如果分页中没有排序,可以直接利用 rownum 的 COUNT STOPKEY 特性来优化分页 SQL。

如果我们想一眼看出分页语句执行计划是正确还是错误的,先看分页语句有没有 ORDER BY,再看执行计划有没有 SORT ORDER BY,如果执行计划中有 SORT ORDER BY,执行计划一般都是错误的。

请大家思考,如下分页语句应该如何建立索引(提示:该 SQL 没有等值过滤)?

```
select *
  from (select *
          from (select a.*, rownum rn
                  from (select *
                          from t_page
                         where owner like 'SYS%'
                           and object_id > 1000
                         order by object_name) a)
         where rownum <= 10)
 where rn >= 1;
```

如果分页语句中排序的表是分区表,这时我们要看分页语句中是否有跨分区扫描,如果有跨分区扫描,创建索引一般都创建为 global 索引,如果不创建 global 索引,就无法保证分页的顺序与索引的顺序一致。如果就只扫描一个分区,这时可以创建 local 索引。

现在我们创建一个根据 object_id 范围分区的分区表 p_test 并且插入测试数据。

```
SQL> create table p_test(
  2   OWNER           VARCHAR2(30),
  3   OBJECT_NAME     VARCHAR2(128),
  4   SUBOBJECT_NAME  VARCHAR2(30),
  5   OBJECT_ID       NUMBER,
  6   DATA_OBJECT_ID  NUMBER,
  7   OBJECT_TYPE     VARCHAR2(19),
  8   CREATED         DATE,
  9   LAST_DDL_TIME   DATE,
 10   TIMESTAMP       VARCHAR2(19),
 11   STATUS          VARCHAR2(7),
 12   TEMPORARY       VARCHAR2(1),
 13   GENERATED       VARCHAR2(1),
 14   SECONDARY       VARCHAR2(1),
 15   NAMESPACE       NUMBER,
 16   EDITION_NAME    VARCHAR2(30)
 17  ) partition by range (object_id)
 18  (
 19  partition p1 values less than (10000),
 20  partition p2 values less than (20000),
 21  partition p3 values less than (30000),
 22  partition p4 values less than (40000),
 23  partition p5 values less than (50000),
 24  partition p6 values less than (60000),
 25  partition p7 values less than (70000),
 26  partition p8 values less than (80000),
 27  partition pmax values less than(maxvalue)
 28  );

Table created.

SQL> insert into p_test select * from dba_objects;
```

```
72662 rows created.

SQL> commit;
```

现有如下分页语句（根据范围分区列排序）。

```
select *
  from (select *
          from (select a.*, rownum rn
                  from (select * from p_test order by object_id) a)
         where rownum <= 10)
 where rn >= 1;
```

该分页语句没有过滤条件，因此会扫描表中所有分区。因为排序列恰好是范围分区列，范围分区每个分区的数据也是递增的，这时我们创建索引可以创建为 local 索引。但是如果将范围分区改成 LIST 分区或者 HASH 分区，这时我们就必须创建 global 索引，因为 LIST 分区和 HASH 分区是无序的。

现在我们创建 local 索引。

```
SQL> create index idx_ptest_id on p_test(object_id,0) local;

Index created.
```

我们查看强制走索引（idx_ptest_id）带有 A-Rows 的执行计划（省略了部分数据）。

```
SQL> select * from table(dbms_xplan.display_cursor(null,null,'ALLSTATS LAST'));

PLAN_TABLE_OUTPUT
--------------------------------------------------------------------------------
SQL_ID  3rp1uz98fgggq, child number 0
--------------------------------------
select *   from (select *          from (select a.*, rownum rn
        from (select /*+ index(p_test idx_ptest_id) */
              *                    from p_test
  order by object_id) a)          where rownum <= 10)  where rn >= 1

Plan hash value: 1636704844

--------------------------------------------------------------------------------
| Id |Operation                            |Name         |Starts|E-Rows|A-Rows|Buffers|
--------------------------------------------------------------------------------
|  0 |SELECT STATEMENT                     |             |    1|      |    10|    5|
|* 1 | VIEW                                |             |    1|    10|    10|    5|
|* 2 |  COUNT STOPKEY                      |             |    1|      |    10|    5|
|  3 |   VIEW                              |             |    1| 51888|    10|    5|
|  4 |    COUNT                            |             |    1|      |    10|    5|
|  5 |     VIEW                            |             |    1| 51888|    10|    5|
|  6 |      PARTITION RANGE ALL            |             |    1| 51888|    10|    5|
|  7 |       TABLE ACCESS BY LOCAL INDEX ROWID|P_TEST    |    1| 51888|    10|    5|
|  8 |        INDEX FULL SCAN              |IDX_PTEST_ID |    1| 51888|    10|    3|
--------------------------------------------------------------------------------

Predicate Information (identified by operation id):
---------------------------------------------------

   1 - filter("RN">=1)
   2 - filter(ROWNUM<=10)
```

8.3 分页语句优化思路

现有如下分页语句(根据 object_name 排序)。

```
select *
  from (select *
          from (select a.*, rownum rn
                  from (select * from p_test order by object_name) a)
         where rownum <= 10)
 where rn >= 1;
```

这时我们就需要创建 global 索引,如果创建 local 索引会导致产生 SORT ORDER BY。

```
SQL> create index idx_ptest_name on p_test(object_name,0) local;

Index created.
```

现在查看强制走索引(idx_ptest_name)带有 A-Rows 的执行计划(省略了部分数据)。

```
SQL> select * from table(dbms_xplan.display_cursor(null,null,'ALLSTATS LAST'));

PLAN_TABLE_OUTPUT
------------------------------------------------------------------------------
SQL_ID  50hgw72gnvs83, child number 0
------------------------------------
select *    from (select *            from (select a.*, rownum rn
         from (select /*+ index(p_test idx_ptest_name) */
           *                       from p_test
    order by object_name) a)        where rownum <= 10)   where rn >=1

Plan hash value: 2548872510

----------------------------------------------------------------------------------------
| Id |Operation                            |Name           |Starts|E-Rows|A-Rows|Buffers |
----------------------------------------------------------------------------------------
|  0 |SELECT STATEMENT                     |               |    1 |      |   10 | 35530 |
|* 1 | VIEW                                |               |    1 |   10 |   10 | 35530 |
|* 2 |  COUNT STOPKEY                      |               |    1 |      |   10 | 35530 |
|  3 |   VIEW                              |               |    1 |51888 |   10 | 35530 |
|  4 |    COUNT                            |               |    1 |      |   10 | 35530 |
|  5 |     VIEW                            |               |    1 |51888 |   10 | 35530 |
|  6 |      SORT ORDER BY                  |               |    1 |51888 |   10 | 35530 |
|  7 |       PARTITION RANGE ALL           |               |    1 |51888 |72662 | 35530 |
|  8 |        TABLE ACCESS BY LOCAL INDEX ROWID|P_TEST     |    9 |51888 |72662 | 35530 |
|  9 |         INDEX FULL SCAN             |IDX_PTEST_NAME |    9 |51888 |72662 |   392 |
----------------------------------------------------------------------------------------

Predicate Information (identified by operation id):
---------------------------------------------------

   1 - filter("RN">=1)
   2 - filter(ROWNUM<=10)
```

现在我们将索引 idx_ptest_name 重建为 global 索引。

```
SQL> drop index idx_ptest_name;

Index dropped.

SQL> create index idx_ptest_name on p_test(object_name,0);

Index created.
```

查看强制走索引（idx_ptest_name）带有 A-Rows 的执行计划（省略了部分数据）。

```
SQL> select * from table(dbms_xplan.display_cursor(null,null,'ALLSTATS LAST'));

PLAN_TABLE_OUTPUT
--------------------------------------------------------------------------------
SQL_ID  50hgw72gnvs83, child number 0
-------------------------------------
select *   from (select *           from (select a.*, rownum rn
          from (select /*+ index(p_test idx_ptest_name) */
                 *                         from p_test
     order by object_name) a)         where rownum <= 10)   where rn >=1

Plan hash value: 4135902528

--------------------------------------------------------------------------------
| Id |Operation                              |Name            |Starts|E-Rows|A-Rows|Buffers|
--------------------------------------------------------------------------------
|  0 |SELECT STATEMENT                       |                |    1|      |    10|    10|
|* 1 | VIEW                                  |                |    1|    10|    10|    10|
|* 2 |  COUNT STOPKEY                        |                |    1|      |    10|    10|
|  3 |   VIEW                                |                |    1| 51888|    10|    10|
|  4 |    COUNT                              |                |    1|      |    10|    10|
|  5 |     VIEW                              |                |    1| 51888|    10|    10|
|  6 |      TABLE ACCESS BY GLOBAL INDEX ROWID|P_TEST          |    1| 51888|    10|    10|
|  7 |       INDEX FULL SCAN                 |IDX_PTEST_NAME  |    1| 51888|    10|     4|
--------------------------------------------------------------------------------

Predicate Information (identified by operation id):
---------------------------------------------------

   1 - filter("RN">=1)
   2 - filter(ROWNUM<=10)
```

8.3.2 多表关联分页优化思路

多表关联分页语句，要利用索引已经排序特性、ROWNUM 的 COUNT STOPKEY 特性以及嵌套循环传值特性来优化。

现在我们创建另外一个测试表 T_PAGE2。

```
SQL> create table t_page2 as select * from dba_objects;

Table created.
```

现有如下分页语句。

```
select *
  from (select *
          from (select a.owner,
                       a.object_id,
                       a.subobject_name,
                       a.object_name,
                       rownum rn
                  from (select t1.owner,
                               t1.object_id,
                               t1.subobject_name,
                               t2.object_name
                          from t_page t1, t_page2 t2
```

```
                        where t1.object_id = t2.object_id
                        order by t2.object_name) a)
           where rownum <= 10)
where rn >= 1;
```

分页语句中排序列是 t_page2 的 object_name，我们需要对其创建一个索引。

```
SQL> create index idx_page2_name on t_page2(object_name,0);

Index created.
```

现在强制 t_page2 走刚才创建的索引并且让其作为嵌套循环驱动表，t_page 作为嵌套循环被驱动表，利用 rownum 的 COUNT STOPKEY 特性，扫描到 10 条数据，SQL 就停止。现在我们查看强制走索引，强制走嵌套循环的 A-ROWS 执行计划。

```
SQL> select * from table(dbms_xplan.display_cursor(null,null,'ALLSTATS LAST'));

PLAN_TABLE_OUTPUT
-----------------------------------
SQL_ID  g0gpgftwrfwzt, child number 0
-----------------------------------
select *    from (select *       from (select
a.owner,a.object_id,a.subobject_name,a.object_name, rownum rn
       from (select /*+ index(t2 idx_page2_name) leading(t2) use_nl(t2,t1) */
t1.owner,t1.object_id,t1.subobject_name,t2.object_name
         from t_page t1, t_page2 t2      where
t1.object_id = t2.object_id    order by
t2.object_name) a)    where rownum <= 10)   where rn >= 1

Plan hash value: 4182646763

---------------------------------------------------------------------------------
| Id |Operation                        |Name           |Starts|E-Rows|A-Rows|Buffers|
---------------------------------------------------------------------------------
|  0 |SELECT STATEMENT                 |               |    1 |      |   10 |   29  |
|* 1 | VIEW                            |               |    1 |   10 |   10 |   29  |
|* 2 |  COUNT STOPKEY                  |               |    1 |      |   10 |   29  |
|  3 |   VIEW                          |               |    1 |61800 |   10 |   29  |
|  4 |    COUNT                        |               |    1 |      |   10 |   29  |
|  5 |     VIEW                        |               |    1 |61800 |   10 |   29  |
|  6 |      NESTED LOOPS               |               |    1 |61800 |   10 |   29  |
|  7 |       TABLE ACCESS BY INDEX ROWID|T_PAGE2       |    1 |66557 |   10 |   10  |
|  8 |        INDEX FULL SCAN          |IDX_PAGE2_NAME |    1 |66557 |   10 |    4  |
|  9 |       TABLE ACCESS BY INDEX ROWID|T_PAGE        |   10 |    1 |   10 |   19  |
|*10 |        INDEX RANGE SCAN         |IDX_PAGE       |   10 |    1 |   10 |   13  |
---------------------------------------------------------------------------------

Predicate Information (identified by operation id):
---------------------------------------------------
   1 - filter("RN">=1)
   2 - filter(ROWNUM<=10)
  10 - access("T1"."OBJECT_ID"="T2"."OBJECT_ID")
```

从执行计划中我们看到，驱动表走的是排序列的索引，扫描了 10 行数据，传值 10 次给被驱动表，然后 SQL 停止运行，逻辑读一共 29 个，该执行计划是正确的，而且是最佳执行计划。

大家思考一下，对于上面的分页语句，能否走 HASH 连接？如果 SQL 走了 HASH 连接，这时两个表关联之后得到的结果无法保证是有序的，这就需要关联完成后再进行一次排序

（SORT ORDER BY），所以不能走 HASH 连接，同理也不能走排序合并连接。

为什么多表关联的分页语句必须走嵌套循环呢？这是因为嵌套循环是驱动表传值给被驱动表，如果驱动表返回的数据是有序的，那么关联之后的结果集也是有序的，这样就可以消除 SORT ORDER BY。

现有如下分页语句（排序列来自两个表）。

```
select *
  from (select *
          from (select a.owner,
                       a.object_id,
                       a.subobject_name,
                       a.object_name,
                       rownum rn
                  from (select t1.owner,
                               t1.object_id,
                               t1.subobject_name,
                               t2.object_name
                          from t_page t1, t_page2 t2
                         where t1.object_id = t2.object_id
                         order by t2.object_name ,t1.subobject_name) a)
         where rownum <= 10)
 where rn >= 1;
```

因为以上分页语句排序列来自多个表，这就需要等两表关联完之后再进行排序，这样无法消除 SORT ORDER BY，所以以上 SQL 语句无法优化，两表之间也只能走 HASH 连接。如果想优化上面分页语句，我们可以与业务沟通，去掉一个表的排序列，这样就不需要等两表关联完之后再进行排序。

现有如下分页语句（根据外连接从表排序）。

```
select *
  from (select *
          from (select a.owner,
                       a.object_id,
                       a.subobject_name,
                       a.object_name,
                       rownum rn
                  from (select t1.owner,
                               t1.object_id,
                               t1.subobject_name,
                               t2.object_name
                          from t_page t1 left join t_page2 t2
                            on t1.object_id = t2.object_id
                         order by t2.object_name) a)
         where rownum <= 10)
 where rn >= 1;
```

两表关联如果是外连接，当两表用嵌套循环进行连接的时候，驱动表只能是主表。这里主表是 t1，但是排序列来自 t2，在分页语句中，对哪个表排序，就应该让其作为嵌套循环驱动表。但是这里相互矛盾。所以该分页语句无法优化，t1 与 t2 只能走 HASH 连接。如果想要优化以上分页语句，我们只能让 t1 表中的列作为排序列。

分页语句中也不能有 distinct、group by、max、min、avg、union、union all 等关键字。因为当分页语句中有这些关键字，我们需要等表关联完或者数据都跑完之后再来分页，这样性能很差。

最后，我们总结一下多表关联分页优化思路。多表关联分页语句，如果有排序，只能对其中一个表进行排序，让参与排序的表作为嵌套循环的驱动表，并且要控制驱动表返回的数据顺序与排序的顺序一致，其余表的连接列要创建好索引。如果有外连接，我们只能选择主表的列作为排序列，语句中不能有 distinct、group by、max、min、avg、union、union all，执行计划中不能出现 SORT ORDER BY。

8.4 使用分析函数优化自连接

现有如下 SQL 及其执行计划。

```
SQL> select ename,deptno,sal
  2    from emp a
  3   where sal = (select max(sal) from emp b where a.deptno = b.deptno);

Execution Plan
----------------------------------------------------------
Plan hash value: 1245077725

--------------------------------------------------------------------------
| Id  | Operation           | Name    | Rows | Bytes | Cost (%CPU)| Time     |
--------------------------------------------------------------------------
|   0 | SELECT STATEMENT    |         |    1 |    39 |     8  (25)| 00:00:01 |
|*  1 |  HASH JOIN          |         |    1 |    39 |     8  (25)| 00:00:01 |
|   2 |   VIEW              | VW_SQ_1 |    3 |    78 |     4  (25)| 00:00:01 |
|   3 |    HASH GROUP BY    |         |    3 |    21 |     4  (25)| 00:00:01 |
|   4 |     TABLE ACCESS FULL| EMP    |   14 |    98 |     3   (0)| 00:00:01 |
|   5 |   TABLE ACCESS FULL | EMP     |   14 |   182 |     3   (0)| 00:00:01 |
--------------------------------------------------------------------------

Predicate Information (identified by operation id):
---------------------------------------------------

   1 - access("SAL"="MAX(SAL)" AND "A"."DEPTNO"="ITEM_1")
```

该 SQL 表示查询员工表中每个部门工资最高的员工的所有信息，访问了 EMP 表两次。我们可以利用分析函数对上面 SQL 进行等价改写，使 EMP 只访问一次。

分析函数的写法如下。

```
SQL> select ename, deptno, sal
  2    from (select a.*, max(sal) over(partition by deptno) max_sal from emp a)
  3   where sal = max_sal;

Execution Plan
----------------------------------------------------------
Plan hash value: 4130734685

--------------------------------------------------------------------------
| Id  | Operation           | Name | Rows | Bytes | Cost (%CPU)| Time     |
--------------------------------------------------------------------------
|   0 | SELECT STATEMENT    |      |   14 |   644 |     4  (25)| 00:00:01 |
|*  1 |  VIEW               |      |   14 |   644 |     4  (25)| 00:00:01 |
|   2 |   WINDOW SORT       |      |   14 |   182 |     4  (25)| 00:00:01 |
|   3 |    TABLE ACCESS FULL| EMP  |   14 |   182 |     3   (0)| 00:00:01 |
```

```
------------------------------------------------
Predicate Information (identified by operation id):
------------------------------------------------

   1 - filter("SAL"="MAX_SAL")
```

使用分析函数改写之后，减少了表扫描次数，EMP 表越大，性能提升越明显。

8.5 超大表与超小表关联优化方法

现有如下 SQL。

```
select * from a,b where a.object_id=b.object_id;
```

表 a 有 30MB，表 b 有 30GB，两表关联后返回大量数据，应该走 HASH 连接，因为 a 是小表所以 a 应该作为 HASH JOIN 的驱动表，大表 b 作为 HASH JOIN 的被驱动表。在进行 HASH JOIN 的时候，驱动表会被放到 PGA 中，这里，因为驱动表 a 只有 30MB，PGA 能够完全容纳下驱动表。因为被驱动表 b 特别大，想要加快 SQL 查询速度，必须开启并行查询。超大表与超小表在进行并行 HASH 连接的时候，可以将小表（驱动表）广播到所有的查询进程，然后对大表进行并行随机扫描，每个查询进程查询部分 b 表数据，然后再进行关联。假设对以上 SQL 启用 6 个并行进程对 a 表的并行广播，对 b 表进行随机并行扫描（每部分记为 b1,b2,b3,b4,b5,b6）其实就相当于将以上 SQL 内部等价改写为下面 SQL。

```
select * from a,b1 where a.object_id=b1.object_id   ---并行进行
union all
select * from a,b2 where a.object_id=b2.object_id   ---并行进行
union all
select * from a,b3 where a.object_id=b3.object_id   ---并行进行
union all
select * from a,b4 where a.object_id=b4.object_id   ---并行进行
union all
select * from a,b5 where a.object_id=b5.object_id   ---并行进行
union all
select * from a,b6 where a.object_id=b6.object_id;  ---并行进行
```

怎么才能让 a 表进行广播呢？我们需要添加 hint: pq_distribute（驱动表 none, broadcast）。

现在我们来查看 a 表并行广播的执行计划（为了方便排版，执行计划中省略了部分数据）。

```
SQL> explain plan for select
 /*+ parallel(6) use_hash(a,b) pq_distribute(a none,broadcast) */
   2    *
   3    from a, b
   4    where a.object_id = b.object_id;

Explained.

SQL> select * from table(dbms_xplan.display);

PLAN_TABLE_OUTPUT
----------------------------------------------------------------------------
```

```
Plan hash value: 3536517442
---------------------------------------------------------------------------------
| Id | Operation              | Name     | Rows  | Bytes |IN-OUT| PQ Distrib |
---------------------------------------------------------------------------------
|  0 | SELECT STATEMENT       |          | 5064K | 1999M |      |            |
|  1 |  PX COORDINATOR        |          |       |       |      |            |
|  2 |   PX SEND QC (RANDOM)  | :TQ10001 | 5064K | 1999M | P->S | QC (RAND)  |
|* 3 |    HASH JOIN           |          | 5064K | 1999M | PCWP |            |
|  4 |     PX RECEIVE         |          | 74893 |   14M | PCWP |            |
|  5 |      PX SEND BROADCAST | :TQ10000 | 74893 |   14M | P->P | BROADCAST  |
|  6 |       PX BLOCK ITERATOR|          | 74893 |   14M | PCWC |            |
|  7 |        TABLE ACCESS FULL| A       | 74893 |   14M | PCWP |            |
|  8 |     PX BLOCK ITERATOR  |          | 5064K |  999M | PCWC |            |
|  9 |      TABLE ACCESS FULL | B        | 5064K |  999M | PCWP |            |
---------------------------------------------------------------------------------

Predicate Information (identified by operation id):
---------------------------------------------------

   3 - access("A"."OBJECT_ID"="B"."OBJECT_ID")
```

如果小表进行了广播，执行计划 Operation 会出现 PX SEND BROADCAST 关键字，PQ Distrib 会出现 BROADCAST 关键字。注意：如果是两个大表关联，千万不能让大表广播。

8.6 超大表与超大表关联优化方法

现有如下 SQL。

```
select * from a,b where a.object_id=b.object_id;
```

表 a 有 4GB，表 b 有 6GB，两表关联后返回大量数据，应该走 HASH 连接。因为 a 比 b 小，所以 a 表应该作为 HASH JOIN 的驱动表。驱动表 a 有 4GB，需要放入 PGA 中。因为 PGA 中 work area 不能超过 2G，所以 PGA 不能完全容纳下驱动表，这时有部分数据会溢出到磁盘（TEMP）进行 on-disk hash join。我们可以开启并行查询加快查询速度。超大表与超大表在进行并行 HASH 连接的时候，需要将两个表根据连接列进行 HASH 运算，然后将运算结果放到 PGA 中，再进行 HASH 连接，这种并行 HASH 连接就叫作并行 HASH HASH 连接。假设对上面 SQL 启用 6 个并行查询，a 表会根据连接列进行 HASH 运算然后拆分为 6 份，记为 a1, a2, a3, a4, a5, a6，b 表也会根据连接列进行 HASH 运算然后拆分为 6 份，记为 b1, b2, b3, b4, b5, b6。那么以上 SQL 开启并行就相当于被改写成如下 SQL。

```
select * from a1,b1 where a1.object_id=b1.object_id   ---并行进行
union all
select * from a2,b2 where a2.object_id=b2.object_id   ---并行进行
union all
select * from a3,b3 where a3.object_id=b3.object_id   ---并行进行
union all
select * from a4,b4 where a4.object_id=b4.object_id   ---并行进行
union all
select * from a5,b5 where a5.object_id=b5.object_id   ---并行进行
union all
select * from a6,b6 where a6.object_id=b6.object_id;  ---并行进行
```

对于上面 SQL，开启并行查询就能避免 on-disk hash join，因为表不是特别大，而且被拆分到内存中了。怎么写 HINT 实现并行 HASH HASH 呢？我们需要添加 hint：`pq_distribute`（被驱动表 hash,hash）。

现在我们来查看并行 HASH HASH 的执行计划（为了方便排版，执行计划中省略了部分数据）。

```
SQL> explain plan for select
/*+ parallel(6) use_hash(a,b) pq_distribute(b hash,hash) */
  2    *
  3    from a, b
  4    where a.object_id = b.object_id;

Explained.

SQL> select * from table(dbms_xplan.display);

PLAN_TABLE_OUTPUT
--------------------------------------------------------------------------------
Plan hash value: 728916813
--------------------------------------------------------------------------------
| Id | Operation              | Name     | Rows  | Bytes |TempSpc|IN-OUT|PQ Distrib|
--------------------------------------------------------------------------------
|  0 | SELECT STATEMENT       |          | 3046M | 1174G |       |      |          |
|  1 |  PX COORDINATOR        |          |       |       |       |      |          |
|  2 |   PX SEND QC (RANDOM)  | :TQ10002 | 3046M | 1174G |       | P->S | QC (RAND)|
|* 3 |    HASH JOIN BUFFERED  |          | 3046M | 1174G |  324M | PCWP |          |
|  4 |     PX RECEIVE         |          | 9323K | 1840M |       | PCWP |          |
|  5 |      PX SEND HASH      | :TQ10000 | 9323K | 1840M |       | P->P | HASH     |
|  6 |       PX BLOCK ITERATOR|          | 9323K | 1840M |       | PCWC |          |
|  7 |        TABLE ACCESS FULL| A       | 9323K | 1840M |       | PCWP |          |
|  8 |     PX RECEIVE         |          |   20M | 4045M |       | PCWP |          |
|  9 |      PX SEND HASH      | :TQ10001 |   20M | 4045M |       | P->P | HASH     |
| 10 |       PX BLOCK ITERATOR|          |   20M | 4045M |       | PCWC |          |
| 11 |        TABLE ACCESS FULL| B       |   20M | 4045M |       | PCWP |          |
--------------------------------------------------------------------------------

Predicate Information (identified by operation id):
--------------------------------------------------

   3 - access("A"."OBJECT_ID"="B"."OBJECT_ID")
```

两表如果进行的是并行 HASH HASH 关联，执行计划 Operation 会出现 PX SEND HASH 关键字，PQ Distrib 会出现 HASH 关键字。

如果表 a 有 20G，表 b 有 30G，即使采用并行 HASH HASH 连接也很难跑出结果，因为要把两个表先映射到 PGA 中，这需要耗费一部分 PGA，之后在进行 HASH JOIN 的时候也需要部分 PGA，此时 PGA 根本就不够用，如果我们查看等待事件，会发现进程一直在做 DIRECT PATH READ/WRITE TEMP。

如何解决超级大表（几十 GB）与超级大表（几十 GB）关联的性能问题呢？我们可以根据并行 HASH HASH 关联的思路，人工实现并行 HASH HASH。下面就是人工实现并行 HASH HASH 的过程。

现在我们创建新表 p1，在表 a 的结构上添加一个字段 HASH_VALUE，同时根据

8.6 超大表与超大表关联优化方法

HASH_VALUE 进行 LIST 分区。

```
SQL> CREATE TABLE P1(
  2  HASH_VALUE NUMBER,
  3  OWNER VARCHAR2(30),
  4  OBJECT_NAME VARCHAR2(128),
  5  SUBOBJECT_NAME VARCHAR2(30),
  6  OBJECT_ID NUMBER,
  7  DATA_OBJECT_ID NUMBER,
  8  OBJECT_TYPE VARCHAR2(19),
  9  CREATED DATE,
 10  LAST_DDL_TIME DATE,
 11  TIMESTAMP VARCHAR2(19),
 12  STATUS VARCHAR2(7),
 13  TEMPORARY VARCHAR2(1),
 14  GENERATED VARCHAR2(1),
 15  SECONDARY VARCHAR2(1),
 16  NAMESPACE NUMBER,
 17  EDITION_NAME VARCHAR2(30)
 18  )
 19      PARTITION BY  list(HASH_VALUE)
 20  (
 21  partition p0 values (0),
 22  partition p1 values (1),
 23  partition p2 values (2),
 24  partition p3 values (3),
 25  partition p4 values (4)
 26  );

Table created.
```

然后我们创建新表 p2，在表 b 的结构上添加一个字段 HASH_VALUE，同时根据 HASH_VALUE 进行 LIST 分区。

```
SQL> CREATE TABLE P2(
  2  HASH_VALUE NUMBER,
  3  OWNER VARCHAR2(30),
  4  OBJECT_NAME VARCHAR2(128),
  5  SUBOBJECT_NAME VARCHAR2(30),
  6  OBJECT_ID NUMBER,
  7  DATA_OBJECT_ID NUMBER,
  8  OBJECT_TYPE VARCHAR2(19),
  9  CREATED DATE,
 10  LAST_DDL_TIME DATE,
 11  TIMESTAMP VARCHAR2(19),
 12  STATUS VARCHAR2(7),
 13  TEMPORARY VARCHAR2(1),
 14  GENERATED VARCHAR2(1),
 15  SECONDARY VARCHAR2(1),
 16  NAMESPACE NUMBER,
 17  EDITION_NAME VARCHAR2(30)
 18  )
 19      PARTITION BY  list(HASH_VALUE)
 20  (
 21  partition p0 values (0),
 22  partition p1 values (1),
 23  partition p2 values (2),
 24  partition p3 values (3),
 25  partition p4 values (4)
 26  );
```

```
Table created.
```

请注意，两个表分区必须一模一样，如果分区不一样，就有数据无法关联上。

我们将 a 表的数据迁移到新表 p1 中。

```
insert into p1
  select ora_hash(object_id, 4), a.* from a;  ---注意排除 object_id 为 null 的数据
commit;
```

然后我们将 b 表的数据迁移到新表 p2 中。

```
insert into p2
  select ora_hash(object_id, 4), b.* from b;  ---注意排除 object_id 为 null 的数据
commit;
```

下面 SQL 就是并行 HASH HASH 关联的人工实现。

```
select *
  from p1, p2
 where p1.object_id = p2.object_id
   and p1.hash_value = 0
   and p2.hash_value = 0;

select *
  from p1, p2
 where p1.object_id = p2.object_id
   and p1.hash_value = 1
   and p2.hash_value = 1;

select *
  from p1, p2
 where p1.object_id = p2.object_id
   and p1.hash_value = 2
   and p2.hash_value = 2;

select *
  from p1, p2
 where p1.object_id = p2.object_id
   and p1.hash_value = 3
   and p2.hash_value = 3;

select *
  from p1, p2
 where p1.object_id = p2.object_id
   and p1.hash_value = 4
   and p2.hash_value = 4;
```

此方法运用了 ora_hash 函数。Oracle 中的 HASH 分区就是利用的 ora_hash 函数。

ora_hash 使用方法如下。

ora_hash(列,HASH 桶)，HASH 桶默认是 4 294 967 295，可以设置 0～4 294 967 295。

ora_hash(object_id,4) 会把 object_id 的值进行 HASH 运算，然后放到 0、1、2、3、4 这些桶里面，也就是说 ora_hash(object_id,4) 只会产生 0、1、2、3、4 这几个值。

将大表（a,b）拆分为分区表（p1, p2）之后，我们只需要依次关联对应的分区，这样就不会出现 PGA 不足的问题，从而解决了超级大表关联查询的效率问题。在实际生产环境中，需要添加多少分区，请自己判断。

8.7 LIKE 语句优化方法

我们先创建测试表 T。

```
SQL> create table t as select * from dba_objects;
Table created.
```

现在有如下语句。

```
select * from t where object_name like '%SEQ%';
```

因为需要对字符串两边进行模糊匹配，而索引根块和分支块存储的是前缀数据（也就是说 object like 'SEQ%' 才能走索引），所以上面 SQL 查询无法走索引。

如果强制走索引，会走 INDEX FULL SCAN。

```
SQL> create index idx_ojbname on t(object_name);
Index created.
```

查看强制走索引的执行计划。

```
SQL> select /*+ index(t) */ * from t where object_name like '%SEQ%';

208 rows selected.

Execution Plan
----------------------------------------------------------
Plan hash value: 3894507753

--------------------------------------------------------------------------------
| Id | Operation                   | Name        | Rows | Bytes | Cost(%CPU)|Time     |
--------------------------------------------------------------------------------
|  0 | SELECT STATEMENT            |             |  219 | 45333 |  2214   (1)|00:00:27|
|  1 |  TABLE ACCESS BY INDEX ROWID| T           |  219 | 45333 |  2214   (1)|00:00:27|
|* 2 |   INDEX FULL SCAN           | IDX_OJBNAME | 3395 |       |   362   (1)|00:00:05|
--------------------------------------------------------------------------------

Predicate Information (identified by operation id):
---------------------------------------------------

   2 - filter("OBJECT_NAME" LIKE '%SEQ%')
```

INDEX FULL SCAN 是单块读，性能不如全表扫描。大家可能会有疑问，可不可以走 INDEX FAST FULL SCAN 呢？答案是不可以，因为 INDEX FAST FULL SCAN 不能回表，而上面 SQL 查询需要回表（select *）。

我们可以创建一个表当索引用，用来代替 INDEX FAST FULL SCAN 不能回表的情况。

```
SQL> create table index_t as select object_name,rowid rid from t;
Table created.
```

现在将 SQL 查询改写为如下 SQL。

```
select *
  from t
 where rowid in (select rid from index_t where object_name like '%SEQ%');
```

改写完 SQL 之后,需要让 index_t 与 t 走嵌套循环,同时让 index_t 作为嵌套循环驱动表,这样就达到了让 index_t 充当索引的目的。

现在我们来对比两个 SQL 的 autotrace 执行计划。

```
SQL> select * from t where object_name like '%SEQ%';

208 rows selected.


Execution Plan
----------------------------------------------------------
Plan hash value: 1601196873

--------------------------------------------------------------------------
| Id  | Operation         | Name | Rows  | Bytes | Cost (%CPU)| Time     |
--------------------------------------------------------------------------
|   0 | SELECT STATEMENT  |      |   135 | 27945 |   235   (1)| 00:00:03 |
|*  1 |  TABLE ACCESS FULL| T    |   135 | 27945 |   235   (1)| 00:00:03 |
--------------------------------------------------------------------------

Predicate Information (identified by operation id):
---------------------------------------------------

   1 - filter("OBJECT_NAME" IS NOT NULL AND "OBJECT_NAME" LIKE '%SEQ%')

Note
-----
   - dynamic sampling used for this statement (level=2)


Statistics
----------------------------------------------------------
          5  recursive calls
          0  db block gets
       1117  consistent gets
          0  physical reads
          0  redo size
      12820  bytes sent via SQL*Net to client
        563  bytes received via SQL*Net from client
         15  SQL*Net roundtrips to/from client
          0  sorts (memory)
          0  sorts (disk)
        208  rows processed

SQL> select /*+ leading(index_t@a) use_nl(index_t@a,t) */
  2         *
  3    from t
  4   where rowid in (select /*+ qb_name(a) */
  5                         rid
  6                    from index_t
  7                   where object_name like '%SEQ%');

208 rows selected.


Execution Plan
```

```
-------------------------------------------------------------------------------
Plan hash value: 2608052908
-------------------------------------------------------------------------------
| Id  | Operation                     | Name    | Rows  | Bytes | Cost (%CPU)| Time     |
-------------------------------------------------------------------------------
|   0 | SELECT STATEMENT              |         |    87 | 25839 |   140   (2)| 00:00:02 |
|   1 |  NESTED LOOPS                 |         |    87 | 25839 |   140   (2)| 00:00:02 |
|   2 |   SORT UNIQUE                 |         |    87 |  6786 |    95   (2)| 00:00:02 |
|*  3 |    TABLE ACCESS FULL          | INDEX_T |    87 |  6786 |    95   (2)| 00:00:02 |
|   4 |   TABLE ACCESS BY USER ROWID  | T       |     1 |   219 |     1   (0)| 00:00:01 |
-------------------------------------------------------------------------------

Predicate Information (identified by operation id):
---------------------------------------------------

   3 - filter("OBJECT_NAME" IS NOT NULL AND "OBJECT_NAME" LIKE '%SEQ%')

Note
-----
   - dynamic sampling used for this statement (level=2)


Statistics
----------------------------------------------------------
          0  recursive calls
          0  db block gets
        499  consistent gets
          0  physical reads
          0  redo size
      12820  bytes sent via SQL*Net to client
        563  bytes received via SQL*Net from client
         15  SQL*Net roundtrips to/from client
          1  sorts (memory)
          0  sorts (disk)
        208  rows processed
```

因为 t 表很小，表字段也不多，所以大家可能感觉性能提升不是特别大。当 t 表越大，性能提升就越明显。采用这个方法还需要对 index_t 进行数据同步，我们可以将 index_t 创建为物化视图，刷新方式采用 on commit 刷新。

8.8 DBLINK 优化

现在有如下两个表，a 表是远端表（1800 万），b 表是本地表（100 行）。

```
SQL> desc a@dblink
 Name                    Null?     Type
 ----------------------- --------- ------
 ID                                NUMBER
 NAME                              VARCHAR2(100)
 ADDRESS                           VARCHAR2(100)

SQL> select count(*) from a@dblink;

  COUNT(*)
----------
  18550272
```

第 8 章　调优技巧

```
SQL> desc b
 Name                 Null?    Type
 -------------------- -------- ------
 ID                            NUMBER
 NAME                          VARCHAR2(100)
 ADDRESS                       VARCHAR2(100)

SQL> select count(*) from b;

  COUNT(*)
----------
       100
```

现有如下 SQL。

select * from a@dblink, b where a.id = b.id;

默认情况下，会将远端表 a 的数据传输到本地，然后再进行关联，autotrace 的执行计划如下。

```
SQL> set timi on
SQL> set autot trace
SQL> select * from a@dblink, b where a.id = b.id;

25600 rows selected.

Elapsed: 00:03:13.80

Execution Plan
----------------------------------------------------------
Plan hash value: 657970699

---------------------------------------------------------------------------------
| Id  | Operation          |Name|Rows | Bytes | Cost (%CPU)| Time     | Inst   |IN-OUT|
---------------------------------------------------------------------------------
|   0 | SELECT STATEMENT   |    |  82 | 19188 |     6  (17)| 00:00:01 |        |      |
|*  1 |  HASH JOIN         |    |  82 | 19188 |     6  (17)| 00:00:01 |        |      |
|   2 |   REMOTE           | A  |  82 |  9594 |     2   (0)| 00:00:01 | DBLINK | R->S |
|   3 |   TABLE ACCESS FULL| B  | 100 | 11700 |     3   (0)| 00:00:01 |        |      |
---------------------------------------------------------------------------------

Predicate Information (identified by operation id):
---------------------------------------------------

   1 - access("A"."ID"="B"."ID")

Remote SQL Information (identified by operation id):
----------------------------------------------------

   2 - SELECT "ID","NAME","ADDRESS" FROM "A" "A" (accessing 'DBLINK' )

Statistics
----------------------------------------------------------
        769  recursive calls
          1  db block gets
         15  consistent gets
      91755  physical reads
        212  redo size
    1477532  bytes sent via SQL*Net to client
      19185  bytes received via SQL*Net from client
       1708  SQL*Net roundtrips to/from client
          0  sorts (memory)
```

```
        0  sorts (disk)
    25600  rows processed
```

远端表 a 很大，对数据进行传输会耗费大量时间，本地表 b 表很小，而且 a 和 b 关联之后返回数据量很少，我们可以将本地表 b 传输到远端，在远端进行关联，然后再将结果集传回本地，这时需要使用 hint:driving_site，下面 SQL 就是将 b 传递到远端关联的示例。

```
select /*+ driving_site(a) */ * from a@dblink, b  where a.id = b.id;
```

autotrace 的执行计划如下。

```
SQL> select /*+ driving_site(a) */ * from a@dblink, b  where a.id = b.id;

25600 rows selected.

Elapsed: 00:00:06.08
Execution Plan
----------------------------------------------------------
Plan hash value: 4284963264

---------------------------------------------------------------------------------
| Id  | Operation              | Name | Rows  | Bytes | Cost (%CPU)| Inst  |IN-OUT|
---------------------------------------------------------------------------------
|   0 | SELECT STATEMENT REMOTE|      | 20931 | 4783K| 25565   (2)|       |      |
|*  1 |  HASH JOIN             |      | 20931 | 4783K| 25565   (2)|       |      |
|   2 |   REMOTE               | B    |    82 | 9594 |     2   (0)|     ! | R->S |
|   3 |   TABLE ACCESS FULL    | A    |   19M | 2173M| 25466   (1)| ORCL  |      |
---------------------------------------------------------------------------------

Predicate Information (identified by operation id):
---------------------------------------------------

   1 - access("A2"."ID"="A1"."ID")

Remote SQL Information (identified by operation id):
----------------------------------------------------

   2 - SELECT "ID","NAME","ADDRESS" FROM "B" "A1" (accessing '!' )


Note
-----
   - fully remote statement


Statistics
----------------------------------------------------------
          6  recursive calls
          0  db block gets
          8  consistent gets
          0  physical reads
          0  redo size
    1428836  bytes sent via SQL*Net to client
      19185  bytes received via SQL*Net from client
       1708  SQL*Net roundtrips to/from client
          0  sorts (memory)
          0  sorts (disk)
      25600  rows processed
```

将本地小表传输到远端关联，再返回结果只需 6 秒，相比将大表传输到本地，在性能上有巨大提升。

现在我们在远端表 a 的连接列建立索引。

```
SQL> create index idx_id on a(id);

Index created.
```

因为 b 表只有 100 行数据，a 表有 1 800 万行数据，两表关联之后返回 2.5 万行数据，我们可以让 a 与 b 走嵌套循环，b 作为驱动表，a 作为被驱动表，而且走连接索引。

```
SQL> select /*+ index(a) use_nl(a,b) leading(b) */ * from a@dblink, b  where a.id = b.id;

25600 rows selected.

Elapsed: 00:00:00.84

Execution Plan
----------------------------------------------------------
Plan hash value: 1489534455

--------------------------------------------------------------------------------
| Id  | Operation          | Name | Rows  | Bytes | Cost (%CPU)| Inst   |IN-OUT|
--------------------------------------------------------------------------------
|   0 | SELECT STATEMENT   |      | 7614K|  1699M| 54680  (100)|        |      |
|   1 |  NESTED LOOPS      |      | 7614K|  1699M| 54680  (100)|        |      |
|   2 |   TABLE ACCESS FULL| B    |   100 | 11700 |     3   (0)|        |      |
|   3 |   REMOTE           | A    | 76146 |  8700K|     3   (0)| DBLINK | R->S |
--------------------------------------------------------------------------------

Remote SQL Information (identified by operation id):
----------------------------------------------------

   3 - SELECT /*+ USE_NL ("A") INDEX ("A") */ "ID","NAME","ADDRESS" FROM "A" "A"
       WHERE "ID"=:1 (accessing 'DBLINK' )


Statistics
----------------------------------------------------------
          0  recursive calls
          0  db block gets
        106  consistent gets
          0  physical reads
          0  redo size
     349986  bytes sent via SQL*Net to client
      19185  bytes received via SQL*Net from client
       1708  SQL*Net roundtrips to/from client
          0  sorts (memory)
          0  sorts (disk)
      25600  rows processed
```

强制 a 表走索引之后，这时我们只需将索引过滤之后的数据传输到本地，而无需将 a 表所有数据传到本地，性能得到极大提升，SQL 耗时不到 1 秒。

现在我们将 b 表传输到远端，强制 b 表作为嵌套循环驱动表。

```
SQL> select /*+ driving_site(a) use_nl(a,b) leading(b) */ * from a@dblink, b  where a.id = b.id;

25600 rows selected.
```

8.8 DBLINK 优化

```
Elapsed: 00:00:02.92

Execution Plan
----------------------------------------------------------
Plan hash value: 557259519
----------------------------------------------------------
| Id  | Operation                   |Name   |Rows  | Bytes | Cost (%CPU)| Inst   |IN-OUT|
----------------------------------------------------------
|  0  | SELECT STATEMENT REMOTE     |       |20931 | 4783K | 20182   (1)|        |      |
|  1  |  NESTED LOOPS               |       |      |       |            |        |      |
|  2  |   NESTED LOOPS              |       |20931 | 4783K | 20182   (1)|        |      |
|  3  |    REMOTE                   |B      |   82 |  9594 |     2   (0)|    !   | R->S |
|* 4  |    INDEX RANGE SCAN         |IDX_ID |  255 |       |     2   (0)| ORCL   |      |
|  5  |   TABLE ACCESS BY INDEX ROWID|A     |  255 | 29835 |   246   (0)| ORCL   |      |
----------------------------------------------------------

Predicate Information (identified by operation id):
---------------------------------------------------

   4 - access("A2"."ID"="A1"."ID")

Remote SQL Information (identified by operation id):
---------------------------------------------------

   3 - SELECT /*+ USE_NL ("A1") */ "ID","NAME","ADDRESS" FROM "B" "A1" (accessing '!' )

Note
-----
   - fully remote statement

Statistics
----------------------------------------------------------
          6  recursive calls
          0  db block gets
          8  consistent gets
          0  physical reads
          0  redo size
     426684  bytes sent via SQL*Net to client
      19185  bytes received via SQL*Net from client
       1708  SQL*Net roundtrips to/from client
          0  sorts (memory)
          0  sorts (disk)
      25600  rows processed
```

该查询耗时 2.9 秒，主要开销耗费在网络传输上，首先我们要将 b 表传输到远端，然后将 a 与 b 的关联结果传输到本地，网络传输耗费了两次。我们可以设置 arraysize 减少网络交互次数，从而减少网络开销，如下所示。

```
SQL> set arraysize 1000
SQL> select /*+ driving_site(a) use_nl(a,b) leading(b) */ * from a@dblink, b  where a
.id = b.id;

25600 rows selected.

Elapsed: 00:00:00.29

Execution Plan
----------------------------------------------------------
Plan hash value: 557259519
```

```
---------------------------------------------------------------------------
| Id  | Operation                    |Name   |Rows  | Bytes  | Cost (%CPU)|Inst  |IN-OUT|
---------------------------------------------------------------------------
|  0  | SELECT STATEMENT REMOTE      |       |20931 | 4783K  | 20182   (1)|      |      |
|  1  |  NESTED LOOPS                |       |      |        |            |      |      |
|  2  |   NESTED LOOPS               |       |20931 | 4783K  | 20182   (1)|      |      |
|  3  |    REMOTE                    |B      |   82 |  9594  |     2   (0)|    ! | R->S |
|* 4  |    INDEX RANGE SCAN          |IDX_ID |  255 |        |     2   (0)| ORCL |      |
|  5  |   TABLE ACCESS BY INDEX ROWID|A      |  255 | 29835  |   246   (0)| ORCL |      |
---------------------------------------------------------------------------

Predicate Information (identified by operation id):
---------------------------------------------------

   4 - access("A2"."ID"="A1"."ID")

Remote SQL Information (identified by operation id):
----------------------------------------------------

   3 - SELECT /*+ USE_NL ("A1") */ "ID","NAME","ADDRESS" FROM "B" "A1" (accessing '!' )

Note
-----
   - fully remote statement

Statistics
----------------------------------------------------------
          3  recursive calls
          0  db block gets
          8  consistent gets
          0  physical reads
          0  redo size
     137698  bytes sent via SQL*Net to client
        694  bytes received via SQL*Net from client
         27  SQL*Net roundtrips to/from client
          0  sorts (memory)
          0  sorts (disk)
      25600  rows processed
```

注意观察执行计划中统计信息栏目 SQL*Net roundtrips 从 1 708 减少到 27。当需要将本地表传输到远端关联、再将关联结果传输到本地的时候，我们可以设置 arraysize 优化 SQL。

如果远端表 a 很大，本地表 b 也很大，两表关联返回数据量多，这时既不能将远端表 a 传到本地，也不能将本地表 b 传到远端，因为无论采用哪种方法，SQL 都很慢。我们可以在本地创建一个带有 dblink 的物化视图，将远端表 a 的数据刷新到本地，然后再进行关联。

如果 SQL 语句中有多个 dblink 源，最好在本地针对每个 dblink 源建立带有 dblink 的物化视图，因为多个 dblink 源之间进行数据传输，网络信息交换会导致严重性能问题。

有时候会使用 dblink 对数据进行迁移，如果要迁移的数据量很大，我们可以使用批量游标进行迁移。以下是使用批量游标迁移数据的示例（将 a@dblink 的数据迁移到 b）。

```
declare
  cursor cur is
    select id, name, address from a@dblink;
  type cur_type is table of cur%rowtype index by binary_integer;
  v_cur cur_type;
begin
  open cur;
```

```
  loop
    fetch cur bulk collect
      into v_cur limit 100000;
    forall i in 1 .. v_cur.count
      insert into b
        (id, name, address)
      values
        (v_cur(i).id, v_cur(i).name, v_cur(i).address);
    commit;
    exit when cur%notfound or cur%notfound is null;
  end loop;
  close cur;
  commit;
end;
```

8.9 对表进行 ROWID 切片

对一个很大的分区表进行 UPDATE、DELETE，想要加快执行速度，可以按照分区，在不同的会话中对每个分区单独进行 UPDATE、DELETE。但是对一个很大的非分区表进行 UPDATE、DELETE，如果只在一个会话里面运行 SQL，很容易引发 UNDO 不够，如果会话连接中断，会导致大量数据从 UNDO 回滚，这将是一场灾难。

对于非分区表，我们可以对表按照 ROWID 切片，然后开启多个窗口同时执行 SQL，这样既能加快执行速度，还能减少对 UNDO 的占用。

Oracle 提供了一个内置函数 DBMS_ROWID.ROWID_CREATE()用于生成 ROWID。对于一个非分区表，一个表就是一个段（Segment），段是由多个区组成，每个区里面的块物理上是连续的。因此，我们可以根据数据字典 DBA_EXTENTS, DBA_OBJECTS 关联，然后再利用生成 ROWID 的内置函数人工生成 ROWID。

例如，我们对 SCOTT 账户下 TEST 表按照每个 Extent 进行 ROWID 切片。

```
select ' and rowid between ' || '''' ||
       dbms_rowid.rowid_create(1,
                               b.data_object_id,
                               a.relative_fno,
                               a.block_id,
                               0) || '''' || ' and ' || '''' ||
       dbms_rowid.rowid_create(1,
                               b.data_object_id,
                               a.relative_fno,
                               a.block_id + blocks - 1,
                               999) || '''' || ';'
  from dba_extents a, dba_objects b
 where a.segment_name = b.object_name
   and a.owner = b.owner
   and b.object_name = 'TEST'
   and b.owner = 'SCOTT'
 order by a.relative_fno, a.block_id;
```

切片后生成的部分数据如下所示。

```
and rowid between 'AAASs5AAEAAB+SIAAA' and 'AAASs5AAEAAB+SPAPn';
and rowid between 'AAASs5AAEAAB+SQAAA' and 'AAASs5AAEAAB+SXAPn';
and rowid between 'AAASs5AAEAAB+SYAAA' and 'AAASs5AAEAAB+SfAPn';
```

```
     and rowid between 'AAASs5AAEAAB+SgAAA' and 'AAASs5AAEAAB+SnAPn';
     and rowid between 'AAASs5AAEAAB+SoAAA' and 'AAASs5AAEAAB+SvAPn';
```

假如要执行 delete test where object_id>50000000，test 表有 1 亿条数据，要删除其中 5 000 万行数据，我们根据上述方法对表按照 ROWID 切片。

```
delete test
 where object_id > 50000000
    and rowid between 'AAASs5AAEAAB+SIAAA' and 'AAASs5AAEAAB+SPAPn';
delete test
 where object_id > 50000000
    and rowid between 'AAASs5AAEAAB+SQAAA' and 'AAASs5AAEAAB+SXAPn';
delete test
 where object_id > 50000000
    and rowid between 'AAASs5AAEAAB+SYAAA' and 'AAASs5AAEAAB+SfAPn';
delete test
 where object_id > 50000000
    and rowid between 'AAASs5AAEAAB+SgAAA' and 'AAASs5AAEAAB+SnAPn';
delete test
 where object_id > 50000000
    and rowid between 'AAASs5AAEAAB+SoAAA' and 'AAASs5AAEAAB+SvAPn';
```

最后，我们将上述 SQL 在不同窗口中执行，这样就能加快 delete 速度，也能减少对 UNDO 的占用。

上述方法需要手动编辑大量 SQL 脚本，如果表的 Extent 很多，这将带来大工作量。我们可以编写存储过程简化上述操作。

因为存储过程需要访问数据字典，我们需要单独授权查询数据字典权限。

```
grant select on dba_extents to scott;

grant select on dba_objects to scott;

CREATE OR REPLACE PROCEDURE P_ROWID(RANGE NUMBER, ID NUMBER) IS
  CURSOR CUR_ROWID IS
    SELECT DBMS_ROWID.ROWID_CREATE(1,
                                   B.DATA_OBJECT_ID,
                                   A.RELATIVE_FNO,
                                   A.BLOCK_ID,
                                   0) ROWID1,
           DBMS_ROWID.ROWID_CREATE(1,
                                   B.DATA_OBJECT_ID,
                                   A.RELATIVE_FNO,
                                   A.BLOCK_ID + BLOCKS - 1,
                                   999) ROWID2
      FROM DBA_EXTENTS A, DBA_OBJECTS B
     WHERE A.SEGMENT_NAME = B.OBJECT_NAME
       AND A.OWNER = B.OWNER
       AND B.OBJECT_NAME = 'TEST'
       AND B.OWNER = 'SCOTT'
       AND MOD(A.EXTENT_ID, RANGE) = ID;
  V_SQL VARCHAR2(4000);
BEGIN
  FOR CUR IN CUR_ROWID LOOP
    V_SQL := 'delete test where object_id > 100 and rowid between :1 and :2';
    EXECUTE IMMEDIATE V_SQL
      USING CUR.ROWID1, CUR.ROWID2;
    COMMIT;
  END LOOP;
```

```
END;
/
```

如果要将表切分为 6 份，我们可以在 6 个窗口中依次执行。

```
begin
  p_rowid(6, 0);
end;
/
begin
  p_rowid(6, 1);
end;
/
begin
  p_rowid(6, 2);
end;
/
begin
  p_rowid(6, 3);
end;
/
begin
  p_rowid(6, 4);
end;
/
begin
  p_rowid(6, 5);
end;
/
```

这样就达到了将表按 ROWID 切片的目的。在工作中，大家可以根据自己的具体需求对存储过程稍作修改（阴影部分）。

8.10 SQL 三段分拆法

如果要优化的 SQL 很长，我们可以将 SQL 拆分为三段，这样就能快速判断 SQL 在写法上是否容易产生性能问题。下面就是 SQL 三段拆分方法。

select ....第一段.... from ....第二段.... where ....第三段....

select 与 from 之间最好不要有标量子查询，也不要有自定义函数。因为有标量子查询或者是自定义函数，会导致子查询或者函数中的表被反复扫描。

from 与 where 之间要关注大表，因为大表很容易引起性能问题；同时要留意子查询和视图，如果有子查询或者视图，要单独运行，看运行得快或是慢，如果运行慢需要单独优化；另外要注意子查询/视图是否可以谓词推入，是否会视图合并；最后还要留意表与表之间是内连接还是外连接，因为外连接会导致嵌套循环无法改驱动表。

where 后面需要特别注意子查询，要能判断各种子查询写法是否可以展开（unnest），同时也要注意 where 过滤条件，尽量不要在 where 过滤列上使用函数，这样会导致列不走索引。

在工作中，我们要养成利用 SQL 三段分拆方法的习惯，这样能大大提升 SQL 优化的速度。

第 9 章 SQL 优化案例赏析

本章将会带大家领略多种多样的 SQL 优化方法,为大家以后优化 SQL 提供宝贵的参考意见。

9.1 组合索引优化案例

2015 年,一位佛山的朋友说某沙发厂 ERP 系统出现大量 read by other session 等待,前台用户卡了一天。数据库版本是 Oracle11gR2,请求协助优化。我们远程连接到朋友电脑之后,利用脚本抓出系统当前正在运行的 SQL,如图 9-1 所示。

图 9-1

从上面查询中我们可以看到,同时运行 SQL:1svyhsn0g56qd,会引发 read by other session 等待,于是从共享池中抓出该 SQL 的执行计划,如图 9-2 所示。

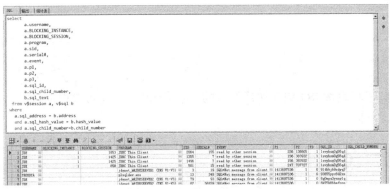

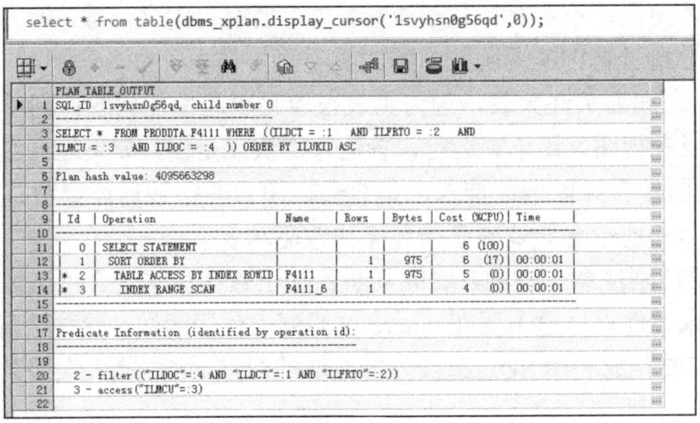

图 9-2

9.1 组合索引优化案例

SQL 文本如下。

```
SELECT *
  FROM PRODDTA.F4111
 WHERE ((ILDCT = :1 AND ILFRTO = :2 AND ILMCU = :3 AND ILDOC = :4))
 ORDER BY ILUKID ASC
```

从执行计划中 Id=3 我们可以看到，该 SQL 走的是 ILMCU 这个列的索引。如图 9-3 所示，表中一共有 2 510 970 行数据。

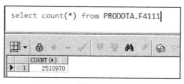

图 9-3

ILMCU 列的数据分布如下，如图 9-4 所示。

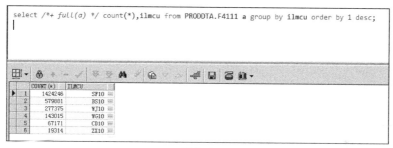

图 9-4

ILMCU 列的数据分布极不均衡。当询问当天做的是不是 SF10 的业务时，朋友确认是做的 SF10 的业务。这就不难解释为什么前台用户抱怨卡了一天。从 2 510 970 条数据中查询 1 424 246 条数据还走索引，这明显大错特错。这个错误的执行计划会导致产生大量的单块读，因为 SQL 执行缓慢，某些耐不住性子的用户可能会多次点击或刷新前台，并且因为做的是 SF10 的业务，前台操作人员可能多达几十位。正是因为有很多人在同时运行该 SQL，而且该 SQL 跑得很慢，又是单块读，所以就发生了多个进程需要同时读取同一个块的情况，这就是产生 read by other session 的原因。

该 SQL 一共有 4 个过滤条件，下面我们分别查看剩余 3 个过滤条件的数据分布，如图 9-5、图 9-6、图 9-7 所示。

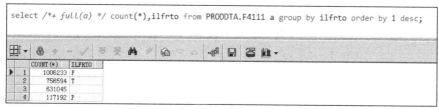

图 9-5

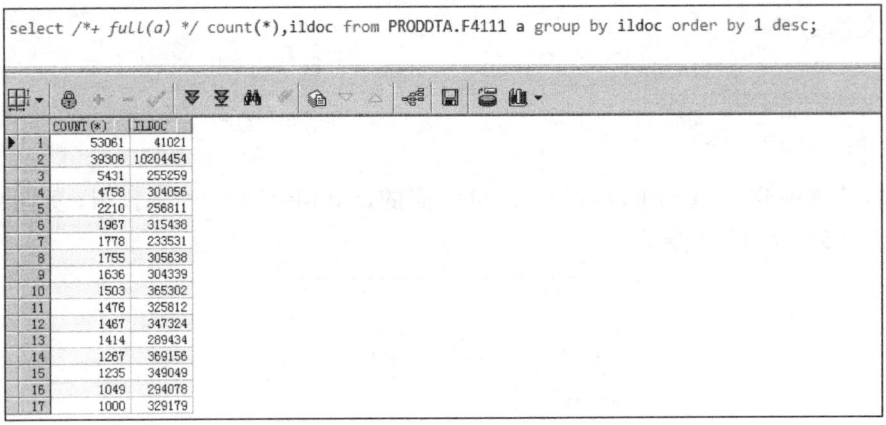

```
select /*+ full(a) */ count(*),ildoc from PRODDTA.F4111 a group by ildoc order by 1 desc;
```

	COUNT(*)	ILDOC
1	53061	41021
2	39306	10204454
3	5431	255259
4	4758	304056
5	2210	256811
6	1967	315438
7	1778	233531
8	1755	305638
9	1636	304339
10	1503	365302
11	1476	325812
12	1467	347324
13	1414	289434
14	1267	369156
15	1235	349049
16	1049	294078
17	1000	329179

图 9-6

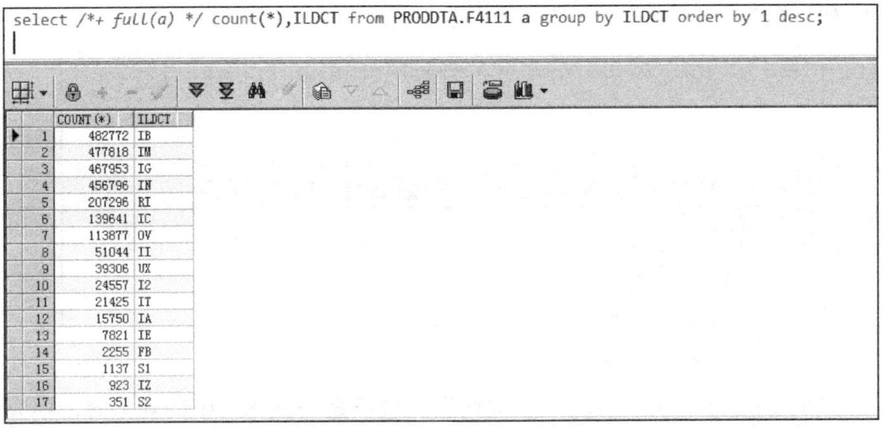

```
select /*+ full(a) */ count(*),ILDCT from PRODDTA.F4111 a group by ILDCT order by 1 desc;
```

	COUNT(*)	ILDCT
1	482772	IB
2	477818	IM
3	467953	IG
4	456796	IN
5	207296	RI
6	139641	IC
7	113877	OV
8	51044	II
9	39306	UX
10	24557	I2
11	21425	IT
12	15750	IA
13	7821	IE
14	2255	FB
15	1137	S1
16	923	IZ
17	351	S2

图 9-7

根据以上查询结果，我们发现，ILDOC 列的数据分布最为均匀，ILDCT 列的数据分布次之，ILMCU 列的数据分布倒数第二，ILFRTO 列的数据分布最不均衡。因为 SQL 都是根据这些列进行等值过滤，于是建立如下组合索引。

```
create index idx_F4111_docdctilmcufrto on F4111(ILDOC,ILDCT,ILMCU,ILFRTO) online nolo
gging;
```

创建完索引之后，系统中的 read other session 等待陆续消失，系统立刻恢复正常，前台用户原本执行了一天还没完成的业务现在可以瞬间完成。

为什么在查询 ILMCU 列的数据分布的时候会使用 HINT:FULL 呢？这是因为原本的 SQL 走该列的索引已经执行不出结果，如果不加 HINT，万一 SQL 查询又使用了该索引，这不是火上浇油吗？至于后面的 HINT，其一是因为复制粘贴，其二是因为表已经全表扫描过了，后面的全表扫描可以直接从 buffer cache 获取数据。

虽然通过创建组合索引优化了该 SQL，但是，在创建组合索引之前，如果优化器能够准确地知道 ILMCU 列的数据分布，那么执行计划也不会走该列的索引而会走其他列的索引（如果存在索引），或者走全表扫描。即使该 SQL 走全表扫描，那也比走索引扫描好太多，至少不会被卡死，不会引发前台用户被卡一天，最多被卡一小会儿。为什么优化器选择了走该列的索引呢？请注意观察执行计划中的 Id=3，Rows=1，优化器认为走 ILMCU 列的索引只返回一行数据。很明显该表统计信息有问题，而且该列很可能没有收集直方图。大家特别是 DBA，一定要重视表的统计信息，另外也要牢牢掌握索引知识，理解透了，就能解决 80%左右的关于 OLTP 的 SQL 性能问题。如果数据库系统不是 OLTP 系统，而是 ERP 系统，或者是 OLAP 中的报表系统、ETL 系统等，只吃透索引没太大帮助，必须精通阅读执行计划、SQL、各种 SQL 等价改写，熟悉分区，同时熟悉系统业务，这样才能游刃有余地进行 SQL 优化。

9.2 直方图优化案例

本案例发生在 2010 年，当时作者罗老师在惠普担任开发 DBA，支撑宝洁公司的数据仓库项目。为了避免泄露信息，他对 SQL 语句做了适当修改。ETL 开发人员发来邮件说有个 long running job，执行了两小时左右还未完成，需要检查一下。收到邮件后，立即检查数据库中正在运行的 SQL，经过与 ETL 开发人员确认，抓出执行计划（为了排版需要，删除了执行计划中非关键部分）。

```
SQL> select * from table(dbms_xplan.display_cursor('gh1hw18uz6dcm',0));

PLAN_TABLE_OUTPUT
------------------------------------------------------------------------------------
SQL_ID  gh1hw18uz6dcm, child number 0
-------------------------------------
create table OPT_REF_BASE_UOM_TEMP_SDIM   parallel 2
  nologging as SELECT PROD_SKID,          RELTV_CURR_QTY,
  STAT_CURR_VAL,          BAR_CURR_CODE      FROM OPT_REF_BASE_UOM_DIM_VW

Plan hash value: 2933813170

------------------------------------------------------------------------------------
| Id  | Operation                       | Name                 | Rows  | Bytes |
------------------------------------------------------------------------------------
|   0 | CREATE TABLE STATEMENT          |                      |       |       |
|   1 |  PX COORDINATOR                 |                      |       |       |
|   2 |   PX SEND QC (RANDOM)           | :TQ10001             |    54 |  2916 |
|   3 |    LOAD AS SELECT               |                      |       |       |
|   4 |     HASH GROUP BY               |                      |    54 |  2916 |
|   5 |      PX RECEIVE                 |                      |    54 |  2916 |
|   6 |       PX SEND HASH              | :TQ10000             |    54 |  2916 |
|   7 |        HASH GROUP BY            |                      |    54 |  2916 |
|   8 |         NESTED LOOPS            |                      |       |       |
|   9 |          NESTED LOOPS           |                      |  3134 |  165K |
|  10 |           PX BLOCK ITERATOR     |                      |       |       |
|* 11 |            TABLE ACCESS FULL    | OPT_REF_UOM_TEMP_SDIM |  3065 |  104K|
|* 12 |           INDEX RANGE SCAN      | PROD_DIM_PK          |     3 |       |
|* 13 |          TABLE ACCESS BY INDEX ROWID| PROD_DIM         |     1 |    19 |
------------------------------------------------------------------------------------

Predicate Information (identified by operation id):
---------------------------------------------------
```

```
   11 - access(:Z>=:Z AND :Z<=:Z)
        filter("UOM"."RELTV_CURR_QTY"=1)
   12 - access("PROD"."PROD_SKID"="UOM"."PROD_SKID")
   13 - filter(("PROD"."BUOM_CURR_SKID" IS NOT NULL AND
               "PROD"."PROD_END_DATE"=TO_DATE(' 9999-12-31 00:00:00', 'syyyy-mm-dd
  hh24:mi:ss') AND "PROD"."CURR_IND"='Y' AND "PROD"."BUOM_CURR_SKID"="UOM"."UOM_SKID"))
```

这个工作很简单，就是 create table .....as select ......。

```
create table OPT_REF_BASE_UOM_TEMP_SDIM  parallel 2  nologging
 as SELECT PROD_SKID, RELTV_CURR_QTY, STAT_CURR_VAL, BAR_CURR_CODE
    FROM OPT_REF_BASE_UOM_DIM_VW;
```

OPT_REF_BASE_UOM_DIM_VW 是一个视图，该视图定义：

```
SELECT UOM.PROD_SKID,
       MAX (UOM.RELTV_CURR_QTY) RELTV_CURR_QTY,
       MAX (UOM.STAT_CURR_VAL) STAT_CURR_VAL,
       MAX (UOM.BAR_CURR_CODE) BAR_CURR_CODE
FROM OPT_REF_UOM_TEMP_SDIM UOM,
     REF_PROD_DIM PROD
WHERE UOM.RELTV_CURR_QTY = 1
      AND PROD.CURR_IND = 'Y'
      AND PROD.PROD_END_DATE = TO_DATE ('31-12-9999', 'dd-mm-yyyy')
      AND PROD.PROD_SKID = UOM.PROD_SKID
      AND PROD.BUOM_CURR_SKID = UOM.UOM_SKID
GROUP BY UOM.PROD_SKID;
```

这个视图的查询效率就直接决定了 ETL JOB 的效率，现在我们查看这个视图的执行计划。

```
SQL> explain plan for SELECT UOM.PROD_SKID,
  2           MAX (UOM.RELTV_CURR_QTY) RELTV_CURR_QTY,
  3           MAX (UOM.STAT_CURR_VAL) STAT_CURR_VAL,
  4           MAX (UOM.BAR_CURR_CODE) BAR_CURR_CODE
  5  FROM OPT_REF_UOM_TEMP_SDIM UOM,
  6       REF_PROD_DIM PROD
  7  WHERE UOM.RELTV_CURR_QTY = 1
  8        AND PROD.CURR_IND = 'Y'
  9        AND PROD.PROD_END_DATE = TO_DATE ('31-12-9999', 'dd-mm-yyyy')
 10        AND PROD.PROD_SKID = UOM.PROD_SKID
 11        AND PROD.BUOM_CURR_SKID = UOM.UOM_SKID
 12  GROUP BY UOM.PROD_SKID;

Explained.

SQL> select * from table(dbms_xplan.display);

PLAN_TABLE_OUTPUT
--------------------------------------------------------------------------------

Plan hash value: 3215660883

--------------------------------------------------------------------------------
| Id  | Operation            | Name                 | Rows  | Bytes | Cost(%CPU)|
--------------------------------------------------------------------------------
|   0 | SELECT STATEMENT     |                      |    78 |  4212 | 15507  (1)|
|   1 |  HASH GROUP BY       |                      |    78 |  4212 | 15507  (1)|
|   2 |   NESTED LOOPS       |                      |       |       |           |
|   3 |    NESTED LOOPS      |                      |  3034 |  159K | 15506  (1)|
|*  4 |     TABLE ACCESS FULL| OPT_REF_UOM_TEMP_SDIM|  2967 |  101K |   650 (14)|
|*  5 |     INDEX RANGE SCAN | PROD_DIM_PK          |     3 |       |     2  (0)|
```

```
|* 6 |      TABLE ACCESS BY INDEX ROWID|PROD_DIM              |    1|    19 |     5   (0)|
---------------------------------------------------------------------------------------------

Predicate Information (identified by operation id):
---------------------------------------------------

   4 - filter("UOM"."RELTV_CURR_QTY"=1)
   5 - access("PROD"."PROD_SKID"="UOM"."PROD_SKID")
   6 - filter("PROD"."BUOM_CURR_SKID" IS NOT NULL AND "PROD"."PROD_END_DATE"=TO_DATE('
              9999-12-31 00:00:00', 'syyyy-mm-dd hh24:mi:ss') AND "PROD"."CURR_IND"='
Y' AND
              "PROD"."BUOM_CURR_SKID"="UOM"."UOM_SKID")

22 rows selected.
```

Id=4 是执行计划的入口，它是嵌套循环的驱动表。CBO 估算 Id=4 返回 2 967 行数据。对于嵌套循环，我们首先要检查驱动表返回的真实行数是否与估算的行数有较大偏差，现在查看驱动表总行数。

```
SQL> select count(*) from OPT_REF_UOM_TEMP_SDIM;

  COUNT(*)
----------
   2137706
```

我们查看驱动表返回的真实行数。

```
SQL> select count(*) from OPT_REF_UOM_TEMP_SDIM where "RELTV_CURR_QTY"=1;

  COUNT(*)
----------
    946432
```

驱动表实际上返回了 94 万行数据，与估算的 2 967 相差巨大。嵌套循环中，驱动表返回多少行数据，被驱动表就会被扫描多少次，这里被驱动表会被扫描 94 万次，这就解释了为什么 SQL 执行了两个小时还没执行成功。显然执行计划是错误的，应该走 HASH 连接。

本案例是因为 Rows 估算有严重偏差，导致走错执行计划。Rows 估算与统计信息有关。Id=4 过滤条件是 RELTV_CURR_QTY = 1，现在我们来查看表和列的统计信息。

```
SQL> select a.table_name name ,a.column_name,b.num_rows,a.num_distinct Cardinality,
  2  a.num_distinct/b.num_rows selectivity,a.histogram from dba_tab_col_statistics a,
  3  dba_tables b where a.owner=b.owner and a.table_name=b.table_name
  4  and a.table_name='OPT_REF_UOM_TEMP_SDIM' and a.column_name='RELTV_CURR_QTY';

NAME                      COLUMN_NAME       NUM_ROWS  CARDINALITY  SELECTIVITY  HISTOGRAM
------------------------- ----------------- --------- ------------ ------------ ---------
OPT_REF_UOM_TEMP_SDIM     RELTV_CURR_QTY    2160000            728     .000337037  NONE
```

统计信息中表总行数有 2 160 000 行数据，与真实的行数（2 137 706）十分接近，这说明表的统计信息没有问题。RELTV_CURR_QTY 列的基数等于 728，没有直方图（HISTOGRAM=NONE）。为什么 Id=4 会估算返回 2 967 行数据呢？正是因为 RELTV_CURR_QTY 列基数太低，而且没有收集直方图，CBO 认为该列数据分布是均衡的，导致在估算 Rows 的时候，直接以表总行数/列基数=216 000/728=2967 来进行估算。所以我们需要对 RELTV_CURR_QTY 列收集直方图。

```
SQL> BEGIN
  2  DBMS_STATS.GATHER_TABLE_STATS(ownname => 'XXXX',
  3  tabname => 'OPT_REF_UOM_TEMP_SDIM',
  4  estimate_percent => 100,
  5  method_opt => 'for columns RELTV_CURR_QTY size skewonly',
  6  degree => DBMS_STATS.AUTO_DEGREE,
  7  cascade=>TRUE
  8  );
  9  END;
 10  /

PL/SQL procedure successfully completed.
```

收集完直方图之后，我们再来查看执行计划。

```
SQL> explain plan for SELECT UOM.PROD_SKID,
  2         MAX (UOM.RELTV_CURR_QTY) RELTV_CURR_QTY,
  3         MAX (UOM.STAT_CURR_VAL) STAT_CURR_VAL,
  4         MAX (UOM.BAR_CURR_CODE) BAR_CURR_CODE
  5  FROM OPT_REF_UOM_TEMP_SDIM UOM,
  6       REF_PROD_DIM PROD
  7  WHERE UOM.RELTV_CURR_QTY = 1
  8    AND PROD.CURR_IND = 'Y'
  9    AND PROD.PROD_END_DATE = TO_DATE ('31-12-9999', 'dd-mm-yyyy')
 10    AND PROD.PROD_SKID = UOM.PROD_SKID
 11    AND PROD.BUOM_CURR_SKID = UOM.UOM_SKID
 12  GROUP BY UOM.PROD_SKID;

Explained.

SQL> select * from table(dbms_xplan.display);

PLAN_TABLE_OUTPUT
--------------------------------------------------------------------------------
Plan hash value: 612020119
--------------------------------------------------------------------------------
| Id | Operation            | Name                  | Rows  | Bytes |TempSpc| Cost(%CPU)|
--------------------------------------------------------------------------------
|  0 | SELECT STATEMENT     |                       | 12097 |  637K |       | 44911  (5)|
|  1 |  HASH GROUP BY       |                       | 12097 |  637K |       | 44911  (5)|
|* 2 |   HASH JOIN          |                       |  951K |   48M |   29M | 44799  (5)|
|* 3 |    TABLE ACCESS FULL | PROD_DIM              |  998K |   18M |       | 43022  (5)|
|* 4 |    TABLE ACCESS FULL | OPT_REF_UOM_TEMP_SDIM |  951K |   31M |       |   654 (15)|
--------------------------------------------------------------------------------

Predicate Information (identified by operation id):
---------------------------------------------------

   2 - access("PROD"."PROD_SKID"="UOM"."PROD_SKID" AND
              "PROD"."BUOM_CURR_SKID"="UOM"."UOM_SKID")
   3 - filter("PROD"."BUOM_CURR_SKID" IS NOT NULL AND "PROD"."PROD_END_DATE"=TO_DATE('
              9999-12-31 00:00:00', 'syyyy-mm-dd hh24:mi:ss') AND "PROD"."CURR_IND"='Y')
   4 - filter("UOM"."RELTV_CURR_QTY"=1)

20 rows selected.
```

现在执行计划自动走了 HASH 连接，这才是正确的执行计划，走了正确的执行计划之后，SQL 能在 8 分钟左右执行完毕。

我们也可以换种思路优化该 SQL。该 SQL 属于 ETL，ETL 一般都需要清洗大量数据，两表关联处理大量数据应该走 HASH 连接，所以我们可以直接让两个表走 HASH 连接。另外该 SQL 有分组汇总（GROUP BY），需要分组汇总的 SQL 一般也是处理大量数据，基于此该 SQL 也应该走 HASH 连接。

9.3 NL 被驱动表不能走 INDEX SKIP SCAN

有如下执行计划（从 AWR 中抓出）。

```
SQL> select * from table(dbms_xplan.display_awr('3m7f7xdpkdrtv', NULL, NULL, 'ALL')) ;
SQL_ID 3m7f7xdpkdrtv
--------------------
select a.int_id,a.zh_label,a0.zh_label from VIEW_RMS_POS_PORT a inner
join (select int_id,zh_label from RMS_LOCALNET_POS where
stateflag=:"SYS_B_00") a0 on to_char(a.up_pos_id)=to_char(a0.int_id)
where    a0.zh_label in (:"SYS_B_01",:"SYS_B_02",:"SYS_B_03",:"SYS_B_04",
:"SYS_B_05",:"SYS_B_06",:"SYS_B_07",:"SYS_B_08",:"SYS_B_09",:"SYS_B_10")
 and   :"SYS_B_11"=:"SYS_B_12" and (a.zh_label in  (:"SYS_B_13")) and
a.stateflag=:"SYS_B_14"

Plan hash value: 494215470

---------------------------------------------------------------------------
| Id  | Operation                     | Name                | Rows | Bytes |
---------------------------------------------------------------------------
|   0 | SELECT STATEMENT              |                     |      |       |
|   1 |  FILTER                       |                     |      |       |
|   2 |   NESTED LOOPS                |                     |    1 |    94 |
|   3 |    INDEX RANGE SCAN           | RMS_JK_POS_PORT_PK  |    1 |    43 |
|   4 |    TABLE ACCESS BY INDEX ROWID| RMS_LOCALNET_POS    |    1 |    51 |
|   5 |     INDEX SKIP SCAN           | RMS_LOCALNET_POS_PUI|   35 |       |
---------------------------------------------------------------------------

Query Block Name / Object Alias (identified by operation id):
---------------------------------------------------------------

   1 - SEL$D26F4AE5
   3 - SEL$D26F4AE5 / RMS_JK_POS_PORT@SEL$2
   4 - SEL$D26F4AE5 / RMS_LOCALNET_POS@SEL$3
   5 - SEL$D26F4AE5 / RMS_LOCALNET_POS@SEL$3
```

该 SQL 在 AWR 中属于 TOP SQL，执行计划走了嵌套循环，被驱动表走了 INDEX SKIP SCAN。在第 5 章中我们讲到，嵌套循环被驱动表只能走 INDEX UNIQUE SCAN 或者 INDEX RANGE SCAN。为什么嵌套循环被驱动表不能走 INDEX SKIP SCAN 呢？这是因为嵌套循环会传值，从驱动表传值给被驱动表，传值相当于过滤条件。有过滤条件但是走了 INDEX SKIP SCAN，很有可能是被驱动表连接列没包含在索引中，或者连接列在索引中放错了位置。

被驱动表连接列是 int_id，现在我们查看索引 RMS_LOCALNET_POS_PUI 具体情况。

```
SQL> SELECT DBMS_METADATA.GET_DDL('INDEX','RMS_LOCALNET_POS_PUI','HBRMW6') FROM DUAL;

  CREATE INDEX "HBRMW6"."RMS_LOCALNET_POS_PUI" ON "HBRMW6"."RMS_LOCALNET_POS" ("PRO_T
ASK_ID", "STATEFLAG")
  PCTFREE 10 INITRANS 2 MAXTRANS 255 COMPUTE STATISTICS
  STORAGE(INITIAL 65536 NEXT 1048576 MINEXTENTS 1 MAXEXTENTS 2147483645
  PCTINCREASE 0 FREELISTS 1 FREELIST GROUPS 1
```

```
BUFFER_POOL DEFAULT FLASH_CACHE DEFAULT CELL_FLASH_CACHE DEFAULT)
TABLESPACE "HBRMW_TBS"
```

被驱动表索引中竟然没有包含连接列。这说明该执行计划是错误的。我们将连接列和过滤列组合起来创建组合索引,从而解决该 SQL 性能问题。

9.4 优化 SQL 需要注意表与表之间关系

一位上海的朋友说以下 SQL 执行不出结果。

```
with tab as (
select bb.card_no cashier_shop_no, aa.card_open_owner merch_id,aa.card_no ,bb.txn_date,bb.last_txn_date,bb.merch_loc_name
 from tb_card aa join
(select t.card_no,a.merch_id,a.merch_loc_name ,t.cust_id txn_date,
to_char(last_day( add_months(to_date(t.cust_id,'yyyymmdd'),1)) ,'yyyymmdd') last_txn_date
from tb_bill_test t join tb_merch a on t.card_no=a.cashier_shop_no
where t.nbr_group='161' and a.merch_id not like '0%' )bb
on aa.card_open_owner=bb.merch_id
where aa.mbr_reg_date between bb.txn_date and  bb.last_txn_date )
select  bb.cashier_shop_no, bb.merch_loc_name,aa.txn_date,
case when aa.merch_id=bb.merch_id and aa.card_no=bb.card_no then '本店会员' else '他店会员' end shop_no,
  count(distinct aa.card_no) card_num,
  sum( case when aa.p_code in ('7646','7686','7208') then -1 else 1 end ) count_num,
  sum(case when  aa.p_code in ('7646','7686','7208') then 0-
    case when aa.txn_amt>aa.earning_amt then aa.txn_amt else aa.earning_amt end
      else   case when aa.txn_amt>aa.earning_amt then aa.txn_amt else aa.earning_amt end end) txn_amt
 from tb_trans aa join tab bb on aa.merch_id=bb.merch_id
where  aa.txn_date between bb.txn_date and bb.last_txn_date and
aa.p_code in ('7647','7687','7207','7646','7686','7208')
and aa.status in ('1','R')
group by bb.cashier_shop_no, bb.merch_loc_name,aa.txn_date,aa.merch_id,
case when aa.merch_id=bb.merch_id and aa.card_no=bb.card_no then '本店会员' else '他店会员' end
order by aa.merch_id, aa.txn_date;
```

执行计划如下。

```
Plan hash value: 4271695044

---------------------------------------------------------------------------------
| Id |Operation                         |Name               |Rows|Bytes|Cost(%CPU)|
---------------------------------------------------------------------------------
|  0 |SELECT STATEMENT                  |                   |  15| 1650|96737  (1)|
|  1 | SORT GROUP BY                    |                   |  15| 1650|96737  (1)|
|  2 |  VIEW                            |VW_DAG_0           |  15| 1650|96735  (1)|
|  3 |   HASH GROUP BY                  |                   |  15| 2430|96735  (1)|
|  4 |    NESTED LOOPS                  |                   |  15|  430|96734  (1)|
|  5 |     NESTED LOOPS                 |                   |2213| 2430|96734  (1)|
|  6 |      NESTED LOOPS                |                   |   1|  103|  542  (1)|
|  7 |       NESTED LOOPS               |                   |   1|   65|   58  (0)|
|  8 |        TABLE ACCESS BY INDEX ROWID|TB_BILL_TEST      |  36|  864|   20  (0)|
|* 9 |         INDEX RANGE SCAN         |TMP_INDEX_BILL_01  |  37|     |    3  (0)|
|*10 |        TABLE ACCESS BY INDEX ROWID|TB_MERCH          |   1|   41|    3  (0)|
|*11 |         INDEX RANGE SCAN         |I1_MERCH           |   1|     |    1  (0)|
```

```
|*12 |      TABLE ACCESS BY INDEX ROWID    |TB_CARD            |    2|  76| 484 (1)|
|*13 |       INDEX RANGE SCAN              |I1_CARD_OPEN_OWNER|3855|    |  24 (0)|
|*14 |      INDEX RANGE SCAN               |I2_TRANS           |2213|    |95972 (1)|
|*15 |     TABLE ACCESS BY GLOBAL INDEX ROWID|TB_TRANS         |   56|3304|96193 (1)|
---------------------------------------------------------------------------------

Predicate Information (identified by operation id):
---------------------------------------------------

   9 - access("T"."NBR_GROUP"='161')
       filter("T"."CARD_NO" IS NOT NULL)
  10 - filter("A"."MERCH_ID" NOT LIKE '0%')
  11 - access("T"."CARD_NO"="A"."CASHIER_SHOP_NO")
  12 - filter("AA"."MBR_REG_DATE">="T"."CUST_ID" AND "AA"."MBR_REG_DATE"<=TO_CHAR(LAS
T_DAY(ADD_MONTHS
       (TO_DATE("T"."CUST_ID",'yyyymmdd'),1)),'yyyymmdd'))
  13 - access("AA"."CARD_OPEN_OWNER"="A"."MERCH_ID")
       filter("AA"."CARD_OPEN_OWNER" IS NOT NULL)
  14 - access("AA"."TXN_DATE">="T"."CUST_ID" AND "AA"."MERCH_ID"="AA"."CARD_OPEN_OWNE
R" AND       "AA"."TXN_DATE"<=TO_CHAR(LAST_DAY(ADD_MONTHS(TO_DATE("T"."CUST_ID",'yyyymm
dd'),1)),'yyyymmdd'))
       filter("AA"."MERCH_ID"="AA"."CARD_OPEN_OWNER")
  15 - filter(("AA"."P_CODE"='7207' OR "AA"."P_CODE"='7208' OR "AA"."P_CODE"='7646' O
R "AA"."P_CODE"='7647'
       OR "AA"."P_CODE"='7686' OR "AA"."P_CODE"='7687') AND ("AA"."STATUS"='1' OR "AA
"."STATUS"='R'))
```

我们拿到一条需要优化的 SQL 语句，怎么入手呢？首先要看 SQL 写法。可以利用 SQL 三段分拆方法，先观察 SQL 语句。该 SQL 语句有个 with as 子句取名为 tab，主查询中就是 tb_trans 与 tab 进行关联。with as 子句一共返回 6 000 多行数据，可以 1 秒内出结果，tb_trans 有两亿条数据。执行计划中，with as 子查询作为一个整体并且作为嵌套循环驱动表，tb_trans 作为嵌套循环被驱动表，乍一看，这也符合嵌套循环关联原则，小表驱动大表，大表走索引。**但是该 SQL 执行不出结果**，最大的可能就是 tab 与 tb_trans 关联之后返回数据量太多，因为返回结果集太多，被驱动表走索引，也就是说该 SQL 可能是被驱动表走索引返回数据量太多导致性能问题。于是检查被驱动表连接列 merch_id 基数，基数很低，tab:tb_trans 是 1 比几十万关系。

因为被驱动表 tb_trans 与 tab 是几十万比 1 的关系，这时就不能走嵌套循环了，只能走 HASH 连接，于是使用 HINT：use_hash(aa,bb)优化 SQL，最终该 SQL 可以在 1 小时左右执行完毕。如果开启并行查询可以更快。

9.5 INDEX FAST FULL SCAN 优化案例

2016 年，北京一位游戏公司的朋友说以下 SQL 最慢的时候要执行 40 分钟，最快的时候只需要几秒至十来秒就可以执行完毕。

```
idle> SELECT COUNT(DISTINCT IDFA)
  FROM SYS_ACTIVATION_SDK_IOS T1
 WHERE CREATE_TIME >= TRUNC(sysdate)
   AND CREATE_TIME < TRUNC(sysdate) + 1
   AND GAME_ID = 153
```

```
          AND NOT EXISTS (SELECT /*+ hash_aj */ IDFA
                FROM SYS_ACTIVATION_SDK_IOS T2
               WHERE CREATE_TIME < TRUNC(sysdate)-1
                 AND T2.GAME_ID = 153
                 AND T1.IDFA = T2.IDFA) ;
```

执行计划如下。

```
Execution Plan
----------------------------------------------------------
Plan hash value: 3686453232
----------------------------------------------------------
| Id  | Operation          | Name                      | Rows | Bytes |
----------------------------------------------------------
|   0 | SELECT STATEMENT   |                           |    1 |   76 |
|   1 |  SORT GROUP BY     |                           |    1 |   76 |
|*  2 |   FILTER           |                           |      |      |
|*  3 |    HASH JOIN ANTI  |                           |   93 | 7068 |
|*  4 |     INDEX RANGE SCAN| SYS_ACTIVATION_SDK_IOS_IDX1 |  304 | 11552 |
|*  5 |     INDEX RANGE SCAN| SYS_ACTIVATION_SDK_IOS_IDX1 | 888K |  32M |
----------------------------------------------------------

Predicate Information (identified by operation id):
---------------------------------------------------

   2 - filter(TRUNC(SYSDATE@!)<TRUNC(SYSDATE@!)+1)
   3 - access("T1"."IDFA"="T2"."IDFA")
   4 - access("GAME_ID"=153 AND "CREATE_TIME">=TRUNC(SYSDATE@!) AND
       "CREATE_TIME"<TRUNC(SYSDATE@!)+1)
   5 - access("T2"."GAME_ID"=153 AND "CREATE_TIME"<TRUNC(SYSDATE@!)-1)
       filter("T2"."IDFA" IS NOT NULL)
```

该 SQL 是一个自关联，SQL 语句里面有 HASH: HASH_AJ 提示 SQL 采用 HASH ANTI JOIN 进行关联。该 SQL 的确走的是 HASH ANTI JOIN，而且都是通过同一个索引访问数据，没有回表。表 SYS_ACTIVATION_SDK_IOS 有 14G，索引 SYS_ACTIVATION_SDK_IOS_IDX1 有 2.5G，根据（game_id,create_time,idfa）创建。

两表关联，我们要搞清楚表大小以及表过滤之后返回的行数。这里表大小已经清楚。

查看 T1 返回行数。

```
SELECT   COUNT(DISTINCT IDFA)
  FROM SYS_ACTIVATION_SDK_IOS T1
 WHERE CREATE_TIME >= TRUNC(sysdate)
   AND CREATE_TIME < TRUNC(sysdate) + 1
   AND GAME_ID = 153;
```

T1 返回 11 799 行数据。我们查看 T2 返回行数。

```
select count(*)
  from (SELECT IDFA
          FROM SYS_ACTIVATION_SDK_IOS T2
         WHERE CREATE_TIME < TRUNC(sysdate) - 1
           AND T2.GAME_ID = 153);
```

T2 返回 1 251 009 行数据。现在我们得到信息，小表 T1（11 799）与较大表 T2（1 251 009）进行关联。一般情况下，小表与大表关联，可以让小表作为 NL 驱动表，大表走连接列索引。在确定能否走 NL 之前，要先检查两个表之间的关系，同时检查表连接列的数据分布，于是我们执行如下 SQL。

```sql
SELECT IDFA, COUNT(*)
  FROM SYS_ACTIVATION_SDK_IOS
 GROUP BY IDFA
 ORDER BY 2 DESC;
```

我们发现 IDFA 基数很低，数据分布不均衡。因为 IDFA 基数很低，所以不能让 T1 与 T2 走嵌套循环，只能走 HASH 连接。执行计划中，T1 与 T2 本来就是走的 HASH 连接，连接方式是正确的，所以问题只能出现在访问路径上。T1 走的是 INDEX RANGE SCAN，返回了 11 799 行数据，T2 走的也是 INDEX RANGE SCAN，返回了 1 251 009 行数据。INDEX RANGE SCAN 是单块读，一般用于返回少量数据，这里返回 1 251 009 行数据显然不合适，因为 INDEX RANGE SCAN 没有回表，所以应该让其走 INDEX FAST FULL SCAN。

```sql
SELECT COUNT(DISTINCT IDFA)
  FROM SYS_ACTIVATION_SDK_IOS T1
 WHERE CREATE_TIME >= TRUNC(sysdate)
   AND CREATE_TIME < TRUNC(sysdate) + 1
   AND GAME_ID = 153
   AND NOT EXISTS (SELECT /*+ hash_aj index_ffs(t2) */
                    IDFA
                     FROM SYS_ACTIVATION_SDK_IOS T2
                    WHERE CREATE_TIME < TRUNC(sysdate) - 1
                      AND T2.GAME_ID = 153
                      AND T1.IDFA = T2.IDFA);
```

最终该 SQL 可以在 1 分钟内执行完毕。该 SQL 跑得慢根本原因就是 INDEX RANGE SCAN 是单块读。

为什么该 SQL 有时要执行 40 多分钟，而有时只需要执行几秒至十来秒呢？原因在于 buffer cache 缓存。当 buffer cache 缓存了索引 SYS_ACTIVATION_SDK_IOS_IDX1，SQL 就能在几秒至十几秒执行完毕；如果 buffer cache 没有缓存 SYS_ACTIVATION_SDK_IOS_IDX1，执行计划中 Id=5 走的是 INDEX RANGE SCAN，导致大量单块读，所以会执行 40 分钟左右。更正了执行计划之后，该 SQL 最慢可以在 1 分钟内执行完毕。

9.6 分页语句优化案例

2013 年一唯品会的朋友有如下语句需要优化。

```sql
select *
  from (select f.*
          from tms.inf_b2c_djwlzt_f f
          inner join tms.orderstatus os on f.transport_code = os.statuscode
         where f.warehouse = 'VIP_BJ'
           and f.is_send = 0
         order by f.created_dtm_loc, os.Sort_No asc)
 where rownum <= 500;
```

该 SQL 类似分页语句，因此我们可以用分页语句优化思路对其进行优化。拿到分页语句，我们首先应该检查分页语句是否符合分页语句编写规范。这里该 SQL 排序列来自两个表，不符合分页语句编写规范。我们在第 8 章中讲到，分页语句只能对一个表的列进行排序。该 SQL 排序列来自 f 和 os，并且显示的时候只有 f 表的数据。因此我们建议去掉 os 表的排序字段，如下所示。

```sql
select *
  from (select f.*
          from tms.inf_b2c_djwlzt_f f
         inner join tms.orderstatus os on f.transport_code = os.statuscode
         where f.warehouse = 'VIP_BJ'
           and f.is_send = 0
         order by f.created_dtm_loc)
 where rownum <= 500;
```

排序列来自 f 表，需要对 f 表创建索引，因为过滤条件是等值访问，我们可以把过滤条件放在前面，排序列放在后面，于是创建如下索引。

```sql
create index idx_f_inf on inf_b2c_djwlzt_f(warehouse,is_send,created_dtm_loc);
```

然后强制 f 表与 os 走嵌套循环，同时让 f 表作为嵌套循环驱动表，走刚才创建的索引。

```sql
select *
  from (select /*+ use_nl(f,os) leading(f) */
        f.*
          from tms.inf_b2c_djwlzt_f f
         inner join tms.orderstatus os on f.transport_code = os.statuscode
         where f.warehouse = 'VIP_BJ'
           and f.is_send = 0
         order by f.created_dtm_loc)
 where rownum <= 500;
```

执行计划如下。

```
----------------------------------------------------------------------------------------
| Id | Operation                       | Name              | Rows  | Bytes | Cost(%CPU)|
----------------------------------------------------------------------------------------
|  0 | SELECT STATEMENT                |                   |   500 |  725K |   754  (1)|
|* 1 |  COUNT STOPKEY                  |                   |       |       |           |
|  2 |   VIEW                          |                   |   502 |  728K |   754  (1)|
|  3 |    NESTED LOOPS                 |                   |   502 |  121K |   754  (1)|
|  4 |     TABLE ACCESS BY INDEX ROWID | INF_B2C_DJWLZT_F  | 2419K |  562M |    71  (0)|
|* 5 |      INDEX RANGE SCAN           | IDX_F_INF         |   502 |       |     5  (0)|
|* 6 |     TABLE ACCESS FULL           | ORDERSTATUS       |     1 |     3 |     1  (0)|
----------------------------------------------------------------------------------------

Predicate Information (identified by operation id):
---------------------------------------------------

   1 - filter(ROWNUM<=500)
   5 - access("F"."WAREHOUSE"='VIP_BJ' AND "F"."IS_SEND"=0)
   6 - filter("F"."TRANSPORT_CODE"="OS"."STATUSCODE")


Statistics
----------------------------------------------------------
          1  recursive calls
          0  db block gets
       1736  consistent gets
          2  physical reads
          0  redo size
      67968  bytes sent via SQL*Net to client
        883  bytes received via SQL*Net from client
         35  SQL*Net roundtrips to/from client
          0  sorts (memory)
          0  sorts (disk)
```

```
       500  rows processed
```

从执行计划中我们看到被驱动表走了全表扫描，嵌套循环被驱动表不能走全表扫描，必须走索引，于是创建如下索引。

```
create index STATUSCODE_IDX on ORDERSTATUS(STATUSCODE);
```

创建索引之后的执行计划如下。

```
---------------------------------------------------------------------------------
| Id |Operation                      | Name             | Rows  | Bytes | Cost(%CPU)|
---------------------------------------------------------------------------------
|  0 |SELECT STATEMENT               |                  |  500  | 725K  |  71   (0)|
|* 1 | COUNT STOPKEY                 |                  |       |       |          |
|  2 |  VIEW                         |                  |  502  | 728K  |  71   (0)|
|  3 |   NESTED LOOPS                |                  |  502  | 121K  |  71   (0)|
|  4 |    TABLE ACCESS BY INDEX ROWID| INF_B2C_DJWLZT_F | 2419K | 562M  |  71   (0)|
|* 5 |     INDEX RANGE SCAN          | IDX_F_INF        |  502  |       |   5   (0)|
|* 6 |    INDEX RANGE SCAN           | STATUSCODE_IDX   |   1   |   3   |   0   (0)|
---------------------------------------------------------------------------------

Predicate Information (identified by operation id):
---------------------------------------------------

  1 - filter(ROWNUM<=500)
  5 - access("F"."WAREHOUSE"='VIP_BJ' AND "F"."IS_SEND"=0)
  6 - access("F"."TRANSPORT_CODE"="OS"."STATUSCODE")

Statistics
----------------------------------------------------------
          1  recursive calls
          0  db block gets
        247  consistent gets
          0  physical reads
          0  redo size
      60433  bytes sent via SQL*Net to client
        883  bytes received via SQL*Net from client
         35  SQL*Net roundtrips to/from client
          0  sorts (memory)
          0  sorts (disk)
        500  rows processed
```

优化完毕之后，该 SQL 逻辑读只有 247 个，最终该 SQL 可以秒杀。

9.7 ORDER BY 取别名列优化案例

2017 年，网络优化班的学生问怎么优化以下语句。

```
select rownum as r, a.*
  from (select npai.AREA_ID,
               npai.PSO_ID,
               npai.RO_ID,
               npai.NO,
               npai.ADDR,
               to_char(npai.CRTD_DT, 'yyyy-mm-dd hH24:mi:ss') as CRTD_DT,
               to_char(npai.CMPLT_DT, 'yyyy-mm-dd hH24:mi:ss') as CMPLT_DT,
               npai.CRM_PROD_ID,
               npai.PROD_SERV_SPEC_ID,
```

```
              npai.PROD_SERV_SPEC_NAME,
              npai.ACTION_TP_ID,
              npai.ACTION_TP_NAME
         from NT_PSO_ARCH_INFO npai
        where npai.crtd_dt >= to_date('2017-01-01', 'yyyy-mm-dd')
          and npai.crtd_dt <= to_date('2017-02-01', 'yyyy-mm-dd')
          and local_area_id = 3
        order by crtd_dt) a
 where rownum <= 20;
```

执行计划如下。

```
Plan hash value: 2467293374

--------------------------------------------------------------------------------
| Id  | Operation                            | Name             | Rows  | Bytes | Cost (%CPU)|
--------------------------------------------------------------------------------
|   0 | SELECT STATEMENT                     |                  |    20 | 28160 |   489K  (1)|
|*  1 |  COUNT STOPKEY                       |                  |       |       |            |
|   2 |   VIEW                               |                  |  950K | 1276M |   489K  (1)|
|*  3 |    SORT ORDER BY STOPKEY             |                  |  950K |   85M |   489K  (1)|
|   4 |     PARTITION LIST SINGLE            |                  |  950K |   85M |   469K  (1)|
|   5 |      TABLE ACCESS BY LOCAL INDEX ROWID| NT_PSO_ARCH_INFO|  950K |   85M |   469K  (1)|
|*  6 |       INDEX RANGE SCAN               | IDX_NTPAI_CRDT   |  950K |       |  2581   (1)|
--------------------------------------------------------------------------------

Predicate Information (identified by operation id):
---------------------------------------------------

   1 - filter(ROWNUM<=20)
   3 - filter(ROWNUM<=20)
   6 - access("NPAI"."CRTD_DT">=TO_DATE(' 2017-01-01 00:00:00', 'syyyy-mm-dd hh24:mi:
ss')
              AND "NPAI"."CRTD_DT"<=TO_DATE('2017-02-01 00:00:00', 'syyyy-mm-dd hh24:
mi:ss'))
```

该 SQL 类似分页语句。拿到分页语句，我们应该先查看分页语句是否符合分页编码规范。这里，SQL 完全符合分页语句编码规范。

该 SQL 排序列是 crtd_dt，执行计划中走的也是 crtd_dt 列的索引。表 nt_pso_arch_info 是 LIST 分区表，分区列是 local_area_id，从执行计划中（Id=4）看到只扫描了一个分区。按道理该 SQL 不应该出现 SORT ORDER BY。为什么执行计划中有 SORT ORDER BY 呢？我们注意观察，SQL 语句中 order by 的列 crtd_dt 在 select 中进行了 to_char 格式化，格式化之后取了别名，但是别名居然与列名一样。正是因为别名与列名一样，才导致无法消除 SORT ORDER BY。

现在我们另外取一个别名（CRTD_DT1）。

```
select rownum as r, a.*
  from (select npai.AREA_ID,
               npai.PSO_ID,
               npai.RO_ID,
               npai.NO,
               npai.ADDR,
               to_char(npai.CRTD_DT, 'yyyy-mm-dd hH24:mi:ss') as CRTD_DT1,
               to_char(npai.CMPLT_DT, 'yyyy-mm-dd hH24:mi:ss') as CMPLT_DT,
               npai.CRM_PROD_ID,
               npai.PROD_SERV_SPEC_ID,
               npai.PROD_SERV_SPEC_NAME,
               npai.ACTION_TP_ID,
```

```
                    npai.ACTION_TP_NAME
             from NT_PSO_ARCH_INFO npai
           where npai.crtd_dt >= to_date('2017-01-01', 'yyyy-mm-dd')
             and npai.crtd_dt <= to_date('2017-02-01', 'yyyy-mm-dd')
             and local_area_id = 3
           order by crtd_dt) a
where rownum <= 20;
```

我们再次查看执行计划。

```
Plan hash value: 3066843972

---------------------------------------------------------------------------------
| Id |Operation                              |Name              |Rows  |Bytes |Cost(%CPU)|
---------------------------------------------------------------------------------
|  0 |SELECT STATEMENT                       |                  |   20 |28160 |  489K (1)|
|* 1 | COUNT STOPKEY                         |                  |      |      |          |
|  2 |  VIEW                                 |                  | 950K |1276M |  489K (1)|
|  3 |   PARTITION LIST SINGLE               |                  | 950K |  85M |  469K (1)|
|  4 |    TABLE ACCESS BY LOCAL INDEX ROWID  |NT_PSO_ARCH_INFO  | 950K |  85M |  469K (1)|
|* 5 |     INDEX RANGE SCAN                  |IDX_NTPAI_CRDT    | 950K |      | 2581  (1)|
---------------------------------------------------------------------------------

Predicate Information (identified by operation id):
---------------------------------------------------

   1 - filter(ROWNUM<=20)
   5 - access("NPAI"."CRTD_DT">=TO_DATE(' 2017-01-01 00:00:00', 'syyyy-mm-dd hh24:mi
:ss') AND
       "NPAI"."CRTD_DT"<=TO_DATE(' 2017-02-01 00:00:00', 'syyyy-mm-dd hh24:mi:ss'))
```

更改别名之后，消除了 SORT ORDER BY，从而达到了优化目的。为什么必须要更改别名呢？这是因为如果不更改别名，order by crtd_dt 就相当于 order by 别名，也就是 order by to_char(npai.CRTD_DT, 'yyyy-mm-dd hH24:mi:ss')，而索引中记录的是 date 类型，现在排序变成了按照 char 类型排序，如果不更改别名执行计划就无法消除 SORT ORDER BY。

在 2014 年的时候也遇到一个类似案例，但是该案例 SQL 和执行计划太长，无法呈现在本书中。大家如有兴趣，可以查看博客：http://blog.csdn.net/robinson1988/ article/details/40870901。

9.8 半连接反向驱动主表案例一

2015 年，网络优化班的学生问如何优化以下 SQL。

```
SQL> explain plan for
  2  select gcode,name,idcode,address,noroom,etime from
  3    LY_T_CHREC t where gcode in (
  4  select gcode from LY_T_CHREC t where name='张三' and bdate ='19941109') a
  5  ;

Explained

SQL> select * from table(dbms_xplan.display(null,null,'ADVANCED -PROJECTION'));

PLAN_TABLE_OUTPUT
---------------------------------------------------------------------------------
Plan hash value: 953100977
---------------------------------------------------------------------------------
```

```
| Id  | Operation                            | Name              | Rows |
----------------------------------------------------------------------------
|   0 | SELECT STATEMENT                     |                   |    2 |
|*  1 |  HASH JOIN RIGHT SEMI                |                   |    2 |
|*  2 |   TABLE ACCESS BY GLOBAL INDEX ROWID | LY_T_CHREC        |    1 |
|*  3 |    INDEX RANGE SCAN                  | IDX_LY_T_CHREC_NAME|  15 |
|   4 |   PARTITION HASH ALL                 |                   | 200M |
|   5 |    TABLE ACCESS FULL                 | LY_T_CHREC        | 200M |
----------------------------------------------------------------------------

Query Block Name / Object Alias (identified by operation id):
------------------------------------------------------------
   1 - SEL$5DA710D3
   2 - SEL$5DA710D3 / T@SEL$2
   3 - SEL$5DA710D3 / T@SEL$2
   5 - SEL$5DA710D3 / T@SEL$1

Outline Data
-------------
  /*+
      BEGIN_OUTLINE_DATA
      SWAP_JOIN_INPUTS(@"SEL$5DA710D3" "T"@"SEL$2")
      USE_HASH(@"SEL$5DA710D3" "T"@"SEL$2")
      LEADING(@"SEL$5DA710D3" "T"@"SEL$1" "T"@"SEL$2")
      INDEX_RS_ASC(@"SEL$5DA710D3" "T"@"SEL$2" ("LY_T_CHREC"."NAME"))
      FULL(@"SEL$5DA710D3" "T"@"SEL$1")
      OUTLINE(@"SEL$2")
      OUTLINE(@"SEL$1")
      UNNEST(@"SEL$2")
      OUTLINE_LEAF(@"SEL$5DA710D3")
      ALL_ROWS
      DB_VERSION('11.2.0.3')
      OPTIMIZER_FEATURES_ENABLE('11.2.0.3')
      IGNORE_OPTIM_EMBEDDED_HINTS
      END_OUTLINE_DATA
  */

Predicate Information (identified by operation id):
---------------------------------------------------
   1 - access("GCODE"="GCODE")
   2 - filter("BDATE"='19941109')
   3 - access("NAME"='张三')
```

朋友提供的信息：子查询返回一个人开房的房号记录，共返回 63 行。该 SQL 就是查与某人相同的房间号的他人的记录。LY_T_CHREC 表有两亿条记录。整个 SQL 执行了 30 分钟还没出结果，子查询可以秒出结果，GCODE、NAME、IDCODE、ADDRESS、NOROOM、ETIME、BDATE 都有索引。

根据以上信息我们得出：该 SQL 主表 LY_T_CHREC 有两亿条数据，没有过滤条件，IN 子查询过滤之后返回 63 行数据，关联列是房间号（GCODE）。LY_T_CHREC 应该存放的是开房记录数据，GCODE 列基数应该比较高。在本书中我们反复强调：小表与大表关联，如果大表连接列基数比较高，可以走嵌套循环，让小表驱动大表，大表走连接列的索引。这里小表就是 IN 子查询，大表就是主表，我们让 IN 子查询作为 NL 驱动表。

```
select /*+ leading(t@a) use_nl(t@a,t) */
 gcode, name, idcode, address, noroom, etime
  from zhxx_lgy.LY_T_CHREC t
 where gcode in (select /*+ qb_name(a) */
```

```
        gcode
  from zhxx_lgy.LY_T_CHREC t
  where name = '张三'
    and bdate = '19941109');
```

最终该 SQL 可以秒出。

9.9 半连接反向驱动主表案例二

2014 年，一位物流行业的朋友说以下 SQL 要执行 4 个多小时。

```
SELECT "VOUCHER".FID "ID",
       "ENTRIES".FID "ENTRIES.ID",
       "ENTRIES".FEntryDC "ENTRIES.ENTRYDC",
       "ACCOUNT".FID "ACCOUNT.ID",
       "ENTRIES".FCurrencyID "CURRENCY.ID",
       "PERIOD".FNumber "PERIOD.NUMBER",
       "ENTRIES".FSeq "ENTRIES.SEQ",
       "ENTRIES".FLocalExchangeRate "LOCALEXCHANGERATE",
       "ENTRIES".FReportingExchangeRate "REPORTINGEXCHANGERATE",
       "ENTRIES".FMeasureUnitID "ENTRYMEASUREUNIT.ID",
       "ASSISTRECORDS".FID "ASSISTRECORDS.ID",
       "ASSISTRECORDS".FSeq "ASSISTRECORDS.SEQ",
       CASE
         WHEN (("ACCOUNT".FCAA IS NULL) AND
               ("ACCOUNT".FhasUserProperty <> 1)) THEN
          "ENTRIES".FOriginalAmount
         ELSE
          "ASSISTRECORDS".FOriginalAmount
       END "ASSISTRECORDS.ORIGINALAMOUNT",
       CASE
         WHEN (("ACCOUNT".FCAA IS NULL) AND
               ("ACCOUNT".FhasUserProperty <> 1)) THEN
          "ENTRIES".FLocalAmount
         ELSE
          "ASSISTRECORDS".FLocalAmount
       END "ASSISTRECORDS.LOCALAMOUNT",
       CASE
         WHEN (("ACCOUNT".FCAA IS NULL) AND
               ("ACCOUNT".FhasUserProperty <> 1)) THEN
          "ENTRIES".FReportingAmount
         ELSE
          "ASSISTRECORDS".FReportingAmount
       END "ASSISTRECORDS.REPORTINGAMOUNT",
       CASE
         WHEN (("ACCOUNT".FCAA IS NULL) AND
               ("ACCOUNT".FhasUserProperty <> 1)) THEN
          "ENTRIES".FQuantity
         ELSE
          "ASSISTRECORDS".FQuantity
       END "ASSISTRECORDS.QUANTITY",
       CASE
         WHEN (("ACCOUNT".FCAA IS NULL) AND
               ("ACCOUNT".FhasUserProperty <> 1)) THEN
          "ENTRIES".FStandardQuantity
         ELSE
          "ASSISTRECORDS".FStandardQuantity
       END "ASSISTRECORDS.STANDARDQTY",
       CASE
```

```sql
                WHEN (("ACCOUNT".FCAA IS NULL) AND
                    ("ACCOUNT".FhasUserProperty <> 1)) THEN
                  "ENTRIES".FPrice
                ELSE
                  "ASSISTRECORDS".FPrice
            END "ASSISTRECORDS.PRICE",
            CASE
                WHEN ("ACCOUNT".FCAA IS NULL) THEN
                  NULL
                ELSE
                  "ASSISTRECORDS".FAssGrpID
            END "ASSGRP.ID"
    FROM T_GL_Voucher "VOUCHER"
    LEFT OUTER JOIN T_BD_Period "PERIOD" ON "VOUCHER".FPeriodID =
                                            "PERIOD".FID
    INNER JOIN T_GL_VoucherEntry "ENTRIES" ON "VOUCHER".FID =
                                              "ENTRIES".FBillID
    INNER JOIN T_BD_AccountView "ACCOUNT" ON "ENTRIES".FAccountID =
                                             "ACCOUNT".FID
    LEFT OUTER JOIN T_GL_VoucherAssistRecord "ASSISTRECORDS" ON "ENTRIES".FID =
                                             "ASSISTRECORDS".FEntryID
WHERE "VOUCHER".FID IN
        (SELECT "VOUCHER".FID "ID"
          FROM T_GL_Voucher "VOUCHER"
          INNER JOIN T_GL_VoucherEntry "ENTRIES" ON "VOUCHER".FID =
                                                    "ENTRIES".FBillID
          INNER JOIN T_BD_AccountView "ACCOUNT" ON "ENTRIES".FAccountID =
                                                   "ACCOUNT".FID
          INNER JOIN t_bd_accountview PAV ON ((INSTR("ACCOUNT".flongnumber,
                                                   pav.flongnumber) = 1 AND
                                              pav.faccounttableid =
                                              "ACCOUNT".faccounttableid) AND
                                              pav.fcompanyid =
                                              "ACCOUNT".fcompanyid)
          WHERE (("VOUCHER".FCompanyID IN ('fSSF82rRSKexM3KKN1d0tMznrtQ=')) AND
                (("VOUCHER".FBizStatus IN (5)) AND
                ((("VOUCHER".FPeriodID IN ('+wQxkBFVRiKnV7OniceMDoI4jEw=')) AND
                "ENTRIES".FCurrencyID =
                'dfd38d11-00fd-1000-e000-1ebdc0a8100dDEB58FDC') AND
                (pav.FID IN ('vyPiKexLRXiyMb41VSVVzJ2pmCY='))))))
ORDER BY "ID" ASC, "ENTRIES.SEQ" ASC, "ASSISTRECORDS.SEQ" ASC;
```

执行计划如下。

```
--------------------------------------------------------------------------------
| Id | Operation                              | Name              | Rows | Bytes | Cost (%CPU) |
--------------------------------------------------------------------------------
|  0 | SELECT STATEMENT                       |                   |   13 |  5733 |   486   (1) |
|  1 |  SORT ORDER BY                         |                   |   13 |  5733 |   486   (1) |
|  2 |   VIEW                                 | VM_NWVW_2         |   13 |  5733 |   486   (1) |
|  3 |    HASH UNIQUE                         |                   |   13 | 11115 |   486   (1) |
|  4 |     NESTED LOOPS OUTER                 |                   |   13 | 11115 |   485   (1) |
|  5 |      NESTED LOOPS                      |                   |    9 |  6606 |   471   (1) |
|  6 |       NESTED LOOPS                     |                   |    9 |  6057 |   467   (1) |
|  7 |        MERGE JOIN OUTER                |                   |    1 |   473 |   459   (1) |
|  8 |         HASH JOIN                      |                   |    1 |   427 |   458   (1) |
|  9 |          NESTED LOOPS                  |                   |      |       |             |
| 10 |           NESTED LOOPS                 |                   |  258 | 83850 |   390   (0) |
| 11 |            NESTED LOOPS                |                   |    6 |  1332 |     3   (0) |
| 12 |             TABLE ACCESS BY INDEX ROWID| T_BD_ACCOUNTVIEW  |    1 |   111 |     2   (0) |
| 13 |              INDEX UNIQUE SCAN         | PK_BD_ACCOUNTVIEW |    1 |       |     1   (0) |
| 14 |              INDEX RANGE SCAN          | IX_BD_ACTCOMLNUM  |    6 |   666 |     1   (0) |
```

```
| 15 |            INDEX RANGE SCAN           |IX_GL_VCHAACCT            |  489|     |   1   (0)|
| 16 |     TABLE ACCESS BY INDEX ROWID       |T_GL_VOUCHERENTRY         |   42| 4326|  65   (0)|
| 17 |            INDEX RANGE SCAN           |IX_GL_VCH_11              | 7536| 750K|  68   (0)|
| 18 |              BUFFER SORT              |                          |    1|   46| 391   (0)|
| 19 |            INDEX RANGE SCAN           |IX_PERIOD_ENC             |    1|   46|   1   (0)|
| 20 |     TABLE ACCESS BY INDEX ROWID       |T_GL_VOUCHERENTRY         |   17| 3400|   8   (0)|
| 21 |            INDEX RANGE SCAN           |IX_GL_VCHENTRYFQ1         |   17|     |   1   (0)|
| 22 |      TABLE ACCESS BY INDEX ROWID      |T_BD_ACCOUNTVIEW          |    1|   61|   1   (0)|
| 23 |            INDEX UNIQUE SCAN          |PK_BD_ACCOUNTVIEW         |    1|     |   1   (0)|
| 24 |      TABLE ACCESS BY INDEX ROWID      |T_GL_VOUCHERASSISTRECORD  |    1|  121|   2   (0)|
| 25 |            INDEX RANGE SCAN           |IX_GL_VCHASSREC_11        |    2|     |   1   (0)|
---------------------------------------------------------------------------------------------

Note
-----
   - 'PLAN_TABLE' is old version
```

执行计划中居然是'PLAN_TABLE' is old version，无法看到谓词信息，这需要重建 PLAN_TABLE。因为没有谓词信息，所以就不打算从执行计划入手优化 SQL 了，而是选择直接分析 SQL，从 SQL 层面优化。

SQL 语句中，select 到 from 之间没有标量子查询，没有自定义函数，from 后面有 5 个表关联，where 条件中只有一个 in（子查询），没有其他过滤条件。SQL 语句中用到的表大小如图 9-8 所示。

#	Table Name	Owner	Num Rows	Table Sample Size	Last Analyzed	Ind
1	T_BD_ACCOUNTVIEW	DEPPON2011	1466547	1466547	04-JAN-14 23:31:43	
2	T_BD_PERIOD	DEPPON2011	134	134	05-JAN-14 00:05:24	
3	T_GL_VOUCHER	DEPPON2011	3578789	3578789	08-JAN-14 23:49:42	
4	T_GL_VOUCHERASSISTRECORD	DEPPON2011	86095467	86095467	05-JAN-14 07:37:22	
5	T_GL_VOUCHERENTRY	DEPPON2011	61165543	61165543	05-JAN-14 08:34:32	

图 9-8

SQL 语句中有 4 个表都是大表，只有一个表 T_BD_PERIOD 是小表，在 SQL 语句中与 T_GL_VOUCHER 外关联，是外连接的从表。如果走嵌套循环，T_BD_PERIOD 只能作为被驱动表，因此排除了让小表 T_BD_PERIOD 作为嵌套循环驱动表的可能性。如果该 SQL 没有过滤条件，以上 SQL 只能走 HASH 连接。

SQL 语句中唯一的过滤条件就是 in（子查询），因此只能把优化 SQL 的希望寄托在子查询身上。in（子查询）与表 T_GL_VOUCHER 进行关联，T_GL_VOUCHER 同时也是外连接的主表，如果 in（子查询）能过滤掉 T_GL_VOUCHER 大量数据，那么可以让 T_GL_VOUCHER 作为嵌套循环驱动表，一直与后面的表 NL 下去，这样或许能优化 SQL。如果 in（子查询）不能过滤掉大量数据，那么 SQL 就无法优化，最终只能全走 HASH。询问 in（子查询）返回多少行，运行多久，得到反馈：in（子查询）返回 16 880 条数据，耗时 23 秒。于是我们将 SQL 改写为 with as 子句，而且固化（/*+ materialize */）with as 子查询，让 with as 子句作为嵌套循环驱动表。

```sql
with x as (
SELECT /*+ materialize */  "VOUCHER".FID "ID"
          FROM T_GL_Voucher "VOUCHER"
          INNER JOIN T_GL_VoucherEntry "ENTRIES" ON "VOUCHER".FID =
                                                   "ENTRIES".FBillID
          INNER JOIN T_BD_AccountView "ACCOUNT" ON "ENTRIES".FAccountID =
                                                   "ACCOUNT".FID
          INNER JOIN t_bd_accountview PAV ON ((INSTR("ACCOUNT".flongnumber,
                                                     pav.flongnumber) = 1 AND
                                              pav.faccounttableid =
                                              "ACCOUNT".faccounttableid) AND
                                              pav.fcompanyid =
                                              "ACCOUNT".fcompanyid)
          WHERE ((("VOUCHER".FCompanyID IN ('fSSF82rRSKexM3KKN1d0tMznrtQ=')) AND
                 (("VOUCHER".FBizStatus IN (5)) AND
                 ((("VOUCHER".FPeriodID IN ('+wQxkBFVRiKnV7OniceMDoI4jEw=')) AND
                 "ENTRIES".FCurrencyID =
                 'dfd38d11-00fd-1000-e000-1ebdc0a8100dDEB58FDC') AND
                 (pav.FID IN ('vyPiKexLRXiyMb41VSVVzJ2pmCY='))))))
)
SELECT "VOUCHER".FID "ID",
       "ENTRIES".FID "ENTRIES.ID",
       "ENTRIES".FEntryDC "ENTRIES.ENTRYDC",
       "ACCOUNT".FID "ACCOUNT.ID",
       "ENTRIES".FCurrencyID "CURRENCY.ID",
       "PERIOD".FNumber "PERIOD.NUMBER",
       "ENTRIES".FSeq "ENTRIES.SEQ",
       "ENTRIES".FLocalExchangeRate "LOCALEXCHANGERATE",
       "ENTRIES".FReportingExchangeRate "REPORTINGEXCHANGERATE",
       "ENTRIES".FMeasureUnitID "ENTRYMEASUREUNIT.ID",
       "ASSISTRECORDS".FID "ASSISTRECORDS.ID",
       "ASSISTRECORDS".FSeq "ASSISTRECORDS.SEQ",
       CASE
         WHEN (("ACCOUNT".FCAA IS NULL) AND
               ("ACCOUNT".FhasUserProperty <> 1)) THEN
            "ENTRIES".FOriginalAmount
         ELSE
            "ASSISTRECORDS".FOriginalAmount
       END "ASSISTRECORDS.ORIGINALAMOUNT",
       CASE
         WHEN (("ACCOUNT".FCAA IS NULL) AND
               ("ACCOUNT".FhasUserProperty <> 1)) THEN
            "ENTRIES".FLocalAmount
         ELSE
            "ASSISTRECORDS".FLocalAmount
       END "ASSISTRECORDS.LOCALAMOUNT",
       CASE
         WHEN (("ACCOUNT".FCAA IS NULL) AND
               ("ACCOUNT".FhasUserProperty <> 1)) THEN
            "ENTRIES".FReportingAmount
         ELSE
            "ASSISTRECORDS".FReportingAmount
       END "ASSISTRECORDS.REPORTINGAMOUNT",
       CASE
         WHEN (("ACCOUNT".FCAA IS NULL) AND
               ("ACCOUNT".FhasUserProperty <> 1)) THEN
            "ENTRIES".FQuantity
         ELSE
            "ASSISTRECORDS".FQuantity
       END "ASSISTRECORDS.QUANTITY",
```

9.9 半连接反向驱动主表案例二

```
      CASE
        WHEN (("ACCOUNT".FCAA IS NULL) AND
              ("ACCOUNT".FhasUserProperty <> 1)) THEN
          "ENTRIES".FStandardQuantity
        ELSE
          "ASSISTRECORDS".FStandardQuantity
      END "ASSISTRECORDS.STANDARDQTY",
      CASE
        WHEN (("ACCOUNT".FCAA IS NULL) AND
              ("ACCOUNT".FhasUserProperty <> 1)) THEN
          "ENTRIES".FPrice
        ELSE
          "ASSISTRECORDS".FPrice
      END "ASSISTRECORDS.PRICE",
      CASE
        WHEN ("ACCOUNT".FCAA IS NULL) THEN
          NULL
        ELSE
          "ASSISTRECORDS".FAssGrpID
      END "ASSGRP.ID"
  FROM T_GL_Voucher "VOUCHER"
  LEFT OUTER JOIN T_BD_Period "PERIOD" ON "VOUCHER".FPeriodID =
                                          "PERIOD".FID
  INNER JOIN T_GL_VoucherEntry "ENTRIES" ON "VOUCHER".FID =
                                          "ENTRIES".FBillID
  INNER JOIN T_BD_AccountView "ACCOUNT" ON "ENTRIES".FAccountID =
                                          "ACCOUNT".FID
  LEFT OUTER JOIN T_GL_VoucherAssistRecord "ASSISTRECORDS" ON "ENTRIES".FID =
                                          "ASSISTRECORDS".FEntryID
WHERE "VOUCHER".FID IN
      (select id from x)
ORDER BY "ID" ASC, "ENTRIES.SEQ" ASC, "ASSISTRECORDS.SEQ" ASC;
```

改写后的执行计划如下。

```
---------------------------------------------------------------------------------------
|Id |Operation                          |Name                        |Rows |Bytes|Cost(%CPU)|
---------------------------------------------------------------------------------------
|  0|SELECT STATEMENT                   |                            |   24|11208| 506   (1)|
|  1| TEMP TABLE TRANSFORMATION         |                            |     |     |          |
|  2|  LOAD AS SELECT                   |SYS_TEMP_0FD9D6853_1AD5C99D |     |     |          |
|  3|   HASH JOIN                       |                            |    1|  415| 458   (1)|
|  4|    NESTED LOOPS                   |                            |     |     |          |
|  5|     NESTED LOOPS                  |                            |  258|83850| 390   (0)|
|  6|      NESTED LOOPS                 |                            |    6| 1332|   3   (0)|
|  7|       TABLE ACCESS BY INDEX ROWID |T_BD_ACCOUNTVIEW            |    1|  111|   2   (0)|
|  8|        INDEX UNIQUE SCAN          |PK_BD_ACCOUNTVIEW           |    1|     |   1   (0)|
|  9|       INDEX RANGE SCAN            |IX_BD_ACTCOMLNUM            |    6|  666|   1   (0)|
| 10|      INDEX RANGE SCAN             |IX_GL_VCHAACCT              |  489|     |   1   (0)|
| 11|     TABLE ACCESS BY INDEX ROWID   |T_GL_VOUCHERENTRY           |   42| 4326|  65   (0)|
| 12|      INDEX RANGE SCAN             |IX_GL_VCH_11                | 7536| 662K|  68   (0)|
| 13|   SORT ORDER BY                   |                            |   24|11208|  48   (5)|
| 14|    NESTED LOOPS OUTER             |                            |   24|11208|  47   (3)|
| 15|     NESTED LOOPS                  |                            |   17| 6086|  21   (5)|
| 16|      NESTED LOOPS                 |                            |   17| 5253|  13   (8)|
| 17|       NESTED LOOPS OUTER          |                            |    1|  121|   5  (20)|
| 18|        NESTED LOOPS               |                            |    1|   87|   4  (25)|
| 19|         VIEW                      |VW_NSO_1                    |    1|   29|   2   (0)|
| 20|          HASH UNIQUE              |                            |    1|   24|          |
| 21|           VIEW                    |                            |    1|   24|   2   (0)|
| 22|            TABLE ACCESS FULL      |SYS_TEMP_0FD9D6853_1AD5C99D |    1|   29|   2   (0)|
| 23|         INDEX RANGE SCAN          |IX_GL_VCH_FIDCMPNUM         |    1|   58|   1   (0)|
```

```
| 24|         INDEX RANGE SCAN        |IX_PERIOD_ENC              |     1|    34|     1   (0)|
| 25|   TABLE ACCESS BY INDEX ROWID   |T_GL_VOUCHERENTRY          |    17|  3196|     8   (0)|
| 26|         INDEX RANGE SCAN        |IX_GL_VCHENTRYFQ1          |    17|      |     1   (0)|
| 27|   TABLE ACCESS BY INDEX ROWID   |T_BD_ACCOUNTVIEW           |     1|    49|     1   (0)|
| 28|         INDEX UNIQUE SCAN       |PK_BD_ACCOUNTVIEW          |     1|      |     1   (0)|
| 29|   TABLE ACCESS BY INDEX ROWID   |T_GL_VOUCHERASSISTRECORD   |     1|   109|     2   (0)|
| 30|         INDEX RANGE SCAN        |IX_GL_VCHASSREC_11         |     2|      |     1   (0)|
```

将 SQL 改写之后，能在 1 分钟内执行完毕，最终 SQL 返回 42 956 条数据。

为什么要将 in 子查询改写为 with as 呢？这是因为原始 SQL 中，in 子查询比较复杂，想直接使用 HINT 让 in 子查询作为嵌套循环驱动表反向驱动主表比较困难，所以将 in 子查询改写为 with as。需要注意的是 with as 子句中必须要添加 HINT:/*+ materialize */，同时主表与子查询关联列必须有索引，如果不添加 HINT: /*+ materialize */，如果主表与子查询关联列没有索引，优化器就不会自动将 with as 作为嵌套循环驱动表。with as 子句添加了 /*+ materialize */ 会生成一个临时表，这时，就将复杂的 in 子查询简单化了，之后优化器会将 with as 子句展开（unnesting），将子查询展开一般是子查询与主表进行 HASH 连接，或者是子查询作为嵌套循环驱动表与主表进行关联，一般不会是主表作为嵌套循环驱动表，因为主表作为嵌套循环驱动表可以直接走 Filter，不用展开。优化器发现 with as 子句数据量较小，而主表较大，而且主表连接列有索引，于是自动让 with as 子句固化的结果作为了嵌套循环驱动表。

9.10 连接列数据分布不均衡导致性能问题

2016 年，一互联网彩票行业的朋友说以下 SQL 要跑几十分钟（数据库环境 Oracle11gR2）。

```
select count(distinct a.user_name), count(distinct a.invest_id)
  from base_data_login_info@agent a
 where a.str_day <= '20160304'
   and a.str_day >= '20160301'
   and a.channel_id in (select channel_rlat
                          from tb_user_channel a, tb_channel_info b
                         where a.channel_id = b.channel_id
                           and a.user_id = 5002)
   and a.platform = a.platform;


Plan hash value: 2367445948

-----------------------------------------------------------------------------------------
| Id  | Operation            | Name                 | Rows  | Bytes | Cost (%CPU)|
-----------------------------------------------------------------------------------------
|   0 | SELECT STATEMENT     |                      |     1 |   130 |   754   (2)|
|   1 |  SORT GROUP BY       |                      |     1 |   130 |            |
|*  2 |   HASH JOIN          |                      |  4067K|  504M |   754   (2)|
|*  3 |    HASH JOIN         |                      | 11535 |  360K |   258   (1)|
|*  4 |     TABLE ACCESS FULL| TB_USER_CHANNEL      | 11535 |  157K |    19   (0)|
|   5 |     TABLE ACCESS FULL| TB_CHANNEL_INFO      | 11767 |  206K |   238   (0)|
|   6 |    REMOTE            | BASE_DATA_LOGIN_INFO |  190K |   17M |   486   (1)|
-----------------------------------------------------------------------------------------

Predicate Information (identified by operation id):
---------------------------------------------------
```

```
    2 - access("A"."CHANNEL_ID"="CHANNEL_RLAT")
    3 - access("A"."CHANNEL_ID"="B"."CHANNEL_ID")
    4 - filter("A"."USER_ID"=5002)

Remote SQL Information (identified by operation id):
----------------------------------------------------

    6 - SELECT "USER_NAME","INVEST_ID","STR_DAY","CHANNEL_ID","PLATFORM" FROM
        "BASE_DATA_LOGIN_INFO" "A" WHERE "STR_DAY"<='20160304' AND "STR_DAY">='20160301'
         AND "PLATFORM" IS NOT NULL (accessing 'AGENT' )
```

想要优化 SQL，必须要知道表大小。TB_USER_CHANNEL 有 1 万行数据，TB_CHANNEL_INFO 有 1 万行左右，BASE_DATA_LOGIN_INFO 有 19 万行，过滤之后剩下 4 万行左右。执行计划走的是 HASH 连接，每个表都只扫描一次，虽然是全表扫描，但是最大表才 19 万行，按道理说不应该执行几十分钟，正常情况下应该可以 1 秒左右出结果。起初我们怀疑是 SQL 中 DBLINK 传输数据导致性能问题，于是在本地创建一个一模一样的表，但是该 SQL 还是执行缓慢。

我们只能一步一步排查 SQL 哪里出了问题，让朋友执行如下 SQL。

```
select count(*)    ---改动了这里
  from base_data_login_info@agent a
 where a.str_day <= '20160304'
   and a.str_day >= '20160301'
   and a.channel_id in (select channel_rlat
                          from tb_user_channel a, tb_channel_info b
                         where a.channel_id = b.channel_id
                           and a.user_id = 5002)
   and a.platform = a.platform;
```

上面 SQL 可以秒出。于是朋友继续执行如下 SQL。

```
select count(a.user_name)    ---改动了这里
  from base_data_login_info@agent a
 where a.str_day <= '20160304'
   and a.str_day >= '20160301'
   and a.channel_id in (select channel_rlat
                          from tb_user_channel a, tb_channel_info b
                         where a.channel_id = b.channel_id
                           and a.user_id = 5002)
   and a.platform = a.platform;
```

上面 SQL 也可以秒出。我们继续排查。

```
select count(a.user_name), count(a.invest_id)    ---改动了这里
  from base_data_login_info@agent a
 where a.str_day <= '20160304'
   and a.str_day >= '20160301'
   and a.channel_id in (select channel_rlat
                          from tb_user_channel a, tb_channel_info b
                         where a.channel_id = b.channel_id
                           and a.user_id = 5002)
   and a.platform = a.platform;
```

以上 SQL 还是可以秒出，我们继续排查。

```
select count(distinct a.user_name), count(a.invest_id)    ---改动了这里
```

```
   from base_data_login_info@agent a
 where a.str_day <= '20160304'
   and a.str_day >= '20160301'
   and a.channel_id in (select channel_rlat
                         from tb_user_channel a, tb_channel_info b
                        where a.channel_id = b.channel_id
                          and a.user_id = 5002)
   and a.platform = a.platform;
```

上面 SQL 依然可以秒出。现在我们找到引起 SQL 慢的原因了，select 中同时 count(distinct a.user_name)，count(distinct a.invest_id)导致 SQL 查询缓慢。

在实际工作中，要优先解决问题，再去查找问题的根本原因。我们将 SQL 进行如下改写。

```
with t1 as
(select /*+ materialize */
 a.user_name, a.invest_id
  from base_data_login_info@agent a
 where a.str_day <= '20160304'
   and a.str_day >= '20160301'
   and a.channel_id in (select channel_rlat
                         from tb_user_channel a, tb_channel_info b
                        where a.channel_id = b.channel_id
                          and a.user_id = 5002)
   and a.platform = a.platform)
select count(distinct user_name) ,count(distinct invest_id) from t1;
```

为什么改写成以上 SQL 能解决性能问题呢？因为在排查问题的时候 count 不加 distinct 是可以秒出的，所以我们先将能秒出的 SQL 放到 with as 子句，通过添加 HINT：/*+ materialize */ 生成临时表，再对临时表进行 count(distinct...),count(distinct)，这样就能解决问题。改写后的 SQL 执行计划如下。

```
Plan hash value: 901326807

---------------------------------------------------------------------------------------------
| Id  | Operation                  | Name                      | Rows  | Bytes | Cost (%CPU)|
---------------------------------------------------------------------------------------------
|   0 | SELECT STATEMENT           |                           |     1 |    54 |  1621   (1)|
|   1 |  TEMP TABLE TRANSFORMATION |                           |       |       |            |
|   2 |   LOAD AS SELECT           | SYS_TEMP_0FD9D6720_EB8EA  |       |       |            |
|*  3 |    HASH JOIN RIGHT SEMI    |                           |  190K |   22M |   744   (1)|
|   4 |     VIEW                   | VW_NSO_1                  | 11535 |  304K |   258   (1)|
|*  5 |      HASH JOIN             |                           | 11535 |  360K |   258   (1)|
|*  6 |       TABLE ACCESS FULL    | TB_USER_CHANNEL           | 11535 |  157K |    19   (0)|
|   7 |       TABLE ACCESS FULL    | TB_CHANNEL_INFO           | 11767 |  206K |   238   (0)|
|   8 |     REMOTE                 | BASE_DATA_LOGIN_INFO      |  190K |   17M |   486   (1)|
|   9 |   SORT GROUP BY            |                           |     1 |    54 |            |
|  10 |    VIEW                    |                           |  190K |    9M |   878   (1)|
|  11 |     TABLE ACCESS FULL      | SYS_TEMP_0FD9D6720_EB8EA  |  190K |    9M |   878   (1)|
---------------------------------------------------------------------------------------------

Predicate Information (identified by operation id):
---------------------------------------------------

   3 - access("A"."CHANNEL_ID"="CHANNEL_RLAT")
   5 - access("A"."CHANNEL_ID"="B"."CHANNEL_ID")
   6 - filter("A"."USER_ID"=5002)

Remote SQL Information (identified by operation id):
```

9.10 连接列数据分布不均衡导致性能问题

```
       8 - SELECT "USER_NAME","INVEST_ID","STR_DAY","CHANNEL_ID","PLATFORM" FROM "BASE_DA
TA_LOGIN_INFO"
          "A" WHERE "STR_DAY"<='20160304' AND "STR_DAY">='20160301'
          AND "PLATFORM" IS NOT NULL (accessing 'AGENT' )
```

解决问题之后，现在我们来查找 SQL 缓慢的根本原因。现在对比缓慢 SQL 的执行计划与秒出 SQL 的执行计划，缓慢 SQL 的执行计划如下。

```
-------------------------------------------------------------------------------
| Id | Operation            | Name                 | Rows  | Bytes | Cost (%CPU)|
-------------------------------------------------------------------------------
|  0 | SELECT STATEMENT     |                      |    1  |  130  |  754   (2)|
|  1 |  SORT GROUP BY       |                      |    1  |  130  |           |
|* 2 |   HASH JOIN          |                      | 4067K |  504M |  754   (2)|
|* 3 |    HASH JOIN         |                      | 11535 |  360K |  258   (1)|
|* 4 |     TABLE ACCESS FULL| TB_USER_CHANNEL      | 11535 |  157K |   19   (0)|
|  5 |     TABLE ACCESS FULL| TB_CHANNEL_INFO      | 11767 |  206K |  238   (0)|
|  6 |    REMOTE            | BASE_DATA_LOGIN_INFO |  190K |   17M |  486   (1)|
-------------------------------------------------------------------------------
```

秒出 SQL 的执行计划如下。

```
-------------------------------------------------------------------------------
| Id  |Operation                   |Name                    | Rows  | Bytes |Cost(%CPU)|
-------------------------------------------------------------------------------
|  0  |SELECT STATEMENT            |                        |    1  |   54  | 1621  (1)|
|  1  | TEMP TABLE TRANSFORMATION  |                        |       |       |          |
|  2  |  LOAD AS SELECT            |SYS_TEMP_0FD9D6720_EB8EA|       |       |          |
|* 3  |   HASH JOIN RIGHT SEMI     |                        |  190K |   22M |  744  (1)|
|  4  |    VIEW                    |VW_NSO_1                | 11535 |  304K |  258  (1)|
|* 5  |     HASH JOIN              |                        | 11535 |  360K |  258  (1)|
|* 6  |      TABLE ACCESS FULL     |TB_USER_CHANNEL         | 11535 |  157K |   19  (0)|
|  7  |      TABLE ACCESS FULL     |TB_CHANNEL_INFO         | 11767 |  206K |  238  (0)|
|  8  |    REMOTE                  |BASE_DATA_LOGIN_INFO    |  190K |   17M |  486  (1)|
|  9  |  SORT GROUP BY             |                        |    1  |   54  |          |
| 10  |   VIEW                     |                        |  190K |    9M |  878  (1)|
| 11  |    TABLE ACCESS FULL       |SYS_TEMP_0FD9D6720_EB8EA|  190K |    9M |  878  (1)|
-------------------------------------------------------------------------------
```

我们注意仔细对比执行计划，缓慢 SQL 执行计划中 Id=2 是 HASH JOIN，而秒出 SQL 的执行计划中 Id=3 是 HASH JOIN RIGHT SEMI。SEMI 是半连接特有关键字，缓慢 SQL 的执行计划中没有 SEMI 关键字，这说明 CBO 将半连接等价改写成了内连接；秒出 SQL 的执行计划有 SEMI 关键字，这说明 CBO 没有将半连接等价改写成内连接。**现在我们得到结论，该 SQL 查询缓慢是因为 CBO 内部将半连接改写为内连接导致。**

大家还记得半连接与内连接接区别吗？半连只返回一个表的数据，关联之后数据量不会翻番，内连接表关联之后数据量可能会翻番。该 SQL 查询缓慢是被改成内连接导致，现在我们有充分理由怀疑内连接关联之后返回的数据量太大，因为如果关联返回的数据量很少是不可能出性能问题的。于是检查两个表连接列的数据分布。

```
select channel_id, count(*)
  from base_data_login_info
 group by channel_id
 order by 2;
```

```
CHANNEL_ID                                         COUNT(*)
-------------------------------------------------- ----------
011a1                                                      2
003a1                                                      3
021a1                                                      3
006a1                                                     12
024h2                                                     16
013a1                                                     19
007a1                                                     24
012a1                                                     25
005a1                                                     27
EPT01                                                     36
028h2                                                    109
008a1                                                    139
029a1                                                    841
009a1                                                    921
014a1                                                   1583
000a1                                                   1975
a0001                                                   2724
004a1                                                   5482
001a1                                                  16329
026h2                                                 160162
```

in 子查询关联列数据分布如下。

```
select channel_rlat, count(*)
  from tb_user_channel a, tb_channel_info b
 where a.channel_id = b.channel_id
   and a.user_id = 5002
 group by channel_rlat
 order by 2 desc;

channel_rlat       count(*)
026h2              10984
024h2              7
002h2              6
023a2              2
007s001022001      1
007s001022002      1
007s001024007      1
007s001024009      1
007s001022009      1
001s001006         1
001s001008         1
001s001001001      1
001s001001003      1
001s001001007      1
001s001001014      1
007s001018003      1
007s001018007      1
007s001019005      1
007s001019008      1
001s001002011      1
001s001011003      1
007s001034         1
007s001023005      1
```

  两表的数据分布果然有问题，其中 026h2 这条数据倾斜特别明显。如果让两表进行内连接，026h2 这条数据关联之后返回结果应该是 160162*10984，现在我们终于发现该 SQL 执行缓慢的根本原因，是因为两个表的连接列中有部分数据倾斜非常严重。

9.10 连接列数据分布不均衡导致性能问题

最初采用的是 with as 子句加/*+ materialize */临时解决 SQL 的性能问题，我们也可以使用 rownum 优化 SQL，rownum 可以让一个查询被当成一个整体。

```
with t1 as
(select
 a.user_name, a.invest_id
  from base_data_login_info@agent a
 where a.str_day <= '20160304'
   and a.str_day >= '20160301'
   and a.channel_id in (select channel_rlat
                         from tb_user_channel a, tb_channel_info b
                        where a.channel_id = b.channel_id
                          and a.user_id = 5002)
   and a.platform = a.platform and rownum>0)
select count(distinct user_name) ,count(distinct invest_id) from t1;
```

如果大家想模拟本案例，可以跟着下面实验步骤执行（请在 11g 中模拟）。

我们先创建如下两个测试表。

```
create table a as select * from dba_objects;
create table b as select * from dba_objects;
```

要执行的缓慢的 SQL 如下。

```
select count(distinct owner), count(distinct object_name)
  from a
 where owner in (select owner from b);
```

优化改写之后的 SQL 如下。

```
with t as(select owner, object_name
            from a
           where owner in (select owner from b)
             and rownum > 0)
select count(distinct owner), count(distinct object_name)
  from t;
```

我们也可以对子查询先去重，将子查询变成 1 的关系，这样也能优化 SQL。

```
select count(distinct owner), count(distinct object_name)
  from a
 where owner in (select owner from b group by owner);
```

请思考为什么 Oracle11g CBO 会将 SQL 改写为内连接？大家是否还记得第 5.6.1 节内容？

select ... from　1 的表　where owner in (select owner from n 的表)改写为内连接，需要加 distinct。

select ... from　n 的表　where owner in (select owner from 1 的表)改写为内连接，不需要加 distinct。

我们的 SQL 是 select count(distinct),count(distinct)，所以 CBO 直接将 SQL 改写为 select count(distinct a.owner),count(distinct object_name) from a,b where a.owner=b.owner; 这个问题在 12c 中已得到纠正。最后我们想说的就是，不管以后优化器进步有多大，我们始终不能依赖优化器，唯一可以依靠的就是自己所掌握的知识。

9.11 Filter 优化经典案例

2012 年，一朋友发来信息说以下 SQL 要跑 5 个小时，请求优化。

```
SELECT
        B.AREA_ID,
        A.PARTY_ID,
        B.AREA_NAME,
        C.NAME           CHANNEL_NAME,
        B.NAME           PARTY_NAME,
        B.ACCESS_NUMBER,
        B.PROD_SPEC,
        B.START_DT,
        A.BO_ACTION_NAME,
        A.SO_STAFF_ID,
        A.ATOM_ACTION_ID,
        A.PROD_ID
   FROM DW_CHANNEL       C,
        DW_CRM_DAY_USER  B,
        DW_BO_ORDER      A
  WHERE A.PROD_ID = B.PROD_ID AND
        A.CHANNEL_ID = C.CHANNEL_ID AND
        A.SO_STAFF_ID LIKE '36%' AND
        A.BO_ACTION_NAME IN ('新装','移机','资费变更') AND
        B.PROD_SPEC IN ('普通电话', 'ADSL','LAN', '手机',
                        'E8 - 2S','E6 移动版', 'E9 版 1M(老版)',
                        '普通 E9','普通新版 E8',
                        '全省_紧密融合型 E9 套餐产品规格',
                        '(新) 全省_紧密融合型 E9 套餐产品规格',
                        '新春欢乐送之 E8 套餐',
                        '新春欢乐送之 E6 套餐') AND
    NOT EXISTS (SELECT *
          FROM DW_BO_ORDER D
         WHERE D.STAFF_ID LIKE '36%' AND
               A.PARTY_ID = D.PARTY_ID AND
               A.BO_ID != D.BO_ID AND
               A.PROD_ID != D.PROD_ID AND
               A.BO_ACTION_NAME IN
               ('新装','移机','资费变更') AND
               A.COMPLETE_DT - INTERVAL '7' DAY < D.COMPLETE_DT);
```

执行计划如下。

```
Plan hash value: 2142862569

-------------------------------------------------------------------------------------
| Id  | Operation                  | Name            | Rows  | Bytes | Cost (%CPU)| Time     |
-------------------------------------------------------------------------------------
|   0 | SELECT STATEMENT           |                 |   905 |  121K|  4152K  (2)| 13:50:32 |
|*  1 |  FILTER                    |                 |       |      |            |          |
|*  2 |   HASH JOIN                |                 |   905 |  121K| 12616   (2)| 00:02:32 |
|*  3 |    HASH JOIN               |                 |   905 | 99550| 12448   (2)| 00:02:30 |
|   4 |     PARTITION RANGE ALL    |                 |  1979 |  108K|  9168   (2)| 00:01:51 |
|*  5 |      TABLE ACCESS FULL     | DW_BO_ORDER     |  1979 |  108K|  9168   (2)| 00:01:51 |
|*  6 |     TABLE ACCESS FULL      | DW_CRM_DAY_USER |  309K |  15M|  3277   (2)| 00:00:40 |
|   7 |    TABLE ACCESS FULL       | DW_CHANNEL      | 48425 | 1276K|   168   (1)| 00:00:03 |
|*  8 |   FILTER                   |                 |       |      |            |          |
|   9 |    PARTITION RANGE ALL     |                 |     1 |    29|  9147   (2)| 00:01:50 |
|* 10 |     TABLE ACCESS FULL      | DW_BO_ORDER     |     1 |    29|  9147   (2)| 00:01:50 |
```

9.11 Filter 优化经典案例

```
--------------------------------------------------------------------------------
Predicate Information (identified by operation id):
--------------------------------------------------------------------------------

   1 - filter( NOT EXISTS (SELECT /*+ */ 0 FROM "DW_BO_ORDER" "D" WHERE (:B1='新装' OR :B2='
              移机'OR :B3='资费变更') AND "D"."PARTY_ID"=:B4 AND TO_CHAR("D"."STAFF_ID")
LIKE
              '36%' AND "D"."COMPLETE_DT">:B5-INTERVAL'+07 00:00:00' DAY(2) TO SECOND
(0) AND
              "D"."PROD_ID"<>:B6 AND "D"."BO_ID"<>:B7))
   2 - access("A"."CHANNEL_ID"="C"."CHANNEL_ID")
   3 - access("A"."PROD_ID"="B"."PROD_ID")
   5 - filter("A"."PROD_ID" IS NOT NULL AND ("A"."BO_ACTION_NAME"='新装' OR
              "A"."BO_ACTION_NAME"='移机' OR "A"."BO_ACTION_NAME"='资费变更') AND
              TO_CHAR("A"."SO_STAFF_ID") LIKE '36%')
   6 - filter("B"."PROD_SPEC"='(新) 全省_紧密融合型 E9 套餐产品规格' OR "B"."PROD_SPEC"
='ADSL'
              OR "B"."PROD_SPEC"='E6移动版' OR "B"."PROD_SPEC"='E8 - 2S' OR
              "B"."PROD_SPEC"='E9 版   1M( 老 版 )' OR "B"."PROD_SPEC"='LAN' OR "B".
"PROD_SPEC"='
              普通 E9' OR "B"."PROD_SPEC"='普通电话' OR "B"."PROD_SPEC"='普通新版 E8' OR
              "B"."PROD_SPEC"='全省_紧密融合型 E9 套餐产品规格' OR "B"."PROD_SPEC"='手机
' OR
              "B"."PROD_SPEC"='新春欢乐送之E6 套餐' OR "B"."PROD_SPEC"='新春欢乐送之E8 套餐')
   8 - filter(:B1='新装' OR :B2='移机' OR :B3='资费变更')
  10 - filter("D"."PARTY_ID"=:B1 AND TO_CHAR("D"."STAFF_ID") LIKE '36%' AND
              "D"."COMPLETE_DT">:B2-INTERVAL'+07 00:00:00' DAY(2) TO SECOND(0) AND
              "D"."PROD_ID"<>:B3 AND "D"."BO_ID"<>:B4)
```

优化 SQL，必须看表大小，表大小信息如下。

```
SQL> select count(*) from dw_bo_order; ----200万行数据

  COUNT(*)
----------
   2282548

SQL> select count(*) from dw_crm_day_user; ----40万行数据

  COUNT(*)
----------
    420918

SQL> select count(*) from dw_channel;  ---4万行数据

  COUNT(*)
----------
     48031
```

SQL 语句中最大表 DW_BO_ORDER 才 200 万行数据，但是 SQL 执行了 5 个多小时，显然执行计划有问题。执行计划中，Id=1 是 Filter，而且 Filter 对应的谓词信息有 EXISTS（子查询:B1），这说明该 Filter 类似嵌套循环。Id=2 和 Id=8 是 Id=1 的儿子，因为这里的 Filter 类似嵌套循环，Id=2 就相当于 NL 驱动表，Id=8 相当于 NL 被驱动表，Id=8 是全表扫描过滤后的数据，所以 Id=8 可以看作全表扫描。本书反复强调过，NL 被驱动表必须走索引。但是 Id=10 并没有走索引。Id=2 估算返回 905 行数据，一般情况下 Rows 会算少，这里就暂且认为 Id=2 返回 905 行数据，那么 Id=8 会被扫描 905 次，也就是说 DW_BO_ORDER 这个 200 万行大表会被扫描 905 次,而且每次都是全表扫描,这就是为什么 SQL 会执行 5 个多小时。

找到 SQL 的性能瓶颈之后,我们就可以想办法优化 SQL。**本案例有两种优化思路,其一是让大表只被扫描一次,其二是不减少扫描次数,但是减少大表每次被扫描的体积。**最优的解决方案是,想办法让 Id=2 和 Id=8 走 HASH 连接消除 Filter,这样只需要扫描 1 次大表,因为当时数据库版本是 Oracle10g,where 子查询中有主表的过滤条件,在 not exists 子查询中添加 HINT:HASH_AJ 无法更改执行计划。我们可以将 not exists 改写为"外连接+子表连接列 is null"的形式,让其走 HASH 连接,但是当时没有采用这种改写方式。因为大表要被扫描 905 次,每次都是全表扫描,如果能减少扫描的体积,也能优化 SQL。我们可以在大表上建立一个组合索引,这样就能避免大表每次全表扫描,从而达到减少扫描体积的目的,但是当时朋友没权限建立索引。最终选择使用 with as 子句优化上述 SQL。

```
SQL> set timi on
SQL> WITH D AS
  2   (SELECT /*+ materialize */
  3     PARTY_ID,
  4     BO_ID,
  5     PROD_ID,
  6     COMPLETE_DT
  7    FROM   DW_BO_ORDER
  8    WHERE  STAFF_ID LIKE '36%' AND
  9           BO_ACTION_NAME IN ('新装',
 10                              '移机',
 11                              '资费变更'))
 12  SELECT
 13              B.AREA_ID,
 14              A.PARTY_ID,
 15              B.AREA_NAME,
 16              C.NAME            CHANNEL_NAME,
 17              B.NAME            PARTY_NAME,
 18              B.ACCESS_NUMBER,
 19              B.PROD_SPEC,
 20              B.START_DT,
 21              A.BO_ACTION_NAME,
 22              A.SO_STAFF_ID,
 23              A.ATOM_ACTION_ID,
 24              A.PROD_ID
 25     FROM     DW_CHANNEL        C,
 26              DW_CRM_DAY_USER   B,
 27              DW_BO_ORDER       A
 28     WHERE    A.PROD_ID = B.PROD_ID AND
 29              A.CHANNEL_ID = C.CHANNEL_ID AND
 30              A.SO_STAFF_ID LIKE '36%' AND
 31              A.BO_ACTION_NAME IN ('新装','移机','资费变更') AND
 32              B.PROD_SPEC IN ('普通电话', 'ADSL','LAN', '手机',
 33                              'E8 - 2S','E6移动版', 'E9版 1M(老版)',
 34                              '普通E9','普通新版 E8',
 35                              '全省 紧密融合型 E9 套餐产品规格',
 36                              '(新) 全省 紧密融合型 E9 套餐产品规格',
 37                              '新春欢乐送之 E8 套餐',
 38                              '新春欢乐送之 E6 套餐') AND
 39              NOT  EXISTS (SELECT  *
 40                FROM   D
 41                WHERE  A.PARTY_ID = D.PARTY_ID AND
 42                       A.BO_ID != D.BO_ID AND
 43                       A.PROD_ID != D.PROD_ID AND
```

44   A.COMPLETE_DT - INTERVAL '7' DAY < D.COMPLETE_DT);

已选择 49 245 行。

已用时间：00: 00: 12.37。

执行计划如下。

```
-------------------------------------------------------------------------------
Plan hash value: 2591883460
-------------------------------------------------------------------------------
| Id  |Operation                    |Name                       |Rows |Bytes|Cost(%CPU)|
-------------------------------------------------------------------------------
|   0 |SELECT STATEMENT             |                           |  905| 121K| 62428  (2)|
|   1 | TEMP TABLE TRANSFORMATION   |                           |     |     |           |
|   2 |  LOAD AS SELECT             |DW_BO_ORDER                |     |     |           |
|   3 |   PARTITION RANGE ALL       |                           | 114K|3228K|  9127  (2)|
|*  4 |    TABLE ACCESS FULL        |DW_BO_ORDER                | 114K|3228K|  9127  (2)|
|*  5 |  FILTER                     |                           |     |     |           |
|*  6 |   HASH JOIN                 |                           |  905| 121K| 12616  (2)|
|*  7 |    HASH JOIN                |                           |  905|99550| 12448  (2)|
|   8 |     PARTITION RANGE ALL     |                           | 1979| 108K|  9168  (2)|
|*  9 |      TABLE ACCESS FULL      |DW_BO_ORDER                | 1979| 108K|  9168  (2)|
|* 10 |     TABLE ACCESS FULL       |DW_CRM_DAY_USER            | 309K|  15M|  3277  (2)|
|  11 |    TABLE ACCESS FULL        |DW_CHANNEL                 |48425|1276K|   168  (1)|
|* 12 |   FILTER                    |                           |     |     |           |
|* 13 |    VIEW                     |                           | 114K|6791K|    90  (3)|
|  14 |     TABLE ACCESS FULL       |SYS_TEMP_0FD9D662E_D625B872| 114K|3228K|    90  (3)|
-------------------------------------------------------------------------------

Predicate Information (identified by operation id):
---------------------------------------------------

   4 - filter(TO_CHAR("STAFF_ID") LIKE '36%')
   5 - filter( NOT EXISTS (SELECT /*+ */ 0 FROM  (SELECT /*+ CACHE_TEMP_TABLE ("T1")
*/ "C0"
              "STAFF_ID","C1" "PARTY_ID","C2" "BO_ID","C3" "PROD_ID","C4" "COMPLETE_D
T" FROM
              "SYS"."SYS_TEMP_0FD9D662E_D625B872" "T1") "D" WHERE (:B1='新装' OR :B2='
移机' OR :B3='
              资费变更') AND TO_CHAR("D"."STAFF_ID") LIKE '36%' AND "D"."PARTY_ID"=:B4
 AND
              "D"."BO_ID"<>:B5 AND "D"."PROD_ID"<>:B6 AND "D"."COMPLETE_DT">:B7-INTER
VAL'+07
              00:00:00' DAY(2) TO SECOND(0)))
   6 - access("A"."CHANNEL_ID"="C"."CHANNEL_ID")
   7 - access("A"."PROD_ID"="B"."PROD_ID")
   9 - filter("A"."PROD_ID" IS NOT NULL AND ("A"."BO_ACTION_NAME"=' 新装 ' OR "A"."BO_
ACTION_NAME"='
              移机' OR "A"."BO_ACTION_NAME"='资费变更') AND TO_CHAR("A"."SO_STAFF_ID")
LIKE '36%')
  10 - filter("B"."PROD_SPEC"='(新) 全省_紧密融合型 E9 套餐产品规格' OR "B"."PROD_SPEC"=
'ADSL' OR
              "B"."PROD_SPEC"='E6 移动版' OR "B"."PROD_SPEC"='E8 - 2S' OR "B"."PROD_
SPEC"='E9 版
              1M(老版)' OR "B"."PROD_SPEC"='LAN' OR "B"."PROD_SPEC"='普通 E9' OR "B".
"PROD_SPEC"='
              普通电话' OR "B"."PROD_SPEC"='普通新版 E8' OR "B"."PROD_SPEC"='全省_紧密融合
型 E9 套餐
              产品规格' OR "B"."PROD_SPEC"='手机' OR "B"."PROD_SPEC"='新春欢乐送之 E6 套餐
' OR
              "B"."PROD_SPEC"='新春欢乐送之 E8 套餐')
  12 - filter(:B1='新装' OR :B2='移机' OR :B3='资费变更')
```

```
    13 - filter(TO_CHAR("D"."STAFF_ID") LIKE '36%' AND "D"."PARTY_ID"=:B1 AND "D"."BO_I
D"<>:B2 AND
             "D"."PROD_ID"<>:B3 AND "D"."COMPLETE_DT">:B4-INTERVAL'+07 00:00:00' DAY
(2) TO
             SECOND(0))


统计信息
----------------------------------------------------------
          2  recursive calls
         29  db block gets
     110506  consistent gets
         22  physical reads
        656  redo size
    2438096  bytes sent via SQL*Net to client
        449  bytes received via SQL*Net from client
         11  SQL*Net roundtrips to/from client
          0  sorts (memory)
          0  sorts (disk)
      49245  rows processed
```

使用 with as 子句将大表要被访问的字段查询出来，一共 4 个字段，然后过滤掉不需要的数据，添加 HINT:MATERIALIZE 将 with as 子句查询结果固化为临时表，这样就达到了减少扫描体积的目的。假设 200 万行的大表 DW_BO_ORDER 有占用 2GB 存储空间，表有 40 个字段，通过 with as 子句改写之后，只需要存储 4 个字段数据，这时只需 200MB 存储空间，而且 with as 子句中还有过滤条件，又可以过滤掉一部分数据，这时 with as 子句可能就只需要几十兆存储空间。虽然被扫描的次数没有改变，但是每次被扫描的体积大大减少，这样就解决了 SQL 查询性能。最终 SQL 可以在 12 秒左右跑完，一共返回 4.9 万行数据。

9.12 树形查询优化案例

2013 年，一朋友咨询如何优化下面树形查询。

```
select rownum, adn, zdn, 'cable'
  from (select distinct connect_by_root(t.tdl_a_dn) adn, t.tdl_z_dn zdn
          from AGGR_1 t
         where t.tdl_operation <> 2
           and exists (select 1
                  from CABLE_1 a
                 where a.tdl_operation <> 2
                   and a.tdl_dn = t.tdl_z_dn)
         start with exists (select 1
                    from RESOURCE_FACING_SERVICE1_1 b
                   where b.tdl_operation <> 2
                     and t.tdl_a_dn = b.tdl_dn)
        connect by nocycle prior t.tdl_z_dn = t.tdl_a_dn);
```

执行计划如下。

```
SQL> select * from table(DBMS_XPLAN.DISPLAY);
Plan hash value: 1439701716

---------------------------------------------------------------------------------
| Id |Operation                                    |Name              |Rows |Bytes|
---------------------------------------------------------------------------------
```

9.12 树形查询优化案例

```
|   0 |SELECT STATEMENT                           |                          |31125|   59M|
|   1 | COUNT                                     |                          |     |      |
|   2 |  VIEW                                     |                          |31125|   59M|
|   3 |   HASH UNIQUE                             |                          |31125|   59M|
|*  4 |    FILTER                                 |                          |     |      |
|*  5 |     CONNECT BY NO FILTERING WITH SW (UNIQUE)|                        |     |      |
|   6 |      TABLE ACCESS FULL                    |AGGR_1                    | 171K| 4353K|
|*  7 |      TABLE ACCESS FULL                    |RESOURCE_FACING_SERVICE1_1|   1 |   18 |
|*  8 |     TABLE ACCESS FULL                     |CABLE_1                   |   1 |   14 |
----------------------------------------------------------------------------------------

Predicate Information (identified by operation id):
---------------------------------------------------

   4 - filter("T"."TDL_OPERATION"<>2 AND  EXISTS (SELECT 0 FROM "CABLE_1" "A" WHERE "
A"."TDL_DN"=:B1
              AND  "A"."TDL_OPERATION"<>2))
   5 - access("T"."TDL_A_DN"=PRIOR "T"."TDL_Z_DN")
              filter( EXISTS (SELECT 0 FROM "RESOURCE_FACING_SERVICE1_1" "B" WHERE "B
"."TDL_DN"=:B1
              AND "B"."TDL_OPERATION"<>2))
   7 - filter("B"."TDL_DN"=:B1 AND "B"."TDL_OPERATION"<>2)
   8 - filter("A"."TDL_DN"=:B1 AND "A"."TDL_OPERATION"<>2)

25 rows selected.
```

阅读过本章 Filter 优化经典案例的读者能很快发现：执行计划中，Id=4 是 Filter，Id=5 和 Id=8 是 Id=4 的儿子，这说明 Id=8 会被多次反复扫描，Id=8 走的是全表扫描，这显然不对。

在进行 SQL 优化的时候，我们需要特别留意执行计划中的谓词过滤信息。执行计划中 Id=7 的谓词过滤中有绑定变量:B1，但是 SQL 语句中并没有绑定变量。大家是否还记得 5.5 节讲到:B1 表示传值。如果 SQL 语句本身没有绑定变量，但是执行计划中谓词过滤信息又有绑定变量（:B1，:B2，B3..），这说明有绑定变量这步需要传值。典型的需要传值的有标量子查询、Filter 以及树形查询中 start with 子查询。**当执行计划中某个步骤需要传值，这个步骤就会被扫描多次**。执行计划中，Id=7 谓词有绑定变量，这说明 Id=7 与 Id=8 一样，要被多次扫描。另外请注意，执行计划中 Id=4 也有绑定变量，但是 Id=4 的绑定变量与 Id=8 是成对出现，Id=5 的绑定变量与 Id=7 也是成对出现，对于成对出现的绑定变量情况，关注有表对应的 Id 即可，这里有表对应的 Id 就是 7 和 8。

通过上面分析，我们知道 SQL 的性能瓶颈在 Id=7 和 Id=8 这两步。对于树形查询，很难通过 SQL 改写减少 start with 子查询中表被多次扫描，所以只能想办法减少表被扫描的体积。我们可以创建下面两个索引来优化 SQL。

```
create index idx_a on CABLE_1(tdl_dn,tdl_operation);
create index idx_b on RESOURCE_FACING_SERVICE1_1(tdl_dn,tdl_operation);
```

本案例也可以使用 with as 子句改写，然后将子查询生成临时表来进行优化，但是 with as 子句改写优化的性能没有创建索引优化的性能高，因为走索引可以进行 INDEX RANGE SCAN，而且不需要回表，而 with as 子句需要对临时表进行全表扫描。本案例的目的是让大家重视执行计划中的谓词信息！

还有一个比较经典的案例，也需要关注谓词信息才能优化 SQL，但是限于 SQL 实在太长，我们无法在书中体现，有兴趣的读者可以查看博客：http://blog.csdn.net/robinson1988/article/details/7002545。

9.13 本地索引优化案例

2012 年，一朋友请求优化下面 SQL，该 SQL 在 RAC 环境中执行，时快时慢，最快时 1 秒，最慢是 3 秒。SQL 语句如下。

```
SELECT /*+INDEX(TMS,IDX1_TB_EVT_DLV_W)*/
 TMS.MAIL_NUM,
 TMS.DLV_BUREAU_ORG_CODE AS DLVORGCODE,
 RO.ORG_SNAME AS DLVORGNAME,
 TMS.DLV_PSEG_CODE AS DLVSECTIONCODE,
 TMS.DLV_PSEG_NAME AS DLVSECTIONNAME,
 TO_CHAR(TMS.DLV_DATE, 'YYYY-MM-DD HH24:MI:SS') AS RECTIME,
 TMS.DLV_STAFF_CODE AS HANDOVERUSERCODE,
 TU2.REALNAME AS HANDOVERUSERNAME,
 DECODE(TMS.DLV_STS_CODE, 'I', '妥投', 'H', '未妥投', TMS.DLV_STS_CODE) AS DLV_STS_CODE,
 CASE
   WHEN TMS.MAIL_NUM LIKE 'EC%' THEN
    '代收'
   WHEN TMS.MAIL_NUM LIKE 'ED%CW' THEN
    '代收'
   WHEN TMS.MAIL_NUM LIKE 'FJ%' THEN
    '代收'
   WHEN TMS.MAIL_NUM LIKE 'GC%' THEN
    '代收'
   ELSE
    '非代收'
 END MAIL_NUM_TYPE
  FROM TB_EVT_DLV_W TMS
  LEFT JOIN RES_ORG RO ON TMS.DLV_BUREAU_ORG_CODE = RO.ORG_CODE
  LEFT JOIN TB_USER TU2 ON TU2.DELVORGCODE = TMS.DLV_BUREAU_ORG_CODE
                       AND TU2.USERNAME = TMS.DLV_STAFF_CODE
 WHERE NOT EXISTS
 (SELECT /*+INDEX(TDW,IDX1_TB_MAIL_SECTION_STORE)*/
         MAIL_NUM
          FROM TB_MAIL_SECTION_STORE TDW
         WHERE TDW.MAIL_NUM = TMS.MAIL_NUM
           AND TDW.DLVORGCODE = TMS.DLV_BUREAU_ORG_CODE
           and TDW.DLVORGCODE = '35000133'
           AND TDW.RECTIME >=
               TO_DATE('2012-11-01 00:00', 'YYYY-MM-DD HH24:MI:SS')
           AND TO_DATE('2012-11-08 15:15', 'YYYY-MM-DD HH24:MI:SS') >=
               TDW.RECTIME
           and rownum = 1)
   AND TMS.DLV_BUREAU_ORG_CODE = '35000133'
   AND TMS.DLV_DATE >= TO_DATE('2012-11-01 00:00', 'YYYY-MM-DD HH24:MI:SS')
   AND TO_DATE('2012-11-08 15:15', 'YYYY-MM-DD HH24:MI:SS') >= TMS.DLV_DATE
   AND ('' IS NULL OR TMS.DLV_STAFF_CODE = '')
   AND ('' IS NULL OR TU2.REALNAME LIKE '%%')
   AND TMS.REC_AVAIL_FLAG = '1';
```

执行计划如下。

```
Plan hash value: 1159587453
```

9.13 本地索引优化案例

```
---------------------------------------------------------------------------------
| Id|Operation                               |Name                   |Rows|Bytes|
---------------------------------------------------------------------------------
|  0|SELECT STATEMENT                        |                       |    |     |
|* 1| FILTER                                 |                       |    |     |
|  2|  NESTED LOOPS OUTER                    |                       | 131|13493|
|* 3|   HASH JOIN RIGHT OUTER                |                       | 129|10191|
|* 4|    TABLE ACCESS BY INDEX ROWID         |EMS_USER               |   6|  120|
|* 5|     INDEX RANGE SCAN                   |EMS_USER_NEW_INX_ORG   |   7|     |
|* 6|    TABLE ACCESS BY GLOBAL INDEX ROWID  |TB_EVT_DLV_W           | 129| 7611|
|* 7|     INDEX RANGE SCAN                   |IDX1_TB_EVT_DLV_W      | 586|     |
|* 8|      COUNT STOPKEY                     |                       |    |     |
|* 9|       FILTER                           |                       |    |     |
| 10|        PARTITION RANGE ITERATOR        |                       |   1|   31|
|*11|         TABLE ACCESS BY LOCAL INDEX ROWID|TB_MAIL_SECTION_STORE|   1|   31|
|*12|          INDEX RANGE SCAN              |IDX1_TB_MAIL_SECTION_STORE| 1|   |
| 13|   TABLE ACCESS BY INDEX ROWID          |RES_ORG                |   1|   24|
|*14|    INDEX RANGE SCAN                    |IDX_RES_ORG            |   1|     |
---------------------------------------------------------------------------------

Predicate Information (identified by operation id):
---------------------------------------------------

   1 - filter(TO_DATE('2012-11-01 00:00','YYYY-MM-DD HH24:MI:SS')<=TO_DATE('2012-11-08
              15:15','YYYY-MM-DD HH24:MI:SS'))
   3 - access("EU"."USERNAME"="TMS"."DLV_STAFF_CODE" AND
              "EU"."DELVORGCODE"="TMS"."DLV_BUREAU_ORG_CODE")
   4 - filter("EU"."POSTMANKIND"<>5)
   5 - access("EU"."DELVORGCODE"='35000133')
   6 - filter(("TMS"."DLV_DATE">=TO_DATE('2012-11-01 00:00','YYYY-MM-DD HH24:MI:SS')
AND
              "TMS"."REC_AVAIL_FLAG"='1' AND "TMS"."DLV_DATE"<=TO_DATE('2012-11-08
              15:15','YYYY-MM-DD HH24:MI:SS')))
   7 - access("TMS"."DLV_BUREAU_ORG_CODE"='35000133')  filter( IS NULL)
   8 - filter(ROWNUM=1)
   9 - filter(TO_DATE('2012-11-01 00:00','YYYY-MM-DD HH24:MI:SS')<=TO_DATE('2012-11-08
              15:15','YYYY-MM-DD HH24:MI:SS') AND :B1='35000133'))
  11 - filter(("TDW"."RECTIME">=TO_DATE('2012-11-01 00:00','YYYY-MM-DD HH24:MI:SS') AND
              "TDW"."RECTIME"<=TO_DATE('2012-11-08 15:15','YYYY-MM-DD HH24:MI:SS')))
  12 - access("TDW"."DLVORGCODE"=:B1 AND "TDW"."MAIL_NUM"=:B2)
  14 - access("TMS"."DLV_BUREAU_ORG_CODE"="RO"."ORG_CODE")
```

首先我们排除执行计划中 Id=8 到 Id=12 会影响 SQL 性能的可能性，因为 Id=8 到 Id=12 只返回 1 行（Id=8，COUNT STOPKEY，ROWNUM=1）数据，返回 1 行数据不可能产生性能问题。执行计划的入口是 Id=5，走的是索引范围扫描，过滤条件是"EU"."DELVORGCODE"='35000133'，Id=4 是 Id=5 中索引范围扫描的回表操作，在回表的时候还进行了过滤"EU"."POSTMANKIND"<>5。如果要追求完美，我们可以将 POSTMANKIND 列放到 Id=5 中的索引中，创建组合索引。Id=5 返回的数据量较少，因此排除了 Id=5 和 Id=4 产生性能问题的可能性。现在我们将目光转移到 Id=7 和 Id=6 上面来。Id=7 走的是索引范围扫描，过滤条件是"TMS"."DLV_BUREAU_ORG_CODE"='35000133'，Id=6 是 Id=7 的索引回表操作，注意 Id=6，Operation 中出现了 GLOBAL 关键字，这说明 TB_EVT_DLV_W 是一个分区表，而且 Id=7 中的索引是全局索引。Id=6 中出现了时间过滤，一般的分区表都是根据时间字段进行分区的。于是我们询问朋友 TB_EVT_DLV_W 是不是根据 DLV_DATE 进行分区的，朋友回答是。得到朋友的肯定回答，我们就知道该 SQL 的性能问题出在何处了，问题出在 Id=7 和 Id=6 上。

我们应该将 Id=7 的全局（global）索引改成本地（local）索引。

```
create index IDX2_TB_EVT_DLV_W on TB_EVT_DLV_W(DLV_BUREAU_ORG_CODE) local;
```

改成本地索引之后，Id=6 就不会再去进行时间过滤了。相比扫描全局索引，扫描本地索引只需要到对应的索引分区中进行扫描，扫描的叶子块数量也大大减少。建立本地索引之后，SQL 多次执行都能稳定在 1 秒内。

如果过滤条件中有分区字段，一般都创建本地索引。

如果过滤条件中没有分区字段，一般都创建 global 索引，如果这时创建成 local 索引，会扫描所有的索引分区，分区数量越多，性能下降越明显。假设有 1 000 个分区，在进行索引扫描的时候会扫描 1 000 个索引分区，此时相比 global 索引，会额外多读取至少 1 000 个索引块。

假设表按月分区，一个月大概几百万行数据，但是只查询几小时的数据，数据也就几千行，这时我们需要将分区列包含在索引中，这样的索引就是有前缀的本地索引。假设表按月分区，但是查询经常按月查询或者跨月查询，这时我们就不需要将分区列包含在索引中，这样创建的本地索引就是非前缀的本地索引。

9.14 标量子查询优化案例

9.14.1 案例一

2011 年，一税务局的朋友请求优化下面 SQL。

```
select *
  from (select t.zxid,
               t.gh,
               t.xm,
               t.bm,
               t.fzjgdm,
               (select count(a.session_id)
                  from test_v a
                 where to_char(t.zxid) = a.ZCRYZH) slzl,
               (select count(a.session_id)
                  from test_v a
                 where to_char(t.zxid) = a.ZCRYZH
                   and a.myd = '0') 无评价,
               (select count(a.session_id)
                  from test_v a
                 where to_char(t.zxid) = a.ZCRYZH
                   and a.myd = '1') 满意,
               (select count(a.session_id)
                  from test_v a
                 where to_char(t.zxid) = a.ZCRYZH
                   and a.myd = '2') 较满意,
               (select count(a.session_id)
                  from test_v a
                 where to_char(t.zxid) = a.ZCRYZH
                   and a.myd = '3') 一般,
               (select count(a.session_id)
                  from test_v a
                 where to_char(t.zxid) = a.ZCRYZH
                   and a.myd = '4') 较不满意,
```

```
                 (select count(a.session_id)
                    from test_v a
                   where to_char(t.zxid) = a.ZCRYZH
                     and a.myd = '5') 不满意
          from CC_ZXJBXX t
         WHERE t.yxbz = 'Y')
where slzl <> 0;
```

该 SQL 有 7 个标量子查询，在 5.5 节中讲到，标量子查询类似嵌套循环，如果主表返回数据很多并且主表连接列基数很高，会导致子查询被多次扫描。该 SQL 竟然有 7 个标量子查询，而且每个标量子查询除了过滤条件不一样，其他都一样，显然我们可以将标量子查询等价改写为外连接，从而优化 SQL，等价改写之后的写法如下。

```
SELECT T.ZXID,
       T.GH,
       T.XM,
       T.BM,
       T.FZJGDM,
       SUM(1) SLZL,
       SUM(DECODE(A.MYD, '0', 1, 0)) 无评价,
       SUM(DECODE(A.MYD, '1', 1, 0)) 满意,
       SUM(DECODE(A.MYD, '2', 1, 0)) 较满意,
       SUM(DECODE(A.MYD, '3', 1, 0)) 一般,
       SUM(DECODE(A.MYD, '4', 1, 0)) 较不满意,
       SUM(DECODE(A.MYD, '5', 1, 0)) 不满意
  FROM CC_ZXJBXX T, test_v A
 where A.ZCRYZH = T.ZXID
   and T.YXBZ = 'Y'
 GROUP BY T.ZXID, T.GH, T.XM, T.BM, T.FZJGDM;
```

SQL 改写之后，因为两表只有关联条件，没有过滤条件，所以两表关联走 HASH 连接，test_v 也只需要被扫描一次，从而大大提升 SQL 性能。

上述 SQL 其实是一个典型的报表开发初学者在刚开始工作的时候编写的，强烈建议大家要加强 SQL 编程技能。

9.14.2 案例二

本案例发生在 2017 年，是一个比较经典的标量子查询改写优化案例。SQL 和执行计划如下。

```
SELECT A.LXR_ID,
       A.SR,
       (SELECT C.JGID || '@' || C.DLS_BM || '@' || C.DLS_MC
          FROM KHGL_DLSJBXX C, KHGL_ZJKJ ZJKJ
         WHERE C.JGID = ZJKJ.JGID
           AND EXISTS (SELECT 1
                  FROM LXR_YH YH
                 WHERE YH.KH_ID = ZJKJ.KJ_ID
                   AND YH.KHLX = '2'
                   AND YH.LXR_ID = A.LXR_ID)) AS ZJJGXX
  FROM LXR_JBXX A
 WHERE A.STATUS = '1'
   AND A.GRDM = :v1
   AND EXISTS (SELECT 1
          FROM LXR_YH YH, KHGL_GRDLXX GRDL
         WHERE YH.FZGS_DM = :v2
           AND YH.KHLX = '2'
```

```
              AND YH.LXR_ID = A.LXR_ID
              AND GRDL.GRDL_ID = YH.KH_ID
              AND GRDL.STATUS = '1')
     AND ROWNUM < 21;
Execution Plan
----------------------------------------------------------
Plan hash value: 704492369

--------------------------------------------------------------------------------------
| Id  | Operation                       | Name                | Rows  | Bytes | Cost (%CPU)| Time     |
--------------------------------------------------------------------------------------
|   0 | SELECT STATEMENT                |                     |     1 |    41 |    12   (0)| 00:00:01 |
|*  1 |  FILTER                         |                     |       |       |            |          |
|*  2 |   HASH JOIN                     |                     | 28114 | 3761K |   156   (2)| 00:00:02 |
|   3 |    TABLE ACCESS FULL            | KHGL_DLSJBXX        | 15342 | 1063K |    89   (2)| 00:00:02 |
|   4 |    TABLE ACCESS FULL            | KHGL_ZJKJ           | 28114 | 1812K |    66   (0)| 00:00:01 |
|*  5 |   INDEX RANGE SCAN              | LXR_YH_ID_IX        |     1 |    68 |     4   (0)| 00:00:01 |
|*  6 |  COUNT STOPKEY                  |                     |       |       |            |          |
|   7 |   NESTED LOOPS SEMI             |                     |     1 |    41 |    12   (0)| 00:00:01 |
|*  8 |    TABLE ACCESS BY INDEX ROWID  | LXR_JBXX            |     1 |    39 |     5   (0)| 00:00:01 |
|*  9 |     INDEX RANGE SCAN            | IDX_LXR_JBXX_GRDM   |     1 |       |     3   (0)| 00:00:01 |
|  10 |    VIEW PUSHED PREDICATE        | VW_SQ_1             |     1 |     2 |     7   (0)| 00:00:01 |
|  11 |     NESTED LOOPS                |                     |     1 |   110 |     7   (0)| 00:00:01 |
|  12 |      TABLE ACCESS BY INDEX ROWID| LXR_YH              |     1 |    75 |     5   (0)| 00:00:01 |
|* 13 |       INDEX RANGE SCAN          | IDX_KHGL_LXRYH_FZGSDM|     1 |       |     4   (0)| 00:00:01 |
|* 14 |      INDEX RANGE SCAN           | IDX_GRDLXX_XZQH_FZGS|     1 |    35 |     2   (0)| 00:00:01 |
--------------------------------------------------------------------------------------

Predicate Information (identified by operation id):
---------------------------------------------------

   1 - filter( EXISTS (SELECT 0 FROM "LXR_YH" "YH" WHERE "YH"."LXR_ID"=:B1 AND "YH"."
KHLX"='2'
              AND "YH"."KH_ID"=:B2))
   2 - access("C"."JGID"="ZJKJ"."JGID")
   5 - access("YH"."KH_ID"=:B1 AND "YH"."KHLX"='2' AND "YH"."LXR_ID"=:B2)
   6 - filter(ROWNUM<21)
   8 - filter("A"."STATUS"='1')
   9 - access("A"."GRDM"=:V1)
  13 - access("YH"."FZGS_DM"=:V2 AND "YH"."LXR_ID"="A"."LXR_ID" AND "YH"."KHLX"='2')
  14 - access("GRDL"."GRDL_ID"="YH"."KH_ID" AND "GRDL"."STATUS"='1')
              filter("GRDL"."STATUS"='1')


Statistics
----------------------------------------------------------
          1  recursive calls
          2  db block gets
     103172  consistent gets
      21144  physical reads
          0  redo size
        533  bytes sent via SQL*Net to client
        472  bytes received via SQL*Net from client
          2  SQL*Net roundtrips to/from client
          0  sorts (memory)
          0  sorts (disk)
          1  rows processed
```

该 SQL 只返回 1 行数据，但是逻辑读为 103 172，显然 SQL 还能进一步优化。从执行计划中可以看到，Id=1 是 Filter，Filter 下面有两个儿子，这属于有害的 Filter。Id=3 和 Id=4 的两个表走的是全表扫描，并且这两个表 Id 前面没有*，也就是说这两个表没有过滤条件。SQL 的

9.14 标量子查询优化案例

逻辑读绝大部分应该是由 Id=1 的 Filter,以及 Id=3 和 Id=4 这两个表贡献而来的。

Id=3 和 Id=4 这两个表来自于标量子查询。注意观察原始 SQL,在标量子查询中,Id=3 与 Id=4 这两个表与主表 LXR_JBXX 没有直接关联,主表是与标量子查询中的半连接进行关联的 (YH.LXR_ID = A.LXR_ID)。

大家还记得标量子查询的原理吗?标量子查询类似嵌套循环,主表通过连接列传值给子查询。因为本案例 SQL 比较特殊,主表是与标量子查询中的半连接的表进行关联的,主表没有直接与标量子查询中 From 后面的表进行关联,这就导致了标量子查询中 From 后面的表没能通过连接列进行传值,从而导致 Id=3 和 Id=4 的表走了全表扫描,也导致了 SQL 使用了 Filter,进而使整个 SQL 运行缓慢。

为了消除 Filter,同时也为了能使 Id=3 和 Id=4 的两个表能走索引,需要对 SQL 进行等价改写,将标量子查询中的半连接改写为内连接就能使 Id=3 和 Id=4 的两个表使用索引了。在标量子查询章节中提到过,标量子查询可以等价改写为外连接。**因为标量子查询中没有聚合函数,因此判断 Id=3 与 Id=4 两表关联之后应该是返回 1 的关系,因为如果两表关联后返回 n 的关系,SQL 会报错**。那么现在只需要考虑将标量子查询的半连接等价改写为内连接即可。因为原始的 SQL 写的是半连接,没有写成内连接,因此我们判断标量子查询中的半连接应该是属于 n 的关系,将半连接改写为内连接,如果半连接属于 n 的关系,要先将半连接变成 1 的关系。所以原始 SQL 可以等价改写为下面 SQL:

```
SELECT A.LXR_ID, A.SR, B.MSG AS ZJJGXX
  FROM LXR_JBXX A,
       (SELECT C.JGID || '@' || C.DLS_BM || '@' || C.DLS_MC AS MSG,
               YH.LXR_ID
          FROM KHGL_DLSJBXX C,
               KHGL_ZJKJ ZJKJ,
               (SELECT LXR_ID, KH_ID
                  FROM LXR_YH
                 WHERE KHLX = '2'
                 GROUP BY LXR_ID, KH_ID) YH  --对连接列分组将 n 的关系变为 1 的关系
         WHERE C.JGID = ZJKJ.JGID
           AND YH.KH_ID = ZJKJ.KJ_ID) B
 WHERE A.LXR_ID = B.LXR_ID(+)
   AND A.STATUS = '1'
   AND A.GRDM = :v1
   AND EXISTS (SELECT 1
          FROM LXR_YH YH, KHGL_GRDLXX GRDL
         WHERE YH.FZGS_DM = :v2
           AND YH.KHLX = '2'
           AND YH.LXR_ID = A.LXR_ID
           AND GRDL.GRDL_ID = YH.KH_ID
           AND GRDL.STATUS = '1')
   AND ROWNUM < 21;
```

改写之后,SQL 的执行计划如下:

```
SQL> /

Elapsed: 00:00:00.01

Execution Plan
----------------------------------------------------------
```

```
Plan hash value: 2638330795

---------------------------------------------------------------------------------------------------
| Id  | Operation                            | Name                  | Rows | Bytes | Cost (%CPU)| Time     |
---------------------------------------------------------------------------------------------------
|   0 | SELECT STATEMENT                     |                       |    1 |   124 |    87   (3)| 00:00:02 |
|*  1 |  COUNT STOPKEY                       |                       |      |       |            |          |
|   2 |   NESTED LOOPS OUTER                 |                       |    1 |   124 |    87   (3)| 00:00:02 |
|   3 |    NESTED LOOPS SEMI                 |                       |    1 |    41 |    12   (0)| 00:00:01 |
|*  4 |     TABLE ACCESS BY INDEX ROWID      | LXR_JBXX              |    1 |    39 |     5   (0)| 00:00:01 |
|*  5 |      INDEX RANGE SCAN                | IDX_LXR_JBXX_GRDM     |    1 |       |     3   (0)| 00:00:01 |
|   6 |     VIEW PUSHED PREDICATE            | VW_SQ_1               |    1 |     2 |     7   (0)| 00:00:01 |
|   7 |      NESTED LOOPS                    |                       |    1 |   110 |     7   (0)| 00:00:01 |
|   8 |       TABLE ACCESS BY INDEX ROWID    | LXR_YH                |    1 |    75 |     5   (0)| 00:00:01 |
|*  9 |        INDEX RANGE SCAN              | IDX_KHGL_LXRYH_FZGSDM |    1 |       |     4   (0)| 00:00:01 |
|* 10 |       INDEX RANGE SCAN               | IDX_GRDLXX_XZQH_FZGS  |    1 |    35 |     2   (0)| 00:00:01 |
|  11 |    VIEW PUSHED PREDICATE             |                       |    1 |    83 |    75   (3)| 00:00:01 |
|  12 |     NESTED LOOPS                     |                       |      |       |            |          |
|  13 |      NESTED LOOPS                    |                       |    1 |   203 |    75   (3)| 00:00:01 |
|* 14 |       HASH JOIN                      |                       |    1 |   132 |    74   (3)| 00:00:01 |
|  15 |        VIEW                          |                       |    1 |    66 |     7  (15)| 00:00:01 |
|  16 |         SORT GROUP BY                |                       |    1 |    68 |     7  (15)| 00:00:01 |
|* 17 |          TABLE ACCESS BY INDEX ROWID | LXR_YH                |    1 |    68 |     6   (0)| 00:00:01 |
|* 18 |           INDEX RANGE SCAN           | IDX_KHGL_LXRYH_LXRID  |    1 |       |     4   (0)| 00:00:01 |
|  19 |        TABLE ACCESS FULL             | KHGL_ZJKJ             |28114 | 1812K |    66   (0)| 00:00:01 |
|* 20 |       INDEX UNIQUE SCAN              | KHGL_DLSJBXX_PK       |    1 |       |     0   (0)| 00:00:01 |
|  21 |      TABLE ACCESS BY INDEX ROWID     | KHGL_DLSJBXX          |    1 |    71 |     1   (0)| 00:00:01 |
---------------------------------------------------------------------------------------------------

Predicate Information (identified by operation id):
---------------------------------------------------

   1 - filter(ROWNUM<21)
   4 - filter("A"."STATUS"='1')
   5 - access("A"."GRDM"=:V1)
   9 - access("YH"."FZGS_DM"=:V2 AND "YH"."LXR_ID"="A"."LXR_ID" AND "YH"."KHLX"='2')
  10 - access("GRDL"."GRDL_ID"="YH"."KH_ID" AND "GRDL"."STATUS"='1')
       filter("GRDL"."STATUS"='1')
  14 - access("YH"."KH_ID"="ZJKJ"."KJ_ID")
  17 - filter("KHLX"='2')
  18 - access("LXR_ID"="A"."LXR_ID")
  20 - access("C"."JGID"="ZJKJ"."JGID")


Statistics
----------------------------------------------------------
          0  recursive calls
          1  db block gets
        400  consistent gets
          0  physical reads
          0  redo size
        533  bytes sent via SQL*Net to client
        472  bytes received via SQL*Net from client
          2  SQL*Net roundtrips to/from client
          1  sorts (memory)
          0  sorts (disk)
          1  rows processed
```

对 SQL 进行等价改写之后，SQL 的逻辑读下降到 400，本次优化也就到此为止。

通过本案例，各位读者应该对 SQL 等价改写引起足够重视，同时也要掌握标量子查询等价改写为外连接，半连接等价改写为内连接，反连接改写为外连接等最基本的 SQL 改写技巧，另外，大家还要对表与表之间关系引起足够重视。

9.15 关联更新优化案例

本案例发生在 2011 年，当时作者罗老师在惠普担任开发 DBA，支撑宝洁公司的数据仓库项目。为了避免泄露信息，他对 SQL 语句做了适当修改。ETL 开发人员发来邮件问能不能想办法提升一下下面 UPDATE 语句性能，该 UPDATE 执行了 30 分钟还没执行完毕，SQL 语句如下。

```
UPDATE OPT_ACCT_FDIM A
   SET ACCT_SKID = (SELECT ACCT_SKID
                      FROM OPT_ACCT_FDIM_BKP B
                     WHERE A.ACCT_ID = B.ACCT_ID);
```

OPT_ACCT_FDIM 有 226 474 行数据，OPT_ACCT_FDIM_BKP 有 227 817 行数据。UPDATE 后面跟子查询类似嵌套循环，它的算法与标量子查询，Filter 一模一样。也就是说 OPT_ACCT_FDIM 表相当于嵌套循环的驱动表，OPT_ACCT_FDIM_BKP 相当于嵌套循环的被驱动表，那么这里表 OPT_ACCT_FDIM_BKP 就会被扫描 20 多万次。OPT_ACCT_FDIM_BKP 是通过 CTAS 创建的备份表，用来备份 OPT_ACCT_FDIM 表的数据。嵌套循环被驱动表应该走索引，但是 OPT_ACCT_FDIM_BKP 是通过 CTAS 创建的，仅仅用于备份，该表上面没有任何索引，这就是说 OPT_ACCT_FDIM_BKP 要被扫描 20 多万次，而且每次都是全表扫描，这就是为什么 UPDATE 执行了 30 分钟还没执行完毕。我们可以创建一个索引（ACCT_ID，ACCT_SKID）从而避免 OPT_ACCT_FDIM_BKP 每次被全表扫描，虽然这种方法能优化该 SQL，但是此时索引会被扫描 20 多万次。如果要更新的表有几千万行甚至上亿行数据，显然不能通过创建索引的方法来优化 SQL。考虑到 ETL 开发人员后续还有类似需求，笔者决定采用存储过程并且利用 ROWID 对关联更新进行优化。存储过程代码如下。

```
SQL> DECLARE
  2     CURSOR CUR_B IS
  3       SELECT
  4        B.ACCT_ID, B.ACCT_SKID, A.ROWID ROW_ID
  5        FROM OPT_ACCT_DIM A, OPT_ACCT_DIM_BKP B
  6       WHERE A.ACCT_ID = B.ACCT_ID
  7       ORDER BY A.ROWID;
  8     V_COUNTER NUMBER;
  9  BEGIN
 10     V_COUNTER := 0;
 11     FOR ROW_B IN CUR_B LOOP
 12       UPDATE OPT_ACCT_DIM
 13          SET ACCT_SKID = ROW_B.ACCT_SKID
 14        WHERE ROWID = ROW_B.ROW_ID;
 15       V_COUNTER := V_COUNTER + 1;
 16       IF (V_COUNTER >= 1000) THEN
 17         COMMIT;
 18         V_COUNTER := 0;
 19       END IF;
 20     END LOOP;
 21     COMMIT;
 22  END;
 23  /
```

```
PL/SQL procedure successfully completed.

Elapsed: 00:01:21.58
```

将关联更新改写成存储过程,利用 ROWID 进行更新只需要 1 分 22 秒就可执行完毕。当时并没有采用批量游标方式进行更新,如果采用批量游标,速度更快。以下是批量游标的 PLSQL 代码。

```
declare
  maxrows          number default 100000;
  rowid_table      dbms_sql.urowid_table;
  acct_skid_table dbms_sql.Number_Table;
  cursor cur_update is
    SELECT B.ACCT_SKID, A.ROWID ROW_ID
      FROM OPT_ACCT_DIM A, OPT_ACCT_DIM_BKP B
      WHERE A.ACCT_ID = B.ACCT_ID
      ORDER BY A.ROWID;
begin
  open cur_update;
  loop
    EXIT WHEN cur_update%NOTFOUND;
    FETCH cur_update bulk collect
      into acct_skid_table, rowid_table limit maxrows;
    forall i in 1 .. rowid_table.count
      update OPT_ACCT_DIM
         set acct_skid = acct_skid_table(i)
       where rowid = rowid_table(i);
    commit;
  end loop;
  close cur_update;
end;
/
```

细心的读者会发现,在游标定义中,我们对要更新的表根据 ROWID 进行了排序操作,这是为什么呢? 同一个块中 ROWID 是连续的,物理上连续的块组成了区,那么同一个区里面 ROWID 也是连续的。**对 ROWID 进行排序是为了保证在更新表的时候,被更新的块尽量不被刷出 buffer cache,从而减少物理 I/O。**假设要被更新的表有 20GB,数据库的 buffer cache 只有 10GB,这时 buffer cache 不能完全容纳要被更新的表,有部分块会被挤压出 buffer cache。这时如果不对 ROWID 进行排序,被更新的块有可能会被反复读入 buffer cache,然后挤压出 buffer cache,然后重复读入、挤压,此时会引发大量的 I/O 读写操作。假设一个块存储 200 行数据,最极端的情况就是每个块要被读入/写出到磁盘 200 次,这样读取的表就不是 20GB,而是(200×20)GB。如果对 ROWID 进行排序,这样就能保证一个块只需被读入 buffer cache 一次,这样就避免了大量的 I/O 读写操作。有读者会问,排序不也耗费资源吗? 这时排序耗费的资源远远低于数据块被反复挤压出 buffer cache 所耗费的开销。如果要被更新的表很小,buffer cache 能完全容纳下要被更新的表,这时就不要对 ROWID 进行排序了,因为 buffer cache 很大,块不会被挤压出 buffer cache,此时对 ROWID 排序反而会影响性能。大家以后遇到类似需求,要先比较被更新的表与 buffer cache 大小,同时也要考虑数据库繁忙程度、buffer cache 还剩余多少空闲块等一系列因素。

下面实验验证如果不对 ROWID 排序,块有可能被反复扫描的观点。

我们先创建两个表，分别取名为 a，b，为了模拟实际情况，将 a，b 中数据随机打乱存储。

```
create table a as select * from dba_objects order by dbms_random.value;
create table b as select * from dba_objects order by dbms_random.value;
```

查看返回结果如下。

```
SQL> select owner,rid as "ROWID",block#
  2    from (SELECT B.owner,
  3                 A.ROWID rid,
  4                 dbms_rowid.rowid_block_number(A.rowid) block#
  5            FROM A, B
  6           WHERE A.object_id = B.object_id)
  7   where rownum <= 10;

OWNER           ROWID              BLOCK#
--------------- ------------------ ----------
PUBLIC          AAAS+CAAEAACEPdAAs  541661
PUBLIC          AAAS+CAAEAACEp2AAP  543350
SYS             AAAS+CAAEAACEgFAAJ  542725
SYS             AAAS+CAAEAACEu9AAc  543677
MDSYS           AAAS+CAAEAACEknAAi  543015
SYS             AAAS+CAAEAACEutAA9  543661
SYS             AAAS+CAAEAACEhRAA4  542801
SYSMAN          AAAS+CAAEAACEvzAAC  543731
PUBLIC          AAAS+CAAEAACElBAAj  543041
PUBLIC          AAAS+CAAEAACEwUAAy  543764
```

从 SQL 查询结果中我们可以看到，返回的数据是无序的。如果关联的两个表连接列本身是有序递增的，比如序列值、时间，这时两表关联返回的结果是部分有序的，可以不用排序，在实际工作中，要具体情况具体分析。

本案例也可以采用 MERGE INTO 对 UPDATE 子查询进行等价改写。

```
merge into OPT_ACCT_FDIM A
 using OPT_ACCT_FDIM_BKP B
on (A.ACCT_ID = B.ACCT_ID)
  when mached then update set a.ACCT_SKID = B.ACCT_SKID;
```

MERGE INTO 可以自由控制走嵌套循环或者走 HASH 连接，而且 MERGE INTO 可以开启并行 DML、并行查询，而采用 PLSQL 更新不能开启并行，所以 MERGE INTO 在速度上有优势。PLSQL 更新可以批量提交，对 UNDO 占用小，而 MERGE INTO 要等提交的时候才会释放 UNDO。采用 PLSQL 更新不需要担心进程突然断开连接，MERGE INTO 更新如果进程断开连接会导致 UNDO 很难释放。所以，如果追求更新速度且被更新的表并发量很小，可以考虑采用 MERGE INTO，如果追求安全、平稳，可以采用 PLSQL 更新。

9.16 外连接有 OR 关联条件只能走 NL

下面 SQL 有 OR 关联条件。

```
SELECT A.CONTRACT_ID, B.BORROWER_ID
  FROM blfct.bl_rtl_con_overdue_fact A
  LEFT JOIN BLpub.Bl_Contract_Dim B ON A.DEALER_ID = B.DEALER_ID
```

```
                       OR A.OVERDUE_DD = B.Overdue_Dd
 WHERE A.ETL_DATE BETWEEN DATE '2016-12-19' AND DATE '2016-12-20';
```

执行计划如下。

```
Plan hash value: 121649910

---------------------------------------------------------------------------------
| Id  | Operation           | Name                    | Rows  | Bytes | Cost (%CPU)|
---------------------------------------------------------------------------------
|  0  | SELECT STATEMENT    |                         |  163M |  5469M|  4421M  (1)|
|  1  |  NESTED LOOPS OUTER |                         |  163M |  5469M|  4421M  (1)|
|* 2  |   TABLE ACCESS FULL | BL_RTL_CON_OVERDUE_FACT |  181K |  3898K|  2192K  (2)|
|  3  |   VIEW              |                         |   903 | 11739 | 24354   (1)|
|* 4  |    TABLE ACCESS FULL| BL_CONTRACT_DIM         |   903 | 12642 | 24354   (1)|
---------------------------------------------------------------------------------

Predicate Information (identified by operation id):
---------------------------------------------------

   2 - filter("A"."ETL_DATE">=TO_DATE(' 2016-12-19 00:00:00', 'syyyy-mm-dd
              hh24:mi:ss') AND "A"."ETL_DATE"<=TO_DATE(' 2016-12-20 00:00:00', 'syyyy-mm-dd
              hh24:mi:ss'))
   4 - filter("A"."OVERDUE_DD"="B"."OVERDUE_DD" OR "A"."DEALER_ID"="B"."DEALER_ID")
```

从执行计划中看到，两表走的是嵌套循环。当两表用外连接进行关联，关联条件中有 OR 关联条件，那么这时只能走嵌套循环，而且驱动表固定为主表，此时不能走 HASH 连接，即使通过 HINT: USE_HASH 也无法修改执行计划。如果主表数据量很大，那么这时就会出现严重性能问题。我们可以将外连接的 OR 关联/过滤条件放到查询中，用 case when 进行过滤，从而让 SQL 可以走 HASH 连接。

```
EXPLAIN PLAN FOR
SELECT A.CONTRACT_ID,
       case
          when A.DEALER_ID = B.DEALER_ID OR A.OVERDUE_DD = B.Overdue_Dd then
            B.BORROWER_ID
          end
  FROM blfct.bl_rtl_con_overdue_fact A
  LEFT JOIN BLpub.Bl_Contract_Dim B ON A.DEALER_ID = B.DEALER_ID
 WHERE A.ETL_DATE BETWEEN DATE '2016-12-19' AND DATE '2016-12-20';
```

执行计划如下。

```
select * from table(dbms_xplan.display());

Plan hash value: 3927476067

------------------------------------------------------------------------------------------
| Id |Operation           | Name                    |Rows | Bytes |TempSpc|Cost(%CPU)|
------------------------------------------------------------------------------------------
|  0 |SELECT STATEMENT    |                         | 57M | 1965M |       | 2218K (2)|
|* 1 | HASH JOIN OUTER    |                         | 57M | 1965M |  6032K| 2218K (2)|
|* 2 |  TABLE ACCESS FULL | BL_RTL_CON_OVERDUE_FACT | 181K| 3898K |       | 2192K (2)|
|  3 |  TABLE ACCESS FULL | BL_CONTRACT_DIM         | 640K| 8763K |       |24349  (1)|
------------------------------------------------------------------------------------------

Predicate Information (identified by operation id):
---------------------------------------------------
```

9.16 外连接有 OR 关联条件只能走 NL

```
     1 - access("A"."DEALER_ID"="B"."DEALER_ID"(+))
     2 - filter("A"."ETL_DATE">=TO_DATE(' 2016-12-19 00:00:00', 'syyyy-mm-dd hh24:mi:
ss') AND
              "A"."ETL_DATE"<=TO_DATE(' 2016-12-20 00:00:00', 'syyyy-mm-dd hh24:mi:
ss'))
```

利用 case when 改写外连接 OR 连接条件有个限制：从表只能是 1 的关系，不能是 n 的关系，从表要展示多少个列，就要写多少个 case when。我们利用 EMP 与 DEPT 进行讲解。EMP 与 DEPT 是 n∶1 关系，现有如下 SQL。

```
select e.*, d.deptno deptno2, d.loc
  from scott.emp e
  left join scott.dept d on d.deptno = e.deptno
                        and (d.deptno >= e.sal and e.sal < 1000 or
                             e.ename like '%O%');
```

执行计划如下。

```
--------------------------------------------------------
Plan hash value: 2962868874

---------------------------------------------------------------------------------------
| Id  | Operation                    | Name    | Rows | Bytes | Cost(%CPU)| Time     |
---------------------------------------------------------------------------------------
|   0 | SELECT STATEMENT             |         |   14 |   826 |    17   (0)| 00:00:01|
|   1 |  NESTED LOOPS OUTER          |         |   14 |   826 |    17   (0)| 00:00:01|
|   2 |   TABLE ACCESS FULL          | EMP     |   14 |   532 |     3   (0)| 00:00:01|
|   3 |   VIEW                       |         |    1 |    21 |     1   (0)| 00:00:01|
|*  4 |    TABLE ACCESS BY INDEX ROWID| DEPT   |    1 |    11 |     1   (0)| 00:00:01|
|*  5 |     INDEX UNIQUE SCAN        | PK_DEPT |    1 |       |     0   (0)| 00:00:01|
---------------------------------------------------------------------------------------

Predicate Information (identified by operation id):
---------------------------------------------------

   4 - filter("E"."ENAME" IS NOT NULL AND "E"."ENAME" IS NOT NULL AND
              "E"."ENAME" LIKE '%O%' OR "D"."DEPTNO">="E"."SAL" AND "E"."SAL"<1000)
   5 - access("D"."DEPTNO"="E"."DEPTNO")
```

执行计划中两表关联走的是嵌套循环，驱动表是主表 EMP。现在我们添加 HINT：USE_HASH 尝试改变表连接方式。

```
SQL> select /*+ use_hash(e,d) */
  2  e.*, d.deptno deptno2, d.loc
  3    from scott.emp e
  4    left join scott.dept d on d.deptno = e.deptno
  5                          and (d.deptno >= e.sal and e.sal < 1000 or
  6                               e.ename like '%O%');

14 rows selected.

Execution Plan
----------------------------------------------------------
Plan hash value: 2962868874

---------------------------------------------------------------------------------------
| Id  | Operation                    | Name    | Rows | Bytes | Cost(%CPU)| Time     |
---------------------------------------------------------------------------------------
|   0 | SELECT STATEMENT             |         |   14 |   826 |    17   (0)| 00:00:01|
|   1 |  NESTED LOOPS OUTER          |         |   14 |   826 |    17   (0)| 00:00:01|
```

```
|   2 |   TABLE ACCESS FULL          | EMP     |    14 |   532 |     3   (0)|00:00:01|
|   3 |   VIEW                       |         |     1 |    21 |     1   (0)|00:00:01|
|*  4 |    TABLE ACCESS BY INDEX ROWID| DEPT   |     1 |    11 |     1   (0)|00:00:01|
|*  5 |     INDEX UNIQUE SCAN        | PK_DEPT |     1 |       |     0   (0)|00:00:01|
---------------------------------------------------------------------------------------

Predicate Information (identified by operation id):
---------------------------------------------------

   4 - filter("E"."ENAME" IS NOT NULL AND "E"."ENAME" IS NOT NULL AND
              "E"."ENAME" LIKE '%O%' OR "D"."DEPTNO">="E"."SAL" AND "E"."SAL"<1000)
   5 - access("D"."DEPTNO"="E"."DEPTNO")
```

添加 HINT 无法更改执行计划。因为 SQL 语句中从表 DEPT 属于 1 的关系，从表 DEPT 要展示两个列，需要对应写上两个 case when。改写的 SQL 如下。

```
select e.*,
       case
         when (d.deptno >= e.sal and e.sal < 1000 or e.ename like '%O%') then
           d.deptno
       end deptno2,
       case
         when (d.deptno >= e.sal and e.sal < 1000 or e.ename like '%O%') then
           d.loc
       end loc
  from scott.emp e
  left join scott.dept d on d.deptno = e.deptno;
```

改写后的执行计划如下。

```
SQL> select e.*,
  2         case
  3           when (d.deptno >= e.sal and e.sal < 1000 or e.ename like '%O%') then
  4             d.deptno
  5         end deptno2,
  6         case
  7           when (d.deptno >= e.sal and e.sal < 1000 or e.ename like '%O%') then
  8             d.loc
  9         end loc
 10    from scott.emp e
 11    left join scott.dept d on d.deptno = e.deptno;

14 rows selected.

Execution Plan
----------------------------------------------------------
Plan hash value: 3387915970

---------------------------------------------------------------------------
| Id  | Operation          | Name | Rows  | Bytes | Cost (%CPU)| Time     |
---------------------------------------------------------------------------
|   0 | SELECT STATEMENT   |      |    14 |   686 |     7  (15)| 00:00:01 |
|*  1 |  HASH JOIN OUTER   |      |    14 |   686 |     7  (15)| 00:00:01 |
|   2 |   TABLE ACCESS FULL| EMP  |    14 |   532 |     3   (0)| 00:00:01 |
|   3 |   TABLE ACCESS FULL| DEPT |     4 |    44 |     3   (0)| 00:00:01 |
---------------------------------------------------------------------------

Predicate Information (identified by operation id):
---------------------------------------------------
```

```
1 - access("D"."DEPTNO"(+)="E"."DEPTNO")
```

用 case when 改写之后，两表自动走了 HASH 连接。

如果主表属于 1 的关系，从表属于 n 的关系，我们就不能用 case when 进行等价改写，例子如下。

```
select d.*, e.deptno deptno2, e.ename, e.sal
  from dept d
  left join emp e on d.deptno = e.deptno
                 and (d.deptno >= e.sal and e.sal < 1000 or
                      e.ename like '%O%');
```

SQL 中 DEPT 是主表，EMP 是从表，DEPT 与 EMP 是 1：n 的关系，此时不能将 SQL 改写为如下写法。

```
select d.*,
       case
         when (d.deptno >= e.sal and e.sal < 1000 or e.ename like '%O%') then
          e.deptno
       end deptno2,
       case
         when (d.deptno >= e.sal and e.sal < 1000 or e.ename like '%O%') then
          e.ename
       end ename,
       case
         when (d.deptno >= e.sal and e.sal < 1000 or e.ename like '%O%') then
          e.sal
       end sal
  from dept d
  left join emp e on d.deptno = e.deptno;
```

我们可以将 SQL 改写为如下写法。

```
select b.*, a.deptno, a.ename, a.sal
  from dept b
  left join (select d.deptno, e.ename, e.sal
               from dept d, emp e
              where d.deptno = e.deptno
                and (d.deptno >= e.sal and e.sal < 1000 or
                     e.ename like '%O%')) a on b.deptno = a.deptno;
```

如果两表是 n：n 关系，这时就无法对 SQL 进行改写了，在日常工作中一般也遇不到 n：n 关系。

9.17 把你脑袋当 CBO

2012 年，一位女性朋友 DBA 请求协助优化如下 SQL。

```
SELECT "A1"."CODE", "A1"."DEVICE_ID", "A1"."SIDEB_PORT_ID", "A1"."VERSION"
  FROM (SELECT
         "A2"."CODE" "CODE",
         "A2"."DEVICE_ID" "DEVICE_ID",
         "A2"."SIDEB_PORT_ID" "SIDEB_PORT_ID",
         "A3"."VERSION" "VERSION",
         ROW_NUMBER() OVER(PARTITION BY "A4"."PROD_ID" ORDER BY "A4"."HIST_TIME" DESC
) "RN"
```

```
              FROM "RM"."H_PROD_2_RF_SERV"              "A4",
                   "RM"."H_RSC_FACING_SERV_LINE_ITEM" "A3",
                   "RM"."CONNECTOR"                    "A2"
             WHERE "A4"."SERV_ID" = "A3"."SERV_ID"
               AND "A3"."LINE_ID" = "A2"."CONNECTOR_ID"
               AND EXISTS (SELECT 0
                    FROM "RM"."DEVICE_ITEM" "A5"
                   WHERE "A5"."DEVICE_ID" = "A2"."DEVICE_ID"
                     AND "A5"."ITEM_SPEC_ID" = 200006
                     AND "A5"."VALUE" ='7')
               AND "A4"."PROD_ID" = 313) "A1"
 WHERE "A1"."RN" = 1;
```

执行计划如下。

```
---------------------------------------------------------------------------------------------
| Id  |Operation                         |Name                        |Rows|Bytes| Cost (%CPU)|
---------------------------------------------------------------------------------------------
|   0 |SELECT STATEMENT                  |                            |  1 | 175 |  20  (10)|
|*  1 | VIEW                             |                            |  1 | 175 |  20  (10)|
|*  2 |  WINDOW SORT PUSHED RANK         |                            |  1 | 109 |  20  (10)|
|   3 |   NESTED LOOPS                   |                            |  1 | 109 |  19   (6)|
|   4 |    NESTED LOOPS                  |                            |  1 |  80 |  17   (6)|
|   5 |     MERGE JOIN CARTESIAN         |                            |  1 |  60 |  13   (8)|
|   6 |      SORT UNIQUE                 |                            |  1 |  36 |   6   (0)|
|*  7 |       TABLE ACCESS BY INDEX ROWID|DEVICE_ITEM                 |  1 |  36 |   6   (0)|
|*  8 |        INDEX RANGE SCAN          |IDX_DEVICE_ITEM_VALE        |  9 |     |   4   (0)|
|   9 |      BUFFER SORT                 |                            |  4 |  96 |   7  (15)|
|  10 |       TABLE ACCESS BY INDEX ROWID|H_PROD_2_RF_SERV            |  4 |  96 |   6   (0)|
|* 11 |        INDEX RANGE SCAN          |IDX_HP2RS_PRODID_SERVID     |  4 |     |   2   (0)|
|  12 |     TABLE ACCESS BY INDEX ROWID  |H_RSC_FACING_SERV_LINE_ITEM |  2 |  40 |   4   (0)|
|* 13 |      INDEX RANGE SCAN            |IDX_HRFSLI_SERV             |  2 |     |   2   (0)|
|* 14 |    TABLE ACCESS BY INDEX ROWID   |CONNECTOR                   |  1 |  29 |   2   (0)|
|* 15 |     INDEX UNIQUE SCAN            |PK_CONNECTOR                |  1 |     |   1   (0)|
---------------------------------------------------------------------------------------------

Predicate Information (identified by operation id):
---------------------------------------------------

   1 - filter("A1"."RN"=1)
   2 - filter(ROW_NUMBER() OVER ( PARTITION BY "A4"."PROD_ID" ORDER BY
              INTERNAL_FUNCTION("A4"."HIST_TIME") DESC )<=1)
   7 - filter("A5"."ITEM_SPEC_ID"=200006)
   8 - access("A5"."VALUE"='7')
  11 - access("A4"."PROD_ID"=313)
  13 - access("A4"."SERV_ID"="A3"."SERV_ID")
  14 - filter("A5"."DEVICE_ID"="A2"."DEVICE_ID")
  15 - access("A3"."LINE_ID"="A2"."CONNECTOR_ID")


Statistics
----------------------------------------------------------
              0  recursive calls
              0  db block gets
        2539920  consistent gets
              0  physical reads
              0  redo size
            735  bytes sent via SQL*Net to client
            492  bytes received via SQL*Net from client
              2  SQL*Net roundtrips to/from client
              3  sorts (memory)
              0  sorts (disk)
              1  rows processed
```

9.17 把你脑袋当 CBO

该 SQL 要执行 9.437 秒，只返回一行数据，其中 A5 有 48 194 511 行数据，A2 有 35 467 304 行数据，其余表都是小表。

首先，笔者运用 SQL 三段分拆方法，检查 SQL 写法，经过检查，SQL 写法没有问题。

其次笔者检查执行计划。执行计划中 Id=5 出现了 MERGE JOIN CARTESIAN，这一般都是统计信息收集不准确，将离 MERGE JOIN CARTESIAN 关键字最近的表(Id=7)Rows 估算为 1 导致。

正常情况下，应该先检查 SQL 中所有表的统计信息是否过期，如果统计信息过期了应该立即收集。因为做了太多的 SQL 优化，遇到 SQL 出现了性能问题，已经形成条件反射想要立刻优化它，所以，当时没有立即对表收集统计信息。

如果想要从执行计划入手优化 SQL，我们一般要从执行计划的入口开始检查，检查 Rows 估算是否准确。当然了，如果执行计划中有明显值得怀疑的地方，我们也可以直接检查值得怀疑之处。

执行计划的入口是 Id=8，Id=8 是索引范围扫描，通过 Id=7 回表。于是让朋友运行如下 SQL。

```
SELECT COUNT(*)
  FROM "RM"."DEVICE_ITEM" "A5"
 WHERE "A5"."ITEM_SPEC_ID" = 200006
   AND "A5"."VALUE" = '7';
```

得到反馈，上面查询返回 68 384 行数据。其次，查询执行计划中 Id=11 和 Id=10 应该返回多少数据（A4），运行如下 SQL。

```
select count(*) from H_PROD_2_RF_SERV where prod_id = 313;
```

得到反馈，上面查询返回 6 行数据。根据以上信息我们知道应该怎么优化上述 SQL 了。我们再来查看原始 SQL 的部分代码。

```
FROM "RM"."H_PROD_2_RF_SERV"             "A4",
     "RM"."H_RSC_FACING_SERV_LINE_ITEM"  "A3",
     "RM"."CONNECTOR"                    "A2"
WHERE "A4"."SERV_ID" = "A3"."SERV_ID"
  AND "A3"."LINE_ID" = "A2"."CONNECTOR_ID"
  AND EXISTS (SELECT 0
                FROM "RM"."DEVICE_ITEM" "A5"
               WHERE "A5"."DEVICE_ID" = "A2"."DEVICE_ID"
                 AND "A5"."ITEM_SPEC_ID" = 200006
                 AND "A5"."VALUE" ='7')
  AND "A4"."PROD_ID" = 313)
```

A4 过滤后只返回 6 行数据，A3 是小表，A2 有 35 467 304 行数据，A5 过滤后返回 6 万行数据，其中 A3，A2 都没有过滤条件。

SQL 语句中 A4 与 A3 进行关联，因为 A4 过滤后返回 6 行数据，A3 是小表，所以让 A4 作为驱动表 leading(a4)，与 A3 使用嵌套循环 use_nl(a4,a3)方式进行关联，关联之后得到一个结果集，因为 A4 与 A3 返回数据量都很小，所以关联之后的结果集也必然很小。

因为 A2 表很大，而且 A2 没有过滤条件，所以我们不能让 A2 走 HASH 连接，因为没有过滤条件，使用 HASH 进行关联只能走全表扫描。如果让 A2 走嵌套循环，作为嵌套循环被驱动表，那么我们可以让 A2 走连接列的索引，这样就避免了大表 A2 因为没有过滤条件而走全表扫描。因此，我们将 A4 与 A3 关联之后的结果集作为嵌套循环驱动表，然后再与 A2 使用嵌套循环进行关联：use_nl(a3,a2)。

因为 A5 过滤后有 6 万行数据，所以我们让 A5 与 A2 进行 HASH 连接，最终添加如下 HINT。

```
SELECT "A1"."CODE", "A1"."DEVICE_ID", "A1"."SIDEB_PORT_ID", "A1"."VERSION"
    FROM (SELECT /*+ leading(a4) use_nl(a4,a3) use_nl(a3,a2) */
            "A2"."CODE" "CODE",
            "A2"."DEVICE_ID" "DEVICE_ID",
            "A2"."SIDEB_PORT_ID" "SIDEB_PORT_ID",
            "A3"."VERSION" "VERSION",
            ROW_NUMBER() OVER(PARTITION BY "A4"."PROD_ID" ORDER BY "A4"."HIST_TIME" DESC) "RN"
            FROM "RM"."H_PROD_2_RF_SERV"              "A4",
                 "RM"."H_RSC_FACING_SERV_LINE_ITEM" "A3",
                 "RM"."CONNECTOR"                   "A2"
            WHERE "A4"."SERV_ID" = "A3"."SERV_ID"
              AND "A3"."LINE_ID" = "A2"."CONNECTOR_ID"
              AND EXISTS (SELECT /*+ hash_sj */ 0
                    FROM "RM"."DEVICE_ITEM" "A5"
                    WHERE "A5"."DEVICE_ID" = "A2"."DEVICE_ID"
                      AND "A5"."ITEM_SPEC_ID" = 200006
                      AND "A5"."VALUE" = '7')
              AND "A4"."PROD_ID" = 313) "A1"
    WHERE "A1"."RN" = 1;
```

执行计划如下。

```
--------------------------------------------------------------------------------------------------
| Id|Operation                        |Name                       |Rows|Bytes|Cost(%CPU)|
--------------------------------------------------------------------------------------------------
|  0|SELECT STATEMENT                 |                           |   1|  175|   40   (3)|
|* 1| VIEW                            |                           |   1|  175|   40   (3)|
|* 2|  WINDOW SORT PUSHED RANK        |                           |   1|  109|   40   (3)|
|* 3|   HASH JOIN SEMI                |                           |   1|  109|   39   (0)|
|  4|    NESTED LOOPS                 |                           |   7|  511|   33   (0)|
|  5|     NESTED LOOPS                |                           |   7|  308|   19   (0)|
|  6|      TABLE ACCESS BY INDEX ROWID|H_PROD_2_RF_SERV           |   4|   96|    7   (0)|
|* 7|       INDEX RANGE SCAN          |IDX_HP2RS_PRODID_SERVID    |   4|     |    3   (0)|
|  8|      TABLE ACCESS BY INDEX ROWID|H_RSC_FACING_SERV_LINE_ITEM|   2|   40|    4   (0)|
|* 9|       INDEX RANGE SCAN          |IDX_HRFSLI_SERV            |   2|     |    2   (0)|
| 10|     TABLE ACCESS BY INDEX ROWID |CONNECTOR                  |   1|   29|    2   (0)|
|*11|      INDEX UNIQUE SCAN          |PK_CONNECTOR               |   1|     |    1   (0)|
|*12|    TABLE ACCESS BY INDEX ROWID  |DEVICE_ITEM                |   1|   36|    6   (0)|
|*13|     INDEX RANGE SCAN            |IDX_DEVICE_ITEM_VALE       |   9|     |    4   (0)|
--------------------------------------------------------------------------------------------------

Predicate Information (identified by operation id):
---------------------------------------------------

   1 - filter("A1"."RN"=1)
   2 - filter(ROW_NUMBER() OVER ( PARTITION BY "A4"."PROD_ID" ORDER BY
              INTERNAL_FUNCTION("A4"."HIST_TIME") DESC )<=1)
   3 - access("A5"."DEVICE_ID"="A2"."DEVICE_ID")
   7 - access("A4"."PROD_ID"=313)
   9 - access("A4"."SERV_ID"="A3"."SERV_ID")
```

```
   11 - access("A3"."LINE_ID"="A2"."CONNECTOR_ID")
   12 - filter("A5"."ITEM_SPEC_ID"=200006)
   13 - access("A5"."VALUE"='7')

Statistics
----------------------------------------------------------
          0  recursive calls
          0  db block gets
      14770  consistent gets
          0  physical reads
          0  redo size
        735  bytes sent via SQL*Net to client
        492  bytes received via SQL*Net from client
          2  SQL*Net roundtrips to/from client
          1  sorts (memory)
          0  sorts (disk)
          1  rows processed
```

最终该 SQL 只需 0.188 秒就能出结果,逻辑读从最开始的 2 539 920 下降到 14 770。

当具备一定优化理论知识之后,我们可以不看执行计划,直接根据 SQL 写法找到 SQL 语句中返回数据量最小的表作为驱动表,然后看它与谁进行关联,根据关联返回的数据量判断走 NL 还是 HASH,然后一直这样进行下去,直到 SQL 语句中所有表都关联完毕。如果大家长期采用此方法进行锻炼,久而久之,你自己的脑袋就是 CBO。

9.18 扩展统计信息优化案例

本案例发生在 2011 年,当时作者罗老师在惠普担任开发 DBA,支撑宝洁公司的数据仓库项目。为了避免泄露信息,他对 SQL 语句做了适当修改。Obiee 终端用户发来邮件说某报表执行了 30 分钟还不出结果,请求协助。通过与 Obiee 开发人员合作,找到报表 SQL 语句如下。

```
select sum(T2083114.MANUL_COST_OVRRD_AMT) as c1,
sum(nvl(T2083114.REVSD_VAR_ESTMT_COST_AMT , 0)) as c2,
T2084525.ACCT_LONG_NAME as c3,
T2084525.NAME as c4,
T2083424.PRMTN_NAME as c5,
T2083424.PRMTN_ID as c6,
case  when case  when T2083424.CORP_PRMTN_TYPE_CODE = 'Target Account'
then 'Corporate' else T2083424.CORP_PRMTN_TYPE_CODE end   is null
then 'Private' else case   when T2083424.CORP_PRMTN_TYPE_CODE = 'Target Account'
then 'Corporate' else T2083424.CORP_PRMTN_TYPE_CODE end   end   as c7,
T2083424.PRMTN_STTUS_CODE as c8,
T2083424.APPRV_BY_DESC as c9,
T2083424.APPRV_STTUS_CODE as c10,
T2083424.AUTO_UPDT_GTIN_IND as c11,
T2083424.CREAT_DATE as c12,
T2083424.PGM_START_DATE as c13,
T2083424.PGM_END_DATE as c14,
nvl(case   when T2083424.PRMTN_STTUS_CODE = 'Confirmed'
then cast(( TRUNC( TO_DATE('2011-06-07' , 'YYYY-MM-DD') ) - TRUNC( T2083424.PGM_END_D
ATE ) ) as  VARCHAR ( 10 ) ) end  , '') as c15,
T2083424.PRMTN_STOP_DATE as c16,
T2083424.SHPMT_START_DATE as c17,
T2083424.SHPMT_END_DATE as c18,
```

```
        T2083424.CNBLN_WK_CNT as c19,
        T2083424.ACTVY_DETL_POP as c20,
        T2083424.CMMNT_DESC as c21,
        T2083424.PRMTN_AVG_POP as c22,
        T2084525.CHANL_TYPE_DESC as c23,
        T2083424.PRMTN_SKID as c24
        from
        OPT_ACCT_FDIM T2084525 /* OPT_ACCT_PRMTN_FDIM */ ,
        OPT_BUS_UNIT_FDIM T2083056,
        OPT_CAL_MASTR_DIM T2083357 /* OPT_CAL_MASTR_DIM01 */ ,
        OPT_PRMTN_FDIM T2083424,
        OPT_ACTVY_FCT T2083114
        where  (T2083056.BUS_UNIT_SKID = T2083114.BUS_UNIT_SKID and T2083114.BUS_UNIT_SKID =
        T2084525.BUS_UNIT_SKID
        and T2083114.DATE_SKID = T2083357.CAL_MASTR_SKID and T2083114.BUS_UNIT_SKID = T208342
        4.BUS_UNIT_SKID
        and T2083114.PRMTN_SKID = T2083424.PRMTN_SKID and T2083056.BUS_UNIT_NAME = 'Chile'
        and T2083114.ACCT_PRMTN_SKID = T2084525.ACCT_SKID and T2083357.FISC_YR_ABBR_NAME = 'F
        Y10/11'
        and T2084525.ACCT_LONG_NAME is not null and (case  when T2083424.CORP_PRMTN_TYPE_CODE
         = 'Target Account'
        then 'Corporate' else T2083424.CORP_PRMTN_TYPE_CODE end  in ('Alternate BDF', 'Corpor
        ate', 'Private'))
        and (T2084525.ACCT_LONG_NAME in ('ADELCO - CHILE - 0066009018', 'ALIMENTOS FRUNA - CH
        ILE - 0066009049',
        'CENCOSUD - CHILE - 0066009007', 'COMERCIAL ALVI - CHILE - 0066009070', 'D&S - CHILE
        - 0066009008',
        'DIPAC - CHILE - 0066009024', 'DIST. COMERCIAL - CHILE - 0066009087', 'DISTRIBUCION L
        AGOS S.A. - CHILE - 2001146505',
        'ECOMMERCE ESCALA 1 - 1900001746', 'EMILIO SANDOVAL - CHILE - 2000402293', 'F. AHUMAD
        A - CHILE - 0066009023',
        'FALABELLA - CHILE - 2000406971', 'FRANCISCO LEYTON - CHILE - 0066009142', 'MAICAO -
        CHILE - 0066009135',
        'MARGARITA UAUY - CHILE - 0066009146', 'PREUNIC - CHILE - 0066009032', 'PRISA DISTRIB
        UCION - CHILE - 2001419970',
        'RABIE - CHILE - 0066009015', 'S Y B FARMACEUTICA S.A. - CHILE - 2000432938',
        'SOC. INV. LA MUNDIAL LTDA - CHILE - 2001270967', 'SOCOFAR - CHILE - 0066009028',
        'SODIMAC - CHILE - 2000402358', 'SOUTHERN CROSS - CHILE - 2002135799',
        'SUPERM. MONSERRAT - CHILE - 0066009120', 'TELEMERCADOS EUROPA - CHILE - 0066009044'))
        and T2083424.PRMTN_LONG_NAME in (select distinct T2083424.PRMTN_LONG_NAME as c1
        from
        OPT_ACCT_FDIM T2084525 /* OPT_ACCT_PRMTN_FDIM */ ,
        OPT_BUS_UNIT_FDIM T2083056,
        OPT_CAL_MASTR_DIM T2083357 /* OPT_CAL_MASTR_DIM01 */ ,
        OPT_PRMTN_FDIM T2083424,
        OPT_PRMTN_PROD_FLTR_LKP T2083698
        where  ( T2083056.BUS_UNIT_SKID = T2083698.BUS_UNIT_SKID and T2083357.CAL_MASTR_SKID
        = T2083698.DATE_SKID
        and T2083698.ACCT_PRMTN_SKID = T2084525.ACCT_SKID and T2083424.PRMTN_SKID = T2083698.
        PRMTN_SKID
        and T2083424.BUS_UNIT_SKID = T2083698.BUS_UNIT_SKID and T2083056.BUS_UNIT_NAME = 'Chi
        le'
        and T2083357.FISC_YR_ABBR_NAME = 'FY10/11' and T2083698.BUS_UNIT_SKID = T2084525.BUS_
        UNIT_SKID
        and (case  when T2083424.CORP_PRMTN_TYPE_CODE = 'Target Account' then 'Corporate'
        else T2083424.CORP_PRMTN_TYPE_CODE end  in ('Alternate BDF', 'Corporate', 'Private'))
        and (T2084525.ACCT_LONG_NAME in ('ADELCO - CHILE - 0066009018',
        'ALIMENTOS FRUNA - CHILE - 0066009049', 'CENCOSUD - CHILE - 0066009007',
        'COMERCIAL ALVI - CHILE - 0066009070', 'D&S - CHILE - 0066009008',
        'DIPAC - CHILE - 0066009024', 'DIST. COMERCIAL - CHILE - 0066009087',
        'DISTRIBUCION LAGOS S.A. - CHILE - 2001146505', 'ECOMMERCE ESCALA 1 - 1900001746',
        'EMILIO SANDOVAL - CHILE - 2000402293', 'F. AHUMADA - CHILE - 0066009023',
```

9.18 扩展统计信息优化案例

```
'FALABELLA - CHILE - 2000406971', 'FRANCISCO LEYTON - CHILE - 0066009142',
'MAICAO - CHILE - 0066009135', 'MARGARITA UAUY - CHILE - 0066009146',
'PREUNIC - CHILE - 0066009032', 'PRISA DISTRIBUCION - CHILE - 2001419970',
'RABIE - CHILE - 0066009015', 'S Y B FARMACEUTICA S.A. - CHILE - 2000432938',
'SOC. INV. LA MUNDIAL LTDA - CHILE - 2001270967', 'SOCOFAR - CHILE - 0066009028',
'SODIMAC - CHILE - 2000402358', 'SOUTHERN CROSS - CHILE - 2002135799',
'SUPERM. MONSERRAT - CHILE - 0066009120', 'TELEMERCADOS EUROPA - CHILE - 0066009044')
) ) )
group by T2083424.PRMTN_SKID, T2083424.PRMTN_ID, T2083424.PRMTN_NAME, T2083424.SHPMT_
END_DATE,
T2083424.SHPMT_START_DATE, T2083424.PRMTN_STTUS_CODE, T2083424.APPRV_STTUS_CODE, T208
3424.CMMNT_DESC,
T2083424.PGM_START_DATE, T2083424.PGM_END_DATE, T2083424.CREAT_DATE, T2083424.APPRV_B
Y_DESC,
T2083424.AUTO_UPDT_GTIN_IND, T2083424.PRMTN_STOP_DATE, T2083424.ACTVY_DETL_POP, T2083
424.CNBLN_WK_CNT,
T2083424.PRMTN_AVG_POP, T2084525.NAME, T2084525.CHANL_TYPE_DESC, T2084525.ACCT_LONG_N
AME,
case  when case  when T2083424.CORP_PRMTN_TYPE_CODE = 'Target Account' then 'Corporate'
else T2083424.CORP_PRMTN_TYPE_CODE end  is null then 'Private' else case
when T2083424.CORP_PRMTN_TYPE_CODE = 'Target Account' then 'Corporate'
else T2083424.CORP_PRMTN_TYPE_CODE  end  end ,
nvl(case  when T2083424.PRMTN_STTUS_CODE = 'Confirmed'
then cast(( TRUNC( TO_DATE('2011-06-07' , 'YYYY-MM-DD') ) - TRUNC( T2083424.PGM_END_D
ATE ) ) as  VARCHAR ( 10 ) ) end , '')
order by c24, c3;
```

该 SQL 是 Obiee 报表工具自动生成的，所以看起来有些凌乱。对于很长的 SQL，我们可以运用 SQL 三段分拆方法，快速查看 SQL 写法有没有性能问题。经过检查，SQL 写法没有任何问题。检查完 SQL 写法之后，我们没有直接检查执行计划，因为执行计划也比较长，因此使用自己编写的脚本抓出该 SQL 要用到的表信息，如下所示。

```
TABLE_NAME                 Size(Mb)    PARTITIONED    DEGREE    NUM_ROWS
----------------------     ---------   -----------    ------    --------
*OPT_BUS_UNIT_FDIM         .001037598  NO             1         16
*OPT_CAL_MASTR_DIM         38.1284523  NO             1         37435
OPT_CAL_MASTR_DIM          38.1284523  NO             1         37435
*OPT_PRMTN_FDIM            74.6365929  YES            1         52140
OPT_PRMTN_FDIM             74.6365929  YES            1         52140
OPT_ACTVY_FCT              19.3430614  YES            1         157230
*OPT_ACCT_FDIM             36.6709185  YES            2         95415
OPT_ACCT_FDIM              36.6709185  YES            2         95415
OPT_PRMTN_PROD_FLTR_LKP    1523.87207  YES            2         30148975
```

"*"号表示该表在执行计划中使用到了索引。一般情况下，只有大表才会引发 SQL 性能问题，SQL 中 OPT_PRMTN_PROD_FLTR_LKP 表走的是全表扫描，有 3 000 万行数据，1.5GB，其他表都是小表。需要说明的是，表 OPT_PRMTN_PROD_FLTR_LKP 大小应该不止 1.5GB，因为当时没有通过 DBA_SEGMENTS 来获取表大小，而是通过 DBA_TABLES 中 NUM_ROWS*AVG_ROW_LEN*估算得来，因为 OPT_PRMTN_PROD_FLTR_LKP 是一个分区表，DBA_TABLES 中的统计不是十分准确。找到大表之后，在我们查看执行计划的时候首先就应该关注大表，SQL 的执行计划如图 9-9 所示（因为执行计划比较长，所以采用截图方式并且省略了谓词）。

```
Id  Operation                                        Name                     Rows    Bytes   Cost (%CPU)  Time       Pstart  Pstop
 0  SELECT STATEMENT                                                            1      352    1551  (17)   00:00:07
 1   SORT GROUP BY                                                              1      352    1551  (17)   00:00:07
 2    VIEW                                           VM_NWVW_2                  1      352    1550  (17)   00:00:07
 3     HASH UNIQUE                                                              1      652    1550  (17)   00:00:07
 4      NESTED LOOPS
 5       NESTED LOOPS                                                           1      652    1549  (17)   00:00:07
 6        NESTED LOOPS                                                          1      639    1548  (17)   00:00:07
 7         NESTED LOOPS                                                         2     1180    1546  (17)   00:00:07
 8          NESTED LOOPS                                                        1      568     130   (5)   00:00:01
 9           NESTED LOOPS                                                       1      509     109   (6)   00:00:01
10            NESTED LOOPS                                                      1      484     108   (6)   00:00:01
*11            HASH JOIN                                                        5      830     103   (6)   00:00:01
12              PARTITION LIST SUBQUERY                                        47     4089      82   (3)   00:00:01  KEY(SQ) KEY(SQ)
13               INLIST ITERATOR
14                TABLE ACCESS BY LOCAL INDEX ROWID  OPT_ACCT_FDIM             47     4089      82   (3)   00:00:01  KEY(SQ) KEY(SQ)
*15                 INDEX RANGE SCAN                 OPT_ACCT_FDIM_NX2         47       43      5    (0)   00:00:01  KEY(SQ) KEY(SQ)
16              NESTED LOOPS                                                10482     808K      20  (15)   00:00:01
17               NESTED LOOPS                                                   1       40      2    (0)   00:00:01
*18                INDEX RANGE SCAN                  OPT_BUS_UNIT_FDIM_UX2      1       26      1    (0)   00:00:01
*19                INDEX RANGE SCAN                  OPT_BUS_UNIT_FDIM_UX2      1       14      1    (0)   00:00:01
20               PARTITION LIST ITERATOR                                    10482    1699K      18  (17)   00:00:01  KEY     KEY
21                TABLE ACCESS FULL                  OPT_ACTVY_FCT           10482    1699K      18  (17)   00:00:01  KEY     KEY
*22           TABLE ACCESS BY GLOBAL INDEX ROWID     OPT_PRMTN_FDIM              1      318      1    (0)   00:00:01  ROWID   ROWID
*23            INDEX UNIQUE SCAN                     OPT_PRMTN_FDIM_PK           1        0      0    (0)   00:00:01
*24          TABLE ACCESS BY INDEX ROWID             OPT_CAL_MASTR_DIM           1       25      1    (0)   00:00:01
*25           INDEX UNIQUE SCAN                      OPT_CAL_MASTR_DIM_PK        1        0      0    (0)   00:00:01
26          PARTITION LIST ALL                                                   1       59     21    (0)   00:00:01    1      17
*27           TABLE ACCESS BY LOCAL INDEX ROWID      OPT_PRMTN_FDIM              1       59     21    (0)   00:00:01    1      17
*28            INDEX RANGE SCAN                      OPT_PRMTN_FDIM_NX3          4       17      1    (0)   00:00:01  KEY     KEY
29         PARTITION LIST ITERATOR                                              39      858   1416  (18)   00:00:07  KEY     KEY
*30          TABLE ACCESS FULL                       OPT_PRMTN_PROD_FLTR_LKP    39      858   1416  (18)   00:00:07  KEY     KEY
*31        TABLE ACCESS BY GLOBAL INDEX ROWID        OPT_ACCT_FDIM               1       49      1    (0)   00:00:01  ROWID   ROWID
*32         INDEX UNIQUE SCAN                        OPT_ACCT_FDIM_PK            1        0      0    (0)   00:00:01
*33       INDEX UNIQUE SCAN                          OPT_CAL_MASTR_DIM_PK        1        0      0    (0)   00:00:01
*34      TABLE ACCESS BY INDEX ROWID                 OPT_CAL_MASTR_DIM           1       13      1    (0)   00:00:01
```

图 9-9 SQL 执行计划

Id=30 就是大表在执行计划中的位置，Id=29 是 Id=30 的父亲，它与 Id=8 对齐。Id=7 是嵌套循环，它是 Id=8 与 Id=29 的父亲。通过分析执行计划，我们发现 OPT_PRMTN_PROD_FLTR_LKP 做了嵌套循环（Id=7）的被驱动表，而且没有走索引，这就是为什么 Obiee 报表执行了 30 分钟还没执行完毕。我们查看 Id=30 的过滤条件如下：

```
30 - filter("T2083056"."BUS_UNIT_SKID"="T2083698"."BUS_UNIT_SKID" AND
            "T2083424"."PRMTN_SKID"="T2083698"."PRMTN_SKID" AND
            "T2083424"."BUS_UNIT_SKID"="T2083698"."BUS_UNIT_SKID")
```

我们根据过滤条件创建索引从而让 NL 被驱动表走索引。

```
SQL> create index OPT_PRMTN_PROD_FLTR_LKP_NX1 ON OPT_PRMTN_PROD_FLTR_LKP(BUS_UNIT_SKI
D,PRMTN_SKID) nologging parallel ;

Index created.

Elapsed: 00:00:33.04
```

创建索引花了 33 分钟，如图 9-10 所示，我们再来看一下 SQL 的执行计划，查看带有 A-TIME 的执行计划。

创建完索引之后，Obiee 报表能在 4 分钟内执行完所有数据。我们注意观察执行计划 Id=11，优化器评估返回 5 行数据，但是实际上返回了 11248 行数据，这导致后续表连接方式全采用了嵌套循环。Id=11 是两表 HASH 连接之后的结果集，如果能够纠正 Id=11 估算 Rows 的误差，那么优化器应该能自我优化该报表。Id=11 是两个表中两个列关联的结果集，优化器一般对多个列进行 Rows 估算的时候通常容易算错，于是对 Id=11 中两个表的连接列收集了扩展统计信息。

9.18 扩展统计信息优化案例

Id	Operation	Name	Starts	E-Rows	A-Rows	A-Time
0	SELECT STATEMENT		1		1324	00:02:42.23
1	SORT GROUP BY		1	1	1324	00:02:42.23
2	VIEW	VM_NWVW_2	1	1	6808	00:02:42.18
3	HASH UNIQUE		1	1	6808	00:02:42.18
4	NESTED LOOPS		1		5220K	00:02:21.06
5	NESTED LOOPS		1	1	5220K	00:02:00.18
6	NESTED LOOPS		1	1	5220K	00:01:49.74
7	NESTED LOOPS		1	2	5220K	00:01:18.42
8	NESTED LOOPS		1	1	6808	00:00:01.62
9	NESTED LOOPS		1	1	6808	00:00:00.54
10	NESTED LOOPS		1		11248	00:00:00.40
* 11	HASH JOIN		1	5	11248	00:00:00.07
12	PARTITION LIST SUBQUERY		1	47	25	00:00:00.01
13	INLIST ITERATOR		1		25	00:00:00.01
14	TABLE ACCESS BY LOCAL INDEX ROWID	OPT_ACCT_FDIM	25	47	25	00:00:00.01
* 15	INDEX RANGE SCAN	OPT_ACCT_FDIM_NX2	25	47	25	00:00:00.01
16	NESTED LOOPS		1	10482	12788	00:00:00.03
17	NESTED LOOPS		1	1	1	00:00:00.01
* 18	INDEX RANGE SCAN	OPT_BUS_UNIT_FDIM_UX2	1	1	1	00:00:00.01
* 19	INDEX RANGE SCAN	OPT_BUS_UNIT_FDIM_UX2	1	1	1	00:00:00.01
20	PARTITION LIST ITERATOR		1	10482	12788	00:00:00.03
* 21	TABLE ACCESS FULL	OPT_ACTVY_FCT	1	10482	12788	00:00:00.03
* 22	TABLE ACCESS BY GLOBAL INDEX ROWID	OPT_PRMTN_FDIM	11248	1	11248	00:00:00.31
* 23	INDEX UNIQUE SCAN	OPT_PRMTN_FDIM_PK	11248	1	11248	00:00:00.12
* 24	TABLE ACCESS BY INDEX ROWID	OPT_CAL_MASTR_DIM	11248	1	6808	00:00:00.14
* 25	INDEX UNIQUE SCAN	OPT_CAL_MASTR_DIM_PK	11248	1	11248	00:00:00.05
26	PARTITION LIST ALL		6808		6808	00:00:01.08
* 27	TABLE ACCESS BY LOCAL INDEX ROWID	OPT_PRMTN_FDIM	115K	1	6808	00:00:01.05
* 28	INDEX RANGE SCAN	OPT_PRMTN_FDIM_NX3	115K	4	6808	00:00:00.78
29	TABLE ACCESS BY GLOBAL INDEX ROWID	OPT_PRMTN_PROD_FLTR_LKP	6808	39	5220K	00:01:19.79
* 30	INDEX RANGE SCAN	OPT_PRMTN_PROD_FLTR_LKP_NX1	6808	3	5220K	00:00:43.96
* 31	TABLE ACCESS BY GLOBAL INDEX ROWID	OPT_ACCT_FDIM	5220K	1	5220K	00:00:23.79
* 32	INDEX UNIQUE SCAN	OPT_ACCT_FDIM_PK	5220K	1	5220K	00:00:08.38
* 33	INDEX UNIQUE SCAN	OPT_CAL_MASTR_DIM_PK	5220K	1	5220K	00:00:07.58
* 34	TABLE ACCESS BY INDEX ROWID	OPT_CAL_MASTR_DIM	5220K	1	5220K	00:00:17.28

Predicate Information (identified by operation id):

11 - access("T2083114"."BUS_UNIT_SKID"="T2084525"."BUS_UNIT_SKID" AND "T2083114"."ACCT_PRMTN_SKID"="T2084525"."ACCT_SKID")

图 9-10

```
SQL> SELECT DBMS_STATS.CREATE_EXTENDED_STATS(USER, 'OPT_ACCT_FDIM', '(BUS_UNIT_SKID,
ACCT_SKID)') FROM DUAL;

DBMS_STATS.CREATE_EXTENDED_STATS(USER,'OPT_ACCT_FDIM','(BUS_UNIT_SKID,ACCT_SKID)')
--------------------------------------------------------------------------------
SYS_STUJ8OD#X2IPA_B9_CH00B046T

SQL> SELECT DBMS_STATS.CREATE_EXTENDED_STATS(USER, 'OPT_ACTVY_FCT', '(BUS_UNIT_SKID,
ACCT_PRMTN_SKID)') FROM DUAL;

DBMS_STATS.CREATE_EXTENDED_STATS(USER,'OPT_ACTVY_FCT','(BUS_UNIT_SKID,ACCT_PRMTN_SKID
)')
--------------------------------------------------------------------------------
SYS_STU#CVQNKK5CCM0W2XEQWSRXSM

SQL> BEGIN
  2    DBMS_STATS.GATHER_TABLE_STATS(ownname => 'XXXXX',   ---为了保密,用户名做了更改
  3    tabname => 'OPT_ACCT_FDIM',
  4    estimate_percent => 20,
  5    method_opt => 'for all columns size auto',
  6    degree => 6,
  7    granularity => 'ALL',
  8    cascade=>TRUE
  9    );
 10  END;
 11  /

PL/SQL procedure successfully completed.

Elapsed: 00:00:57.76
```

```
SQL> BEGIN
  2    DBMS_STATS.GATHER_TABLE_STATS(ownname => 'XXXX',    ---为了保密,用户名做了更改
  3    tabname => 'OPT_ACTVY_FCT',
  4    estimate_percent => 20,
  5    method_opt => 'for all columns size auto',
  6    degree => 6,
  7    granularity => 'ALL',
  8    cascade=>TRUE
  9    );
 10   END;
 11  /

PL/SQL procedure successfully completed.

Elapsed: 00:01:15.10
```

收集完扩展统计信息之后,SQL 能在 1 秒左右执行完毕,带有 A-Time 的执行计划如图 9-11 所示。

Id	Operation	Name	Starts	E-Rows	A-Rows	A-Time	Buffers
0	SELECT STATEMENT		1		1324	00:00:01.85	210K
1	SORT GROUP BY		1	1	1324	00:00:01.85	210K
* 2	FILTER		1		6808	00:00:01.84	210K
3	NESTED LOOPS		1		6808	00:00:00.04	52722
4	NESTED LOOPS		1	4	11248	00:00:00.03	41474
5	NESTED LOOPS		1	12	11248	00:00:00.02	30247
* 6	HASH JOIN		1	403	11248	00:00:00.01	172
7	PARTITION LIST SUBQUERY		1	47	25	00:00:00.01	50
8	INLIST ITERATOR				25	00:00:00.01	47
9	TABLE ACCESS BY LOCAL INDEX ROWID	OPT_ACCT_FDIM	25	47	25	00:00:00.01	47
* 10	INDEX RANGE SCAN	OPT_ACCT_FDIM_NX2	25	47	25	00:00:00.01	27
11	NESTED LOOPS		1	10508	12788	00:00:00.01	122
* 12	INDEX RANGE SCAN	OPT_BUS_UNIT_FDIM_UX2	1	1	1	00:00:00.01	1	0
13	PARTITION LIST ITERATOR		1	10508	12788	00:00:00.01	121
* 14	TABLE ACCESS FULL	OPT_ACTVY_FCT	1	10508	12788	00:00:00.01	121
15	TABLE ACCESS BY GLOBAL INDEX ROWID	OPT_PRMTN_FDIM	11248	1	11248	00:00:00.01	30075
* 16	INDEX UNIQUE SCAN	OPT_PRMTN_FDIM_PK	11248	1	11248	00:00:00.01	11250
* 17	INDEX UNIQUE SCAN	OPT_CAL_MASTR_DIM_PK	11248	1	11248	00:00:00.01	11227
* 18	TABLE ACCESS BY INDEX ROWID	OPT_CAL_MASTR_DIM	11248	1	6808	00:00:00.01	11248
19	NESTED LOOPS		6206		6206	00:00:01.79	158K
20	NESTED LOOPS		6206	1	6206	00:00:01.79	151K
21	NESTED LOOPS		6206	1	6206	00:00:01.79	145K
22	NESTED LOOPS		6206	5	6206	00:00:01.79	128K
23	NESTED LOOPS		6206	1	6206	00:00:00.09	103K
* 24	INDEX RANGE SCAN	OPT_BUS_UNIT_FDIM_UX2	6206	1	6206	00:00:00.01	6206
25	PARTITION LIST ALL		6206	1	6206	00:00:00.09	97324
* 26	TABLE ACCESS BY LOCAL INDEX ROWID	OPT_PRMTN_FDIM	49648	1	6206	00:00:00.09	97324
* 27	INDEX RANGE SCAN	OPT_PRMTN_FDIM_NX3	49648	4	6206	00:00:00.01	86887
28	TABLE ACCESS BY GLOBAL INDEX ROWID	OPT_PRMTN_PROD_FLTR_LKP	6206	39	6206	00:00:01.69	24825
* 29	INDEX RANGE SCAN	OPT_PRMTN_PROD_FLTR_LKP_NX1	6206	3	6206	00:00:01.53	18618
30	TABLE ACCESS BY GLOBAL INDEX ROWID	OPT_ACCT_FDIM	6206	1	6206	00:00:00.01	17241
* 31	INDEX UNIQUE SCAN	OPT_ACCT_FDIM_PK	6206	1	6206	00:00:00.01	11035
* 32	INDEX UNIQUE SCAN	OPT_CAL_MASTR_DIM_PK	6206	1	6206	00:00:00.01	6211
* 33	TABLE ACCESS BY INDEX ROWID	OPT_CAL_MASTR_DIM	6206	1	6206	00:00:00.01	6206

图 9-11

大家在工作中如果遇到多列过滤或者多列关联 Rows 估算出现较大偏差的时候,不妨收集扩展统计信息试一试。

其实当时是项目经理找到作者罗老师来优化 SQL 的,当时他应该是被美国宝洁的客户批评了。客户的原话是说:"我已经抽完一支烟了,报表还没打开,我原本以为当我抽完第二支烟的时候报表能打开,谁知当我抽完第三支烟的时候报表还没打开!"罗老师优化完报表之后,幽默地说了句,现在客户可以在掏打火机、烟还没点燃之前就能打开报表了。

9.19 使用 LISGAGG 分析函数优化 WMSYS.WM_CONCAT

2016 年，在上周末优化班的时候，一个同学请求现场优化如下 SQL。

```
with temp as
      (select sgd.detail_id id,
              wmsys.wm_concat(distinct(sg.gp_name)) groupnames,
              wmsys.wm_concat(distinct(su.user_name)) usernames
         from  sgd
         left join   sg
           on sg.id = sgd.gp_id
         left join   sug
           on sg.id = sug.gp_id
         left join   su
           on sug.user_id = su.id
        group by sgd.detail_id)
    select zh.id,
           zh.id detailid,
           zh.name detailname,
           zh.p_level hospitallevel,
           zh.type hospitaltype,
           dza.name region,
           temp.groupnames,
           temp.usernames,
           (case
             when gd.gp_id is null then
              0
             else
              1
           end) isalloted
      from  zh
      left join   dza
        on zh.area_id = dza.id
      left join temp
        on zh.id = temp.id
   left join (select gp_id, detail_id from sys_gp_detail where gp_Id = :0) gd
        on zh.id = gd.detail_id order by length(id),zh.id asc;
```

该 SQL 返回 20 779 行数据，要执行 4 分 32 秒。该执行计划中全是 HASH JOIN，这里就不贴执行计划了。

首先这条 SQL 最终返回 20 779 行数据，该 SQL 语句最后部分没有 GROUP BY，表与表之间关联全是外连接，主表 zh 没有过滤条件，因此判断 zh 表最多 20 779 行数据，因为它是外连接的主表，不管关联有没有关联上，zh 会返回表中全部数据，如果 zh 与 dza 是 1:n 关系，那么 zh 表总行数还将少于 20 779 行数据。同时也判定 dza，TEMP 数据量都不大，因为所有表关联完只返回 20779 行数据。既然都是小表，为什么最终要执行 4 分 32 秒呢？遇到此类问题，我们需要将 SQL 拆开，分步执行，这样就能判断 SQL 中哪一步是性能瓶颈。9.10 节中案例也是采用分步执行方法找到问题根本原因。

SQL 语句中有个 with as 子句，对其单独执行，发现要执行两分钟左右。with as 子句中有两个列转行函数：wmsys.wm_concat，将其注释之后 with as 子句能秒出。现在我们定位到，SQL 性能问题是由 wmsys.wm_concat 导致。对于列转行，Oracle 还提供了 Listagg 分析函数，

wmsys.wm_concat 从 Oracle11g 之后返回的是 Clob 类型,而 Listagg 返回的是 varchar2 类型。因此我们尝试对 with as 子句进行等价改写,利用分析函数 Listagg 代替 wmsys.wm_concat,以验证改写之后是否还会出现性能问题。with as 子句原始 SQL 如下。

```sql
select sgd.detail_id id,
       wmsys.wm_concat(distinct(sg.gp_name)) groupnames,
       wmsys.wm_concat(distinct(su.user_name)) usernames
  from sgd
  left join sg on sg.id = sgd.gp_id
  left join sug on sg.id = sug.gp_id
  left join su on sug.user_id = su.id
 group by sgd.detail_id;
```

因为 with as 子句中有两个 wmsys.wm_concat,而且 wmsys.wm_concat 中有 distinct,而 Listagg 不支持 distinct,所以我们只能一个一个去掉 wmsys.wm_concat。现在将 with as 子句中 wmsys.wm_concat(distinct(su.user_name)) usernames 去掉,只保留 wmsys.wm_concat(distinct(sg.gp_name)) groupnames。因为 usernames 关联了 su、sug,而现在只保留 groupnames,所以我们需要将 su、sug 去掉,去掉 usernames 的 SQL 如下。

```sql
select sgd.detail_id id, wmsys.wm_concat(distinct(sg.gp_name)) groupnames
  from sys_gp_detail sgd
  left join sys_gp sg on sg.id = sgd.gp_id
 group by sgd.detail_id;
```

其执行计划如下。

```
已用时间: 00: 00: 58.04.
执行计划
----------------------------------------------------------
Plan hash value: 3491823204

--------------------------------------------------------------------------------
| Id  | Operation                    | Name          | Rows  | Bytes |TempSpc| Cost (%CPU)|
--------------------------------------------------------------------------------
|   0 | SELECT STATEMENT             |               | 20584 |  824K |       |  1308   (8)|
|   1 |  SORT GROUP BY               |               | 20584 |  824K |   15M |  1308   (8)|
|*  2 |   HASH JOIN RIGHT OUTER      |               |  313K |   12M |       |   449   (6)|
|   3 |    TABLE ACCESS FULL         | SYS_GP        |     3 |    69 |       |     3   (0)|
|   4 |    TABLE ACCESS FULL         | SYS_GP_DETAIL |  313K |  5518K|       |   438   (5)|
--------------------------------------------------------------------------------

Predicate Information (identified by operation id):
---------------------------------------------------

   2 - access("SG"."ID"(+)="SGD"."GP_ID")


统计信息
----------------------------------------------------------
          1  recursive calls
     249348  db block gets
      44447  consistent gets
          0  physical reads
          0  redo size
    9993548  bytes sent via SQL*Net to client
    6067828  bytes received via SQL*Net from client
      83118  SQL*Net roundtrips to/from client
```

9.19 使用 LISGAGG 分析函数优化 WMSYS.WM_CONCAT

```
        1  sorts (memory)
        0  sorts (disk)
```

执行计划中的 db block gets 来自于 Clob。因为 Listagg 不支持 distinct，所以我们需要先去重，再采用 Listagg，Listagg 改写的 SQL 如下。

```
select detail_id, listagg(gp_name, ',') within
 group(
 order by null)
  from (select sgd.detail_id, sg.gp_name
          from sys_gp_detail sgd
          left join sys_gp sg on sg.id = sgd.gp_id
          group by sgd.detail_id, sg.gp_name)
 group by detail_id;
```

改写后的执行计划如下。

```
已用时间: 00: 00: 01.12
执行计划
----------------------------------------------------------
Plan hash value: 147456425

----------------------------------------------------------------------------------------
| Id  | Operation                     | Name          | Rows  | Bytes |TempSpc| Cost(%CPU)|
----------------------------------------------------------------------------------------
|   0 | SELECT STATEMENT              |               | 20584 | 1547K|        | 1467   (7)|
|   1 |  SORT GROUP BY                |               | 20584 | 1547K|        | 1467   (7)|
|   2 |   VIEW                        | VM_NWVW_0     | 43666 | 3283K|        | 1467   (7)|
|   3 |    HASH GROUP BY              |               | 43666 | 1748K|   15M | 1467   (7)|
|*  4 |     HASH JOIN RIGHT OUTER|    |               |  313K |  12M |        |  449   (6)|
|   5 |      TABLE ACCESS FULL        | SYS_GP        |     3 |   69 |        |    3   (0)|
|   6 |      TABLE ACCESS FULL        | SYS_GP_DETAIL |  313K | 5518K|        |  438   (5)|
----------------------------------------------------------------------------------------

Predicate Information (identified by operation id):
---------------------------------------------------

   4 - access("SG"."ID"(+)="SGD"."GP_ID")


统计信息
----------------------------------------------------------
          1  recursive calls
          0  db block gets
       2775  consistent gets
          0  physical reads
          0  redo size
     450516  bytes sent via SQL*Net to client
      15595  bytes received via SQL*Net from client
       1387  SQL*Net roundtrips to/from client
          1  sorts (memory)
          0  sorts (disk)
      20779  rows processed
```

使用 Listagg 改写之后，SQL 能在 1 秒执行完毕，而采用 wmsys.wm_concat 需要 58 秒，这说明采用 Listagg 代替 wmsys.wm_concat 能达到优化目的。

下面我们改写另外一个 wmsys.wm_concat，改写的思路一模一样，先去重，再使用 Listagg。

```
select detail_id, listagg(user_name, ',') within
 group(
```

```
        order by null)
  from (select sgd.detail_id id, su.user_name
          from sgd
          left join sg on sg.id = sgd.gp_id
          left join sug on sg.id = sug.gp_id
          left join su on sug.user_id = su.id
        group by sgd.detail_id, su.user_name)
 group by detail_id;
```

最终的 with as 子句如下。

```
select a.detail_id id , a.groupnames, b.usernames
  from (select detail_id, listagg(gp_name, ',') within
         group(
         order by null) groupnames
          from (select sgd.detail_id, sg.gp_name
                 from sys_gp_detail sgd
                 left join sys_gp sg on sg.id = sgd.gp_id
                group by sgd.detail_id, sg.gp_name)
        group by detail_id) a,
       (select detail_id, listagg(user_name, ',') within
         group(
         order by null) usernames
          from (select sgd.detail_id, su.user_name
                 from sgd
                 left join sg on sg.id = sgd.gp_id
                 left join sug on sg.id = sug.gp_id
                 left join su on sug.user_id = su.id
               group by sgd.detail_id, su.user_name)
        group by detail_id) b
 where a..detail_id = b.detail_id;
```

用改写后的 with as 子句替换原始 SQL 中的 with as 子句，最终 SQL 能在两秒左右执行完毕。

在工作中尽量使用 Listagg 代替 wmsys.wm_concat。

9.20 INSTR 非等值关联优化案例

2014 年，曾遇到一个 INSTR 优化案例。因为当初 SQL 代码并非运行在 Oracle 中，所以，在 Oracle 中创建测试数据以便演示该案例，不管是 Oracle 数据库还是其他数据库，优化的思想都是一样的。

需求是这样的：查找事实表中 URL 字段包含了维度表中 URL 的记录，然后进行汇总统计。

创建事实表如下。

```
create table T_FACT
(msisdn number(11),
url varchar2(50)
);
```

插入测试数据。

```
insert into T_FACT
  select '139' || chr(dbms_random.value(48, 57)) ||
         chr(dbms_random.value(48, 57)) || chr(dbms_random.value(48, 57)) ||
         chr(dbms_random.value(48, 57)) || chr(dbms_random.value(48, 57)) ||
```

9.20 INSTR 非等值关联优化案例

```
            chr(dbms_random.value(48, 57)) || chr(dbms_random.value(48, 57)),
            lpad(chr(dbms_random.value(97, 122)),
                 dbms_random.value(1, 20),
                 chr(dbms_random.value(97, 122))) ||
            lpad(chr(dbms_random.value(97, 122)) ||
                 chr(dbms_random.value(97, 122)) ||
                 chr(dbms_random.value(97, 122)) ||
                 chr(dbms_random.value(97, 122)),
                 dbms_random.value(4, 20),
                 chr(dbms_random.value(97, 122)))
     from dual
   connect by rownum <= 10000;
```

反复插入数据,直到表中一共有 128 万条数据。

```
begin
  for i in 1..7 loop
    insert into T_FACT
      select * from T_FACT;
    commit;
  end loop;
end;
```

在实际案例中事实表有上亿条数据,演示只取 100 万条数据。

创建维度表如下。

```
create table T_DIM as
    select cast(rownum as number(6)) code,cast(c1 as varchar2(50)) url
from (
      select distinct substr(url, -dbms_random.value(2, length(url) - 3)) c1
  from T_FACT);
```

创建汇总统计表。

```
create table T_RESULT
(
msisdn number(11),
code number(6),
url varchar2(50),
cnt number(6)
);
```

现在我们要执行如下 SQL,统计 T_FACT 表中 URL 包含了 T_DIM 的记录。

```
insert into T_RESULT
  (msisdn, code, url, cnt)
  select t1.msisdn, t2.code, t2.url, sum(1)
    from T_FACT t1
   inner join T_DIM t2 on instr(t1.url, t2.url) > 0
   group by t1.msisdn, t2.code, t2.url;
```

因为 SQL 中关联条件是 instr,这时只能走嵌套循环,不能走 HASH 连接,也不能走排序合并连接,排序合并连接一般用于>=, >, <, <=。以上 SQL 执行计划如下。

```
SQL> select * from table(dbms_xplan.display);

PLAN_TABLE_OUTPUT
--------------------------------------------------------------------------------
Plan hash value: 2285685195
```

```
---------------------------------------------------------------------------
| Id  | Operation                 | Name     | Rows  | Bytes | Cost (%CPU)| Time       |
---------------------------------------------------------------------------
|   0 | INSERT STATEMENT          |          |  10G  |  798G |  134M  (3)|448:22:51 |
|   1 |  LOAD TABLE CONVENTIONAL  | T_RESULT |       |       |           |          |
|   2 |   HASH GROUP BY           |          |  10G  |  798G |  134M  (3)|448:22:51 |
|   3 |    NESTED LOOPS           |          |  10G  |  798G |  133M  (3)|445:43:32 |
|   4 |     TABLE ACCESS FULL     | T_FACT   | 1192K |   45M |  1363  (1)| 00:00:17 |
|*  5 |     TABLE ACCESS FULL     | T_DIM    |  8993 |  351K |   112  (2)| 00:00:02 |
---------------------------------------------------------------------------

Predicate Information (identified by operation id):
---------------------------------------------------

   5 - filter(INSTR("T1"."URL","T2"."URL")>0)
```

本书反复强调，嵌套循环被驱动表必须走索引。但是，如果执行计划是因为 INSTR、LIKE、REGEXP_LIK 等而导致的嵌套循环，这时被驱动表反而不能走索引。INSTR、LIKE、REGEXP_LIKE 会匹配所有数据，走索引的访问路径只能是 INDEX FULL SCAN，而 INDEX FULL SCAN 是单块读，全表扫描是多块读。如果 INDEX FULL SCAN 需要回表，这时效率远远不如全表扫描效率高。如果被驱动表走 INDEX FULL SCAN 不回表，这时我们也可以根据索引中的索引列，建立一个临时表，将需要的列包含在临时表中，用临时表代替 INDEX FULL SCAN，因为临时表不像索引那样需要存储根、分支、叶子节点，临时表相比索引体积反而更小，这样可以减少被驱动表每次被扫描的体积。被驱动表因为要被反复扫描多次，buffer cache 最好要有足够的空间用于存放被驱动表，从而避免被驱动表每次被扫描都需要物理 I/O。

经过上面分析，如果从执行计划方向入手，我们无法优化 SQL。我们再来看一下原始 SQL 语句。

```
insert into T_RESULT
   (msisdn, code, url, cnt)
   select t1.msisdn, t2.code, t2.url, sum(1)
     from T_FACT t1
     inner join T_DIM t2 on instr(t1.url, t2.url) > 0
     group by t1.msisdn, t2.code, t2.url;
```

SQL 语句中有 GROUP BY（汇总），事实表与维度表一般都是 N:1 关系，因为 SQL 语句中有汇总，我们可以先对事实表进行汇总，去掉重复数据，然后再与维度表关联。因为执行计划中事实表是驱动表，维度表是被驱动表，将事实表提前汇总可以将数据量大大减少，这样我们就可以减少嵌套循环的循环次数，从而达到优化目的。

事实表原始数据为 128 万行，我们对事实表提前汇总。

```
create table T_MIDDLE as select
     msisdn,url,sum(1) cnt
     from  T_FACT group by msisdn,url;
```

提前汇总之后，数据从 128 万行减少到 1 万行。

```
SQL> select count(*) from  T_MIDDLE;

  COUNT(*)
----------
     10000
```

改写后的 SQL 如下。

```
insert into T_RESULT
  (msisdn, code, url, cnt)
 select t1.msisdn, t2.code, t2.url, sum(cnt)
   from T_MIDDLE t1
   inner join T_DIM t2 on instr(t1.url, t2.url) > 0
  group by t1.msisdn, t2.code, t2.url;
```

对数据进行提前汇总之后，被驱动表 T_DIM 只需要循环 1 万次，而之前需要循环 128 万次，性能得到极大提升。

如果想要最大程度优化 INSTR，LIKE，REGEXP_LIKE 等非等值关联，我们只能从业务角度入手，设法从业务本身、数据本身着手，使其进行等值连接，从而可以走 HASH 连接。

如果业务手段无法优化，除了上面讲到的提前汇总数据，我们还可以开启并行查询（并行广播），从而优化 SQL。如果不想开启并行查询，我们可以对表进行拆分（类似并行广播），人工模拟并行查询，从而优化 SQL。我们可以对驱动表进行拆分，也可以对被驱动表进行拆分，但是最好不要同时拆分驱动表和被驱动表，因为连接条件是非等值连接，同时拆分驱动表和被驱动表会导致交叉关联（将驱动表和被驱动表都拆分为 6 份，会关联 36 次）。如果表是非分区表，我们可以利用 ROWID 进行拆分。如果表是分区表，我们可以针对分区进行拆分。关于具体的拆分方法，请大家阅读 8.5 节内容。

9.21 REGEXP_LIKE 非等值关联优化案例

本案例为好友南京越烟（QQ: 843999405）分享。

一个存储过程从周五晚上执行了到了周一还没有执行完，存储过程代码如下。

```
declare
  isMatch   Boolean := false;
  dealPnCnt number(10) := 0;
begin
  for c_no_data in (select nbn.no, 69 as partition_id
                      from TMP_NBR_NO_XXXX nbn
                     where nbn.level_id = 1
                       and length(nbn.no) = 8) loop
    dealPnCnt := dealPnCnt + 1;
    for c_data in (select nli.*, nl.nbr_level_id
                     from tmp_xxx_item nli,
                          a_level_item nl2i,
                          b_level_item nl,
                          c_level_item ns2l
                    where nli.nbr_level_item_id = nl2i.nbr_level_item_id
                      and nl2i.nbr_level_id = nl.nbr_level_id
                      and nl.nbr_level_id = ns2l.nbr_level_id
                      and ns2l.area_id = c_no_data.partition_id
                      and ns2l.res_spec_id = 6039
                      and ns2l.nbr_level_id between 201 and 208
                    order by nl2i.priority) loop
      if (regexp_like(c_no_data.no, c_data.expression)) then
        update TMP_NBR_NO_XXXX n
           set n.level_id = c_data.nbr_level_id
         where n.no = c_no_data.no;
```

```
          exit;
        end if;
      end loop;
      if mod(dealPnCnt, 5000) = 0 then
        commit;
      end if;
    end loop;
end;
```

TMP_NBR_NO_XXXX 共有 400w 行数据，180MB。

```
select nli.*, nl.nbr_level_id
  from tmp_xxx_item nli,
       a_level_item nl2i,
       b_level_item nl,
       c_level_item ns2l
 where nli.nbr_level_item_id = nl2i.nbr_level_item_id
   and nl2i.nbr_level_id = nl.nbr_level_id
   and nl.nbr_level_id = ns2l.nbr_level_id
   and ns2l.area_id = c_no_data.partition_id
   and ns2l.res_spec_id = 6039
   and ns2l.nbr_level_id between 201 and 208
 order by nl2i.priority;
```

上面 SQL 查询返回 43 行数据。

在 5.1 节提到过，嵌套循环就是一个 LOOP 循环，LOOP 套 LOOP 相当于笛卡儿积。该 PLSQL 代码中有 LOOP 套 LOOP 的情况，这就导致 UPDATE TMP_NBR_NO_XXXX 要执行（400 万*43）次，TMP_NBR_NO_XXXX.no 列没有索引，TMP_NBR_NO_XXXX 每次更新都要进行全表扫描。这就是为什么存储过程从周五执行到周一还没执行完。

大家可能会问，为什么不用 MERGE INTO 对 PLSQL 代码进行改写呢？PLSQL 代码中是用 regexp_like(c_no_data.no, c_data.expression)进行关联的，使用 like、regexp_like 关联，无法走 HASH 连接，也无法走排序合并连接，两表只能走嵌套循环并且被驱动表无法走索引。如果强行使用 MERGE INTO 进行改写，因为该 SQL 执行时间很长，会导致 UNDO 不释放，所以，我们没有采用 MERGE INTO 对代码进行改写。

大家可能也会问，为什么不对 TMP_NBR_NO_XXXX.no 建立索引呢？这是因为关联更新可以采用 ROWID 批量更新，所以没有采用建立索引方法优化。

下面我们采用 ROWID 批量更新方法改写上面 PLSQL，为了方便大家阅读 PLSQL 代码，先创建一个临时表用于存储 43 记录。

```
create table TMP_DATE_TEST
(
  expression   VARCHAR2(255) not null,
  nbr_level_id NUMBER(9) not null,
  priority     NUMBER(8) not null
);

insert into TMP_DATE_TEST
    select nli.expression, nl.nbr_level_id, priority  from tmp_xxx_item  nli,
                             a_level_item nl2i,
                             b_level_item       nl,
                             c_level_item ns2l
                    where nli.nbr_level_item_id = nl2i.nbr_level_item_id
```

9.21 REGEXP_LIKE 非等值关联优化案例

```
                   and nl2i.nbr_level_id = nl.nbr_level_id
                   and nl.nbr_level_id = ns21.nbr_level_id
                   and ns21.area_id = 69
                   and ns21.res_spec_id = 6039
                   and ns21.nbr_level_id between 201 and 208;
```

我们创建另外一个临时表,用于存储要被更新的表的 ROWID 以及过滤条件字段。

```
create table TMP_NBR_NO_XXXX_TEXT
(
  rid   ROWID,
  no    VARCHAR2(255),
);

 insert into  TMP_NBR_NO_XXXX_TEXT
   select rowid rid,  nbn.no, from   TMP_NBR_NO_XXXX nbn where  nbn.level_id=1 and length(nbn.no)= 8 ;
```

改写之后的 PLSQL 代码如下。

```
declare
  type rowid_table_type is table of rowid index by pls_integer;
  updateCur sys_refcursor;
  v_rowid    rowid_table_type;
  v_rowid2   rowid_table_type;
begin
  for c_no_data in (select t.expression, t.nbr_level_id, t.priority
                      from TMP_DATE_TEST t
                     order by 3) loop
    open updateCur for
      select rid
        from TMP_NBR_NO_XXXX_TEXT nbn
       where regexp_like(nbn.no, c_no_data.expression);
    loop
      fetch updateCur bulk collect
        into v_rowid LIMIT 20000;
        forall i in v_rowid.FIRST .. v_rowid.LAST
        update TMP_NBR_NO_XXXX
           set level_id = c_no_data.nbr_level_id
         where rowid = v_rowid(i);
      commit;
      exit when updateCur%notfound;
    end loop;
    CLOSE updateCur;
  end loop;
end;
```

改写后的 PLSQL 能在 4 小时左右执行完。有没有什么办法进一步优化呢?单个进程能在 4 小时左右执行完,如果开启 8 个并行进程,那应该能在 30 分钟左右执行完。但是 PLSQL 怎么开启并行呢?正常情况下 PLSQL 是无法开启并行的,如果我们直接在多个窗口中执行同一个 PLSQL 代码,会遇到锁争用,如果能解决锁争用,在多个窗口中执行同一个 PLSQL 代码,这样就变相实现了 PLSQL 开并行功能。在第 8 章提到过,可以利用 ROWID 切片变相实现并行。

```
select DBMS_ROWID.ROWID_CREATE(1, c.oid, e.RELATIVE_FNO, e.BLOCK_ID, 0) minrid,
       DBMS_ROWID.ROWID_CREATE(1,
                               c.oid,
                               e.RELATIVE_FNO,
                               e.BLOCK_ID + e.BLOCKS - 1,
```

```
                                 10000) maxrid
  from dba_extents e,
       (select max(data_object_id) oid
          from dba_objects
         where object_name = upper('TMP_NBR_NO_XXXX_TEXT')
           and owner = upper('RESCZ2')
           and data_object_id is not null) c
 where e.segment_name = 'TMP_NBR_NO_XXXX_TEXT'
   and e.owner = 'RESCZ2';
```

但是这时我们发现，切割出来的数据分布严重不均衡，这是因为创建表空间的时候没有指定 uniform size 的 Extent。于是我们新建一个表空间，指定采用 uniform size 方式管理 Extent。

```
create tablespace TBS_BSS_FIXED  datafile
           '/oradata/osstest2/tbs_bss_fixed_500.dbf'
     size 500M extent management local uniform size 128k;
```

我们重建一个表用来存储要被更新的 ROWID。

```
create table RID_TABLE
(
  rowno   NUMBER,
  minrid VARCHAR2(18),
  maxrid VARCHAR2(18)
) ;
```

我们将 ROWID 插入到新表中。

```
insert into rid_table
  select rownum rowno,
         DBMS_ROWID.ROWID_CREATE(1, c.oid, e.RELATIVE_FNO, e.BLOCK_ID, 0) minrid,
         DBMS_ROWID.ROWID_CREATE(1,
                                 c.oid,
                                 e.RELATIVE_FNO,
                                 e.BLOCK_ID + e.BLOCKS - 1,
                                 10000) maxrid
    from dba_extents e,
         (select max(data_object_id) oid
            from dba_objects
           where object_name = upper('TMP_NBR_NO_XXXX_TEXT')
             and owner = upper('RESCZ2')
             and data_object_id is not null) c
   where e.segment_name = 'TMP_NBR_NO_XXXX_TEXT'
     and e.owner = 'RESCZ2';
```

这样 RID_TABLE 中每行指定的数据都很均衡，大概 4 035 条数据。最终更改的 PLSQL 代码如下。

```
create or replace  procedure  pro_phone_grade(flag_num in number)
as
 type rowid_table_type is table of  rowid index  by  pls_integer;
  updateCur  sys_refcursor;
 v_rowid  rowid_table_type;
 v_rowid2  rowid_table_type;
begin
for  rowid_cur in (select  *  from  rid_table  where mod(rowno, 8)=flag_num
  loop
     for c_no_data in (select t.expression, t.nbr_level_id, t.priority  from TMP_DATE_TEST t order by 3 )
       loop
```

```
            open  updateCur  for  select rid,rowid  from TMP_NBR_NO_XXXX_TEXT  nbn
              where rowid between rowid_cur.minrid and rowid_cur.maxrid
              and regexp_like(nbn.no, c_no_data.expression);
            loop
              fetch updateCur  bulk collect  into  v_rowid, v_rowid2 LIMIT 20000;
                 forall i in v_rowid.FIRST ..v_rowid.LAST
                 update TMP_NBR_NO_XXXX  set  level_id = c_no_data.nbr_level_id  where rowid = v_rowid(i);
                 commit;
                 exit when  updateCur%notfound;
            end loop;
            CLOSE updateCur;
         end loop;
      end loop;
end;
```

然后我们在 8 个窗口中同时运行以上 PLSQL 代码。

```
begin
pro_phone_grade(0);
end;

begin
pro_phone_grade(1);
end;

begin
pro_phone_grade(2);
end;

……

begin
pro_phone_grade(7);
end;
```

最终我们能在 29 分钟左右执行完所有存储过程。本案例经典之处就在于 ROWID 切片实现并行，同时考虑到了数据分布对并行的影响，其次还使用了 ROWID 关联更新技巧。

9.22 ROW LEVEL SECURITY 优化案例

在做报表开发的时候，有时我们会遇到这样的需求：不同权限的账户各自对应不同的权限，从而看到不同的数据，这时我们一般会采用 Row Level Security 实现这样的需求。2011 年，作者罗老师在惠普的时候，遇到过多起 Row Level Security 引发的 SQL 性能问题。

Obiee 报表开发人员发来邮件反映，使用权限较低的账户打开报表非常缓慢，报表运行了 15 分钟还没响应；而使用权限最高的账户，报表可以在 16 秒内执行完毕。执行缓慢的 SQL 代码如下。

```
select sum(nvl(T1796547.ACTL_GIV_AMT , 0)) as c1,
T1792779.ACCT_LONG_NAME as c2,
T1792779.NAME as c3,
T1796631.PRMTN_NAME as c4,
T1796631.PRMTN_ID as c5,
case  when case  when T1796631.CORP_PRMTN_TYPE_CODE = 'Target Account' then 'Corporate' else T1796631.CORP_PRMTN_TYPE_CODE end  is null then 'Private' else
```

第 9 章　SQL 优化案例赏析

```sql
      case when T1796631.CORP_PRMTN_TYPE_CODE = 'Target Account' then 'Corporate' else T17
96631.CORP_PRMTN_TYPE_CODE end   end   as c6,
      T1796631.PRMTN_STTUS_CODE as c7,
      T1796631.APPRV_BY_DESC as c8,
      T1796631.APPRV_STTUS_CODE as c9,
      T1796631.AUTO_UPDT_GTIN_IND as c10,
      T1796631.CREAT_DATE as c11,
      T1796631.PGM_START_DATE as c12,
      T1796631.PGM_END_DATE as c13,
      nvl(case  when T1796631.PRMTN_STTUS_CODE = 'Confirmed' then cast(( TRUNC( TO_DATE('20
11-04-26' , 'YYYY-MM-DD') ) - TRUNC( T1796631.PGM_END_DATE ) ) as
VARCHAR ( 10 ) ) end  , '') as c14,
      T1796631.PRMTN_STOP_DATE as c15,
      T1796631.SHPMT_START_DATE as c16,
      T1796631.SHPMT_END_DATE as c17,
      T1796631.CNBLN_WK_CNT as c18,
      T1796631.ACTVY_DETL_POP as c19,
      T1796631.CMMNT_DESC as c20,
      T1796631.PRMTN_AVG_POP as c21,
      T1792779.CHANL_TYPE_DESC as c22,
      T1796631.PRMTN_SKID as c23
from
      OPT_ACCT_FDIM T1792779 /* OPT_ACCT_PRMTN_FDIM */ ,
      OPT_BUS_UNIT_FDIM T1796263,
      OPT_CAL_MASTR_DIM T1796564 /* OPT_CAL_MASTR_DIM01 */ ,
      OPT_PRMTN_FDIM T1796631,
      OPT_BASLN_FCT T1796547
where  ( T1792779.ACCT_SKID = T1796547.ACCT_SKID
      and T1792779.BUS_UNIT_SKID = T1796547.BUS_UNIT_SKID
      and T1796263.BUS_UNIT_SKID = T1796547.BUS_UNIT_SKID
      and T1796547.WK_SKID = T1796564.CAL_MASTR_SKID
      and T1796547.BUS_UNIT_SKID = T1796631.BUS_UNIT_SKID
      and T1792779.ACCT_LONG_NAME = 'FN-AEON_GROUP(JUSCOJ4)(C005) - 1900001326'
      and T1796263.BUS_UNIT_NAME = 'Japan'
      and T1796547.PRMTN_SKID = T1796631.PRMTN_SKID
      and T1796564.FISC_YR_ABBR_NAME = 'FY10/11'
      and T1792779.ACCT_LONG_NAME is not null
      -- add RLS
      and T1796547.acct_skid IN (select  org.org_skid from (SELECT DISTINCT ap.org_skid
            FROM opt_acct_postn_lkp ap, opt_party_persn_lkp pp, opt_user_lkp u
            WHERE ap.postn_id = pp.party_id
              AND pp.persn_id = u.user_id
              AND u.login_name = 'BT0016'
          union select 0 as org_skid
          from sys.dual) org
        )
      and T1792779.bus_unit_skid IN (0,11769,11772,11774,11777,11779,11780,14329,14334,1433
9,14340,14341,14350,14800,14801)
      and T1796547.bus_unit_skid IN (0,11769,11772,11774,11777,11779,11780,14329,14334,1433
9,14340,14341,14350,14800,14801)
      and T1796263.bus_unit_skid IN (0,11769,11772,11774,11777,11779,11780,14329,14334,1433
9,14340,14341,14350,14800,14801)
      and T1796631.bus_unit_skid IN (0,11769,11772,11774,11777,11779,11780,14329,14334,1433
9,14340,14341,14350,14800,14801)
      -- end RLS
      and (case when T1796631.CORP_PRMTN_TYPE_CODE = 'Target Account' then 'Corporate'
      else T1796631.CORP_PRMTN_TYPE_CODE end  in ('Corporate', 'Planned', 'Private'))
      and T1796631.PRMTN_LONG_NAME in (select distinct T1796631.PRMTN_LONG_NAME as c1
from
      OPT_ACCT_FDIM T1792779 /* OPT_ACCT_PRMTN_FDIM */ ,
      OPT_BUS_UNIT_FDIM T1796263,
      OPT_CAL_MASTR_DIM T1796564 /* OPT_CAL_MASTR_DIM01 */ ,
```

9.22 ROW LEVEL SECURITY 优化案例

```
     OPT_PRMTN_FDIM T1796631,
     OPT_PRMTN_PROD_FLTR_LKP T1796906
where ( T1792779.ACCT_SKID = T1796906.ACCT_PRMTN_SKID
  and T1792779.BUS_UNIT_SKID = T1796906.BUS_UNIT_SKID
  and T1796263.BUS_UNIT_SKID = T1796906.BUS_UNIT_SKID
  and T1796564.CAL_MASTR_SKID = T1796906.DATE_SKID
  and T1796631.PRMTN_SKID = T1796906.PRMTN_SKID
  and T1792779.ACCT_LONG_NAME = 'FN-AEON_GROUP(JUSCOJ4)(C005) - 1900001326'
  and T1796263.BUS_UNIT_NAME = 'Japan' and T1796564.FISC_YR_ABBR_NAME = 'FY10/11'
  and T1796631.BUS_UNIT_SKID = T1796906.BUS_UNIT_SKID
  and (case when T1796631.CORP_PRMTN_TYPE_CODE = 'Target Account' then 'Corporate' els
e T1796631.CORP_PRMTN_TYPE_CODE end  in ('Corporate', 'Planned',
'Private')) and ROWNUM >= 1 ) ) )
group by T1792779.NAME, T1792779.CHANL_TYPE_DESC,
T1792779.ACCT_LONG_NAME, T1796631.PRMTN_SKID,
T1796631.PRMTN_ID, T1796631.PRMTN_NAME,
T1796631.SHPMT_END_DATE, T1796631.SHPMT_START_DATE,
T1796631.PRMTN_STTUS_CODE, T1796631.APPRV_STTUS_CODE,
T1796631.CMMNT_DESC, T1796631.PGM_START_DATE,
T1796631.PGM_END_DATE, T1796631.CREAT_DATE,
T1796631.APPRV_BY_DESC, T1796631.AUTO_UPDT_GTIN_IND,
T1796631.PRMTN_STOP_DATE, T1796631.ACTVY_DETL_POP,
T1796631.CNBLN_WK_CNT, T1796631.PRMTN_AVG_POP,
case  when case
when T1796631.CORP_PRMTN_TYPE_CODE = 'Target Account'
then 'Corporate' else T1796631.CORP_PRMTN_TYPE_CODE end  is null
then 'Private' else case  when
T1796631.CORP_PRMTN_TYPE_CODE = 'Target Account' then 'Corporate'
else T1796631.CORP_PRMTN_TYPE_CODE end   end , nvl(case  when
T1796631.PRMTN_STTUS_CODE = 'Confirmed'
then cast(( TRUNC( TO_DATE('2011-04-26' , 'YYYY-MM-DD') ) - TRUNC( T1796631.PGM_END_D
ATE ) ) as  VARCHAR ( 10 ) )
end , '')
order by c23, c2;
```

执行缓慢的 SQL 与执行较快的 SQL 相比，缓慢的 SQL 在 where 条件中多了以下部分代码。

```
-- add RLS
and T1796547.acct_skid IN (select  org.org_skid from (SELECT DISTINCT ap.org_skid
    FROM opt_acct_postn_lkp ap, opt_party_persn_lkp pp, opt_user_lkp u
     WHERE ap.postn_id = pp.party_id
       AND pp.persn_id = u.user_id
       AND u.login_name = 'BT0016'
    union select 0 as org_skid
    from sys.dual) org
   )
and T1792779.bus_unit_skid IN (0,11769,11772,11774,11777,11779,11780,14329,14334,1433
9,14340,14341,14350,14800,14801)
and T1796547.bus_unit_skid IN (0,11769,11772,11774,11777,11779,11780,14329,14334,1433
9,14340,14341,14350,14800,14801)
and T1796263.bus_unit_skid IN (0,11769,11772,11774,11777,11779,11780,14329,14334,1433
9,14340,14341,14350,14800,14801)
and T1796631.bus_unit_skid IN (0,11769,11772,11774,11777,11779,11780,14329,14334,1433
9,14340,14341,14350,14800,14801)
-- end RLS
```

这部分代码就是实现 Row Level Security 功能的代码，对于权限较低的账户过滤掉一部分数据，而对于权限最高的账号不做过滤。如果不加 RLS 代码，报表能在 16 秒内执行完毕，但是增加了 RLS 代码，报表执行了 15 分钟不出结果。通过以上信息，我们判断是由于增加了 RLS 代码，导致执行计划发生了变化，从而导致 SQL 性能问题。

RLS 代码中有一个 in 子查询，in 子查询中有 union 关键字。在第 7 章中讲到过子查询非嵌套，当 where 条件中有子查询，优化器会尝试将子查询展开，从而消除 Filter。in 子查询中有 union 是可以展开的（unnest），而 exists 子查询中有 union 是不可以展开的。如果 where 条件中的子查询不能展开（no_unnest），执行计划中会出现 Filter，Filter 一般是在 SQL 的最后阶段执行。如果 where 条件中的子查询展开了，子查询会与主表提前关联。

因为增加了 RLS 代码导致 SQL 产生了性能问题，RLS 代码中有 in 子查询，因为 in 子查询可以展开（unnest），所以我们推断是优化器的子查询非嵌套（Subquery Unnesting）导致产生的性能问题，让 Obiee 开发人员在 in 子查询中添加 HINT：NO_UNNEST，让子查询不展开。子查询不展开，执行计划中就会出现 Filter，但是 Filter 是在最后进行过滤，子查询不展开就不会干扰原始的（跑得快的）执行计划，只是在跑得快的执行计划的最后一步添加 Filter 过滤而已。添加完 HINT 之后，SQL 能在 12 秒内执行完毕。

因为子查询中有 union，这里也可以不添加 HINT：NO_UNNEST，将 in 改写为 exists，这时优化器会自动走 Filter，也能达到优化目的。需要提醒大家的是，千万不要因为我们将 in 改写为 exists、exists 执行快就说 exists 性能比 in 高。如果有谁遇到本案例，将 in 改写为 exists，然后发布博客说今天又用 exists 优化了 in 子查询，这只会让人贻笑大方。

罗老师的个人技术博客中还记录了另一个 RLS 引发的性能问题，大家如有兴趣也可以查看网页：http://blog.csdn.net/robinson1988/article/details/8644565。

9.23 子查询非嵌套优化案例一

2011 年，一位朋友请求优化如下 SQL。

```
select tpc.policy_id,
       tcm.policy_code,
       tpf.organ_id,
       to_char(tpf.insert_time, 'YYYY-MM-DD') As insert_time,
       tpc.change_id,
       d.policy_code,
       e.company_name,
       f.real_name,
       tpf.fee_type,
       sum(tpf.pay_balance) as pay_balance,
       c.actual_type,
       tpc.notice_code,
       d.policy_type,
       g.mode_name as pay_mode
  from t_policy_change     tpc,
       t_contract_master   tcm,
       t_policy_fee        tpf,
       t_fee_type          c,
       t_contract_master   d,
       t_company_customer  e,
       t_customer          f,
       t_pay_mode          g
 where tpc.change_id = tpf.change_id
   and tpf.policy_id = d.policy_id
   and tcm.policy_id = tpc.policy_id
```

9.23 子查询非嵌套优化案例一

```
       and tpf.receiv_status = 1
       and tpf.fee_status = 1
       and tpf.payment_id is null
       and tpf.fee_type = c.type_id
       and tpf.pay_mode = g.mode_id
       and d.company_id = e.company_id(+)
       and d.applicant_id = f.customer_id(+)
       and tpf.organ_id in
           (select
              organ_id
               from t_company_organ
             start with organ_id = '101'
           connect by prior organ_id = parent_id)
 group by tpc.policy_id,
          tpc.change_id,
          tpf.fee_type,
          to_char(tpf.insert_time, 'YYYY-MM-DD'),
          c.actual_type,
          d.policy_code,
          g.mode_name,
          e.company_name,
          f.real_name,
          tpc.notice_code,
          d.policy_type,
          tpf.organ_id,
          tcm.policy_code
 order by change_id, fee_type;
```

执行计划如下。

```
SQL> select * from table(dbms_xplan.display);

PLAN_TABLE_OUTPUT
---------------------------------------------------------------------------------
| Id|Operation                          | Name                   |Rows  |Bytes|TempSpc| Cost (%CPU)|
---------------------------------------------------------------------------------
|  0|SELECT STATEMENT                   |                        |45962 | 11M |       |45650   (0)|
|  1| SORT GROUP BY                     |                        |45962 | 11M | 23M | 45650   (0)|
|* 2|  HASH JOIN                        |                        |45962 | 11M |     | 43908   (0)|
|  3|   INDEX FULL SCAN                 |T_FEE_TYPE_IDX_003      |  106 | 636 |     |     1   (0)|
|  4|   NESTED LOOPS OUTER              |                        |45962 | 11M |     | 43906   (0)|
|* 5|    HASH JOIN                      |                        |45962 |7271K|6824K| 43905   (0)|
|  6|     NESTED LOOPS                  |                        |45961 |6283K|     | 42312   (0)|
|* 7|      HASH JOIN SEMI               |                        |45961 |5655K| 50M | 33120   (1)|
|* 8|       HASH JOIN OUTER             |                        | 400K | 45M | 44M | 32315   (1)|
|* 9|        HASH JOIN                  |                        | 400K | 39M | 27M | 26943   (0)|
|*10|         HASH JOIN                 |                        | 400K | 23M |     | 16111   (0)|
| 11|          TABLE ACCESS FULL        |T_PAY_MODE              |   25 | 525 |     |     2   (0)|
|*12|          TABLE ACCESS FULL        |T_POLICY_FEE            | 400K | 15M |     | 16107   (0)|
| 13|         TABLE ACCESS FULL         |T_CONTRACT_MASTER       |1136K | 46M |     |  9437   (0)|
| 14|        VIEW                       |index_join_007          |2028K | 30M |     |       |
|*15|         HASH JOIN                 |                        | 400K | 45M | 44M | 32315   (1)|
| 16|          INDEX FAST FULL SCAN     |PK_T_CUSTOMER           |2028K | 30M |     |   548   (0)|
| 17|          INDEX FAST FULL SCAN     |IDX_CUSTOMER__BIR_REAL_GEN|2028K | 30M |     |   548   (0)|
| 18|       VIEW                        |VW_NSO_1                |    7 |  42 |     |       |
|*19|        CONNECT BY WITH FILTERING  |                        |      |     |     |       |
| 20|         NESTED LOOPS              |                        |      |     |     |       |
|*21|          INDEX UNIQUE SCAN        |PK_T_COMPANY_ORGAN      |    1 |   6 |     |       |
| 22|          TABLE ACCESS BY USER ROWID|T_COMPANY_ORGAN        |      |     |     |       |
| 23|         NESTED LOOPS              |                        |      |     |     |       |
| 24|          BUFFER SORT              |                        |    7 |  70 |     |       |
| 25|           CONNECT BY PUMP         |                        |      |     |     |       |
```

```
|*26|          INDEX RANGE SCAN          |T_COMPANY_ORGAN_IDX_002 |    7|   70|         |    1   (0)|
| 27| TABLE ACCESS BY INDEX ROWID        |T_POLICY_CHANGE         |    1|   14|         |    2  (50)|
|*28|         INDEX UNIQUE SCAN          |PK_T_POLICY_CHANGE      |    1|     |         |    1   (0)|
| 29|       INDEX FAST FULL SCAN         |IDX1_ACCEPT_DATE        |1136K|  23M|         |  899   (0)|
| 30| TABLE ACCESS BY INDEX ROWID        |T_COMPANY_CUSTOMER      |    1|   90|         |    2  (50)|
|*31|         INDEX UNIQUE SCAN          |PK_T_COMPANY_CUSTOMER   |    1|     |         |            |
-----------------------------------------------------------------------------------------------------

Predicate Information (identified by operation id):
---------------------------------------------------

   2 - access("TPF"."FEE_TYPE"="C"."TYPE_ID")
   5 - access("TCM"."POLICY_ID"="TPC"."POLICY_ID")
   7 - access("TPF"."ORGAN_ID"="VW_NSO_1"."$nso_col_1")
   8 - access("D"."APPLICANT_ID"="F"."CUSTOMER_ID"(+))
   9 - access("TPF"."POLICY_ID"="D"."POLICY_ID")
  10 - access("TPF"."PAY_MODE"="G"."MODE_ID")
  12 - filter("TPF"."CHANGE_ID" IS NOT NULL AND TO_NUMBER("TPF"."RECEIV_STATUS")=1 AN
D "TPF"."FEE_STATUS"=1 AND
              "TPF"."PAYMENT_ID" IS NULL)
  15 - access("indexjoin_alias_012".ROWID="indexjoin_alias_011".ROWID)
  19 - filter("T_COMPANY_ORGAN"."ORGAN_ID"='101')
  21 - access("T_COMPANY_ORGAN"."ORGAN_ID"='101')
  26 - access("T_COMPANY_ORGAN"."PARENT_ID"=NULL)
  28 - access("TPC"."CHANGE_ID"="TPF"."CHANGE_ID")
  31 - access("D"."COMPANY_ID"="E"."COMPANY_ID"(+))

55 rows selected

Statistics
----------------------------------------------------------
         21  recursive calls
          0  db block gets
     125082  consistent gets
      21149  physical reads
          0  redo size
       2448  bytes sent via SQL*Net to client
        656  bytes received via SQL*Net from client
          2  SQL*Net roundtrips to/from client
          4  sorts (memory)
          0  sorts (disk)
         11  rows processed
```

上述 SQL 要执行 12 秒左右,逻辑读 12 万。该 SQL 中,t_policy_fee tpf 有 400 万行,t_contract_master tcm 有 1000 万行。其余表都是小表。

根据 SQL 三段分拆方法首先检查了 SQL 写法,SQL 写法没有明显不妥之处。然后开始检查执行计划。我们注意观察执行计划的统计信息(Statistics),该 SQL 最终只返回 11 行数据(11 rows processed)。**SQL 中有 13 个 GROUP BY 字段,一般而言,GROUP BY 字段越少,去重能力越强;GROUP BY 字段越多,去重能力越弱。因此,我们判断该 SQL 在 GROUP BY 之前只返回少量数据,返回少量数据应该走嵌套循环,而不是走 HASH 连接。既然推断出该 SQL 最终返回数据量较少,那么 SQL 中的大表都应该走索引,但是 SQL 语句中的两个大表 t_policy_fee tpf 与 t_contract_master tcm 都是走的全表扫描,这显然不对。它们应该走索引,或者作为嵌套循环的被驱动表。**

根据上面分析,我们将注意力集中在了大表(Id=12 和 Id=13)上,同时也将注意力集中在了 HASH 连接上。执行计划中 Id=12 有 TO_NUMBER("TPF"."RECEIV_STATUS")=1,开发

9.23 子查询非嵌套优化案例一

人员少写了引号，这可能导致 SQL 不走索引。Id=13 前面没有"*"号，这说明 T_CONTRACT_MASTER 没有过滤条件，如果走 HASH 连接，那么该表只能走全表扫描。但是该表有 1000 万条数据，所以只能让它作为嵌套循环被驱动表，然后走连接列的索引。

SQL 语句中有个 in 子查询，并且子查询中有固化子查询关键字 start with，在 7.1 节中讲到，in 子查询中有固化子查询关键字，子查询可以展开（unnest）。这个 in 子查询只返回 1 行数据，在执行计划中它属于 Id=18，然后它与 Id=8 进行的是 HASH 连接。Where 子查询 unnest 之后，一般都会打乱执行计划，也就是说 Id=8，Id=9，Id=10，Id=11，Id=12，Id=13，Id=14 的执行计划都会因为子查询被展开而在一起关联的。

我们再回去看原始 SQL，原始 SQL 中只有 tpf 表有过滤条件，其他表均无过滤条件。而 tpf 表的过滤条件要么是状态字段过滤（tpf.receiv_status = 1 and tpf.fee_status = 1），要么是组织编号过滤 tpf.organ_id in（子查询）。因此判断这些过滤条件并不能过滤掉大部分数据。SQL 中有两处外链接，d.company_id = e.company_id(+)，d.applicant_id = f.customer_id(+)，如果走嵌套循环，外连接无法更改驱动表。如果走 HASH 连接，外连接可以更改驱动表。

因为 SQL 最终只返回少量数据，我们判断执行计划应该走嵌套循环。走嵌套循环首先要确定好谁做驱动表。根据上面的分析 e,f 首先被排除掉做驱动表的可能性，因为它们是外连接的从表，tpf, tcm 也被排除掉作为驱动表的可能性，因为它们是大表。现在只剩下 tpc, c 和 g 可以作为驱动表候选，tpc, c,g 都是与 tpf 关联的，只需要看谁最小，谁就作为驱动表。而在原始执行计划中，因为 in 子查询被展开了，扰乱了执行计划，导致 Id=11，Id=12，Id=13 走了 HASH 连接，所以笔者对子查询添加了 HINT：NO_UNNEST，让子查询不展开，从而不去干扰执行计划，添加 HINT 后的 SQL 如下。

```
select tpc.policy_id,
       tcm.policy_code,
       tpf.organ_id,
       to_char(tpf.insert_time, 'YYYY-MM-DD') As insert_time,
       tpc.change_id,
       d.policy_code,
       e.company_name,
       f.real_name,
       tpf.fee_type,
       sum(tpf.pay_balance) as pay_balance,
       c.actual_type,
       tpc.notice_code,
       d.policy_type,
       g.mode_name as pay_mode
  from t_policy_change     tpc,
       t_contract_master   tcm,
       t_policy_fee        tpf,
       t_fee_type          c,
       t_contract_master   d,
       t_company_customer  e,
       t_customer          f,
       t_pay_mode          g
 where tpc.change_id = tpf.change_id
   and tpf.policy_id = d.policy_id
   and tcm.policy_id = tpc.policy_id
   and tpf.receiv_status = '1'    ---这里原来没引号，是开发搞忘了写''
   and tpf.fee_status = 1
```

```
           and tpf.payment_id is null
           and tpf.fee_type = c.type_id
           and tpf.pay_mode = g.mode_id
           and d.company_id = e.company_id(+)
           and d.applicant_id = f.customer_id(+)
           and tpf.organ_id in
              (select /*+ no_unnest */
                organ_id
                 from t_company_organ
                start with organ_id = '101'
              connect by prior organ_id = parent_id)
  group by tpc.policy_id,
           tpc.change_id,
           tpf.fee_type,
           to_char(tpf.insert_time, 'YYYY-MM-DD'),
           c.actual_type,
           d.policy_code,
           g.mode_name,
           e.company_name,
           f.real_name,
           tpc.notice_code,
           d.policy_type,
           tpf.organ_id,
           tcm.policy_code
  order by change_id, fee_type
```

执行计划如下。

```
SQL> select * from table(dbms_xplan.display);

PLAN_TABLE_OUTPUT
-------------------------------------------------------------------------------------------------
| Id|Operation                           | Name                       |Rows |Bytes| Cost (%CPU)|
-------------------------------------------------------------------------------------------------
|  0|SELECT STATEMENT                    |                            |20026|4928K| 68615  (30)|
|  1| SORT GROUP BY                      |                            |20026|4928K| 28563   (0)|
|* 2|  FILTER                            |                            |     |     |            |
|  3|   NESTED LOOPS                     |                            |20026|4928K| 27812   (0)|
|  4|    NESTED LOOPS                    |                            |20026|4498K| 23807   (0)|
|  5|     NESTED LOOPS OUTER             |                            |20026|4224K| 19802   (0)|
|  6|      NESTED LOOPS OUTER            |                            |20026|3911K| 15797   (0)|
|  7|       NESTED LOOPS                 |                            |20026|2151K| 15796   (0)|
|* 8|        HASH JOIN                   |                            |20026|1310K| 11791   (0)|
|  9|         INDEX FULL SCAN            |T_FEE_TYPE_IDX_003          |  106|  636|     1   (0)|
|*10|         HASH JOIN                  |                            |20026|1192K| 11789   (0)|
| 11|          TABLE ACCESS FULL         |T_PAY_MODE                  |   25|  525|     2   (0)|
|*12|          TABLE ACCESS BY INDEX ROWID|T_POLICY_FEE               |20026| 782K| 11786   (0)|
|*13|           INDEX RANGE SCAN         |IDX_POLICY_FEE_RECEIV_STATUS|1243K|     | 10188   (0)|
| 14|        TABLE ACCESS BY INDEX ROWID |T_CONTRACT_MASTER           |    1|   43|     2  (50)|
|*15|         INDEX UNIQUE SCAN          |PK_T_CONTRACT_MASTER        |    1|     |     1   (0)|
| 16|       TABLE ACCESS BY INDEX ROWID  |T_COMPANY_CUSTOMER          |    1|   90|     2  (50)|
|*17|        INDEX UNIQUE SCAN           |PK_T_COMPANY_CUSTOMER       |    1|     |            |
| 18|      TABLE ACCESS BY INDEX ROWID   |T_CUSTOMER                  |    1|   16|     2  (50)|
|*19|       INDEX UNIQUE SCAN            |PK_T_CUSTOMER               |    1|     |     1   (0)|
| 20|     TABLE ACCESS BY INDEX ROWID    |T_POLICY_CHANGE             |    1|   14|     2  (50)|
|*21|      INDEX UNIQUE SCAN             |PK_T_POLICY_CHANGE          |    1|     |     1   (0)|
| 22|    TABLE ACCESS BY INDEX ROWID     |T_CONTRACT_MASTER           |    1|   22|     2  (50)|
|*23|     INDEX UNIQUE SCAN              |PK_T_CONTRACT_MASTER        |    1|     |     1   (0)|
|*24|  FILTER                            |                            |     |     |            |
|*25|   CONNECT BY WITH FILTERING        |                            |     |     |            |
| 26|    NESTED LOOPS                    |                            |     |     |            |
|*27|     INDEX UNIQUE SCAN              |PK_T_COMPANY_ORGAN          |    1|    6|            |
```

```
|  28|        TABLE ACCESS BY USER ROWID   |T_COMPANY_ORGAN        |    |    |          |
|  29|         NESTED LOOPS                |                       |    |    |          |
|  30|          BUFFER SORT                |                       |  7 | 70 |          |
|  31|           CONNECT BY PUMP           |                       |    |    |          |
|*32|           INDEX RANGE SCAN           |T_COMPANY_ORGAN_IDX_002|  7 | 70 |   1  (0) |
---------------------------------------------------------------------------------------

Predicate Information (identified by operation id):
---------------------------------------------------

   2 - filter( EXISTS (SELECT /*+ NO_UNNEST */ 0 FROM "T_COMPANY_ORGAN" "T_COMPANY_OR
GAN" WHERE
              "T_COMPANY_ORGAN"."PARENT_ID"=NULL AND ("T_COMPANY_ORGAN"."ORGAN_ID"=:B
1)))
   8 - access("SYS_ALIAS_1"."FEE_TYPE"="C"."TYPE_ID")
  10 - access("SYS_ALIAS_1"."PAY_MODE"="G"."MODE_ID")
  12 - filter("SYS_ALIAS_1"."CHANGE_ID" IS NOT NULL AND "SYS_ALIAS_1"."FEE_STATUS"=1
AND
              "SYS_ALIAS_1"."PAYMENT_ID" IS NULL)
  13 - access("SYS_ALIAS_1"."RECEIV_STATUS"='1')
  15 - access("SYS_ALIAS_1"."POLICY_ID"="D"."POLICY_ID")
  17 - access("D"."COMPANY_ID"="E"."COMPANY_ID"(+))
  19 - access("D"."APPLICANT_ID"="F"."CUSTOMER_ID"(+))
  21 - access("TPC"."CHANGE_ID"="SYS_ALIAS_1"."CHANGE_ID")
  23 - access("TCM"."POLICY_ID"="TPC"."POLICY_ID")
  24 - filter("T_COMPANY_ORGAN"."ORGAN_ID"=:B1)
  25 - filter("T_COMPANY_ORGAN"."ORGAN_ID"='101')
  27 - access("T_COMPANY_ORGAN"."ORGAN_ID"='101')
  32 - access("T_COMPANY_ORGAN"."PARENT_ID"=NULL)

58 rows selected.

Statistics
----------------------------------------------------------
          0  recursive calls
          0  db block gets
       2817  consistent gets
          0  physical reads
          0  redo size
       2268  bytes sent via SQL*Net to client
        656  bytes received via SQL*Net from client
          2  SQL*Net roundtrips to/from client
         40  sorts (memory)
          0  sorts (disk)
          9  rows processed
```

添加完 HINT 之后，SQL 能在 1 秒内执行完毕，逻辑读也降低到 2 817。如果不想添加 HINT，我们可以将 in 改成 exists，因为子查询中有固化子查询关键字，这时 SQL 不能展开，会自动走 Filter，也能达到添加 HINT：NO_UNNEST 的效果，但是，这并不是说 exists 比 in 性能好！

我们推荐大家在 Oracle 中使用 in 而不是使用 exists。因为 exists 子查询中有固化子查询关键字会自动走 Filter，想要消除 Filter 只能改写 SQL。in 可以控制走 Filter 或者不走，in 执行计划可控，而 exists 执行计划不可控。

对于 in 子查询，我们一定要搞清楚 in 子查询返回多少数据，究竟能起到多大过滤作用。如果 in 子查询能过滤掉主表大量数据，这时我们一定要让 in 子查询展开并且作为 NL 驱动表反向驱动主表，主表作为 NL 被驱动表，走连接列索引。如果 in 子查询不能过滤掉主表大量数据，这时要检查 in 子查询返回数据量多少，如果返回数据量很少，in 子查询即使不展开，走 Filter 也不大会影响 SQL 性能。如果 in 子查询返回数据量很多，但是并不能过滤掉主表大量数据，这时一定要让 in 子查询展开并且与主表走 HASH 连接。

本案例中，in 子查询返回数据量很少，只有 1 行数据，但是主表并不能用子查询过滤大量数据，因为过滤条件是 tpf.organ_id，组织关系 id 这种列一般基数很低。其实原始 SQL 相当于如下写法：

```sql
select tpc.policy_id,
       tcm.policy_code,
       tpf.organ_id,
       to_char(tpf.insert_time, 'YYYY-MM-DD') As insert_time,
       tpc.change_id,
       d.policy_code,
       e.company_name,
       f.real_name,
       tpf.fee_type,
       sum(tpf.pay_balance) as pay_balance,
       c.actual_type,
       tpc.notice_code,
       d.policy_type,
       g.mode_name as pay_mode
  from t_policy_change    tpc,
       t_contract_master  tcm,
       t_policy_fee       tpf,
       t_fee_type         c,
       t_contract_master  d,
       t_company_customer e,
       t_customer         f,
       t_pay_mode         g
 where tpc.change_id = tpf.change_id
   and tpf.policy_id = d.policy_id
   and tcm.policy_id = tpc.policy_id
   and tpf.receiv_status = 1
   and tpf.fee_status = 1
   and tpf.payment_id is null
   and tpf.fee_type = c.type_id
   and tpf.pay_mode = g.mode_id
   and d.company_id = e.company_id(+)
   and d.applicant_id = f.customer_id(+)
   and tpf.organ_id in ('xxx')   ---将子查询换成具体值，这样就不会干扰执行计划
 group by tpc.policy_id,
          tpc.change_id,
          tpf.fee_type,
          to_char(tpf.insert_time, 'YYYY-MM-DD'),
          c.actual_type,
          d.policy_code,
          g.mode_name,
          e.company_name,
          f.real_name,
          tpc.notice_code,
          d.policy_type,
          tpf.organ_id,
          tcm.policy_code
 order by change_id, fee_type;
```

因为原始 SQL 本意相当于以上 SQL，子查询只起过滤作用，所以使用 HINT:NO_UNNEST，让子查询不去干扰正常执行计划，从而达到优化目的。

9.24 子查询非嵌套优化案例二

本案例与上一个案例是同一个人的优化请求，SQL 语句如下。

```
select distinct decode(length(a.category_id),
                       5,
                       decode(a.origin_type, 801, 888888, 999999),
                       a.category_id) category_id,
            a.notice_code,
            a.treat_status,
            lr.real_name as receiver_name,
            f.send_code,
            f.policy_code,
            g.real_name agent_name,
            f.organ_id,
            f.dept_id,
            a.policy_id,
            a.change_id,
            a.case_id,
            a.group_policy_id,
            a.fee_id,
            a.auth_id,
            a.pay_id,
            cancel_appoint.appoint_time cancel_appoint_time,
            a.insert_time,
            a.send_time,
            a.end_time,
            f.agency_code,
            a.REPLY_TIME,
            a.REPLY_EMP_ID,
            a.FIRST_DUTY,
            a.NEED_SEND_PRINT,
            11 source
  from t_policy_problem       a,
       t_policy               f,
       t_agent                g,
       t_letter_receiver      lr,
       t_problem_category     pc,
       t_policy_cancel_appoint cancel_appoint
 where f.agent_id = g.agent_id(+)
   and a.policy_id = f.policy_id(+)
   and lr.main_receiver = 'Y'
   and a.category_id = pc.category_id
   and a.item_id = lr.item_id
   and a.policy_id = cancel_appoint.policy_id(+)
   And a.Item_Id = (Select Max(item_id)
                      From t_Policy_Problem
                     Where notice_code = a.notice_code)
   and a.policy_id is not null
   and a.notice_code is not null
   and a.change_id is null
   and a.case_id is null
   and a.group_policy_id is null
   and a.origin_type not in (801, 802)
   and a.pay_id is null
   and a.category_id not in (130103, 130104, 130102, 140102, 140101)
   and f.policy_type = '1'
   and (a.fee_id is null or (a.fee_id is not null and a.origin_type = 701))
   and exists((select 1
```

```
                from t_dept
                where f.dept_id = dept_id
                start with dept_id = '1020200028'
                connect by parent_id = prior dept_id))
    and exists (select 1
                from T_COMPANY_ORGAN
                where f.organ_id = organ_id
                start with organ_id = '10202'
                connect by parent_id = prior organ_id)
    and pc.NEED_PRITN = 'Y';
```

朋友说这个 SQL 执行不出结果。执行计划如下。

```
PLAN_TABLE_OUTPUT
--------------------------------------------------------------------------------
| Id  |Operation                        | Name                      |Rows |Bytes|Cost(%CPU)|
--------------------------------------------------------------------------------
|  0  |SELECT STATEMENT                 |                           |  1| 236|  741   (1)|
|  1  | SORT UNIQUE                     |                           |  1| 236|  681   (0)|
|* 2  |  FILTER                         |                           |   |    |           |
|  3  |   NESTED LOOPS                  |                           |  1| 236|  666   (1)|
|  4  |    NESTED LOOPS OUTER           |                           |  1| 219|  665   (1)|
|  5  |     NESTED LOOPS                |                           |  1| 203|  664   (1)|
|  6  |      NESTED LOOPS OUTER         |                           |  1| 196|  663   (1)|
|  7  |       NESTED LOOPS              |                           |  1| 182|  662   (1)|
|* 8  |        TABLE ACCESS FULL        |T_POLICY_PROBLEM           |  1| 107|  660   (0)|
|* 9  |        TABLE ACCESS BY INDEX ROWID|T_POLICY                 |  1|  75|    2  (50)|
|*10  |         INDEX UNIQUE SCAN       |PK_T_POLICY                |  1|    |    1   (0)|
| 11  |       TABLE ACCESS BY INDEX ROWID|T_POLICY_CANCEL_APPOINT   |  1|  14|    2  (50)|
|*12  |        INDEX UNIQUE SCAN        |UK1_POLICY_CANCEL_APPOINT  |  1|    |           |
|*13  |      TABLE ACCESS BY INDEX ROWID|T_PROBLEM_CATEGORY         |  1|   7|    2  (50)|
|*14  |       INDEX UNIQUE SCAN         |PK_T_PROBLEM_CATEGORY      |  1|    |           |
| 15  |     TABLE ACCESS BY INDEX ROWID |T_AGENT                    |  1|  16|    2  (50)|
|*16  |      INDEX UNIQUE SCAN          |PK_T_AGENT                 |  1|    |           |
|*17  |    INDEX RANGE SCAN             |T_LETTER_RECEIVER_IDX_001  |  1|  17|    2   (0)|
| 18  |  SORT AGGREGATE                 |                           |  1|  21|           |
| 19  |   TABLE ACCESS BY INDEX ROWID   |T_POLICY_PROBLEM           |  1|  21|    2  (50)|
|*20  |    INDEX RANGE SCAN             |IDX_POLICY_PROBLEM__N_CODE |  1|    |    3   (0)|
|*21  |  FILTER                         |                           |   |    |           |
|*22  |   CONNECT BY WITH FILTERING     |                           |   |    |           |
| 23  |    NESTED LOOPS                 |                           |   |    |           |
|*24  |     INDEX UNIQUE SCAN           |PK_T_DEPT                  |  1|  17|    1   (0)|
| 25  |     TABLE ACCESS BY USER ROWID  |T_DEPT                     |   |    |           |
| 26  |    HASH JOIN                    |                           |   |    |           |
| 27  |     CONNECT BY PUMP             |                           |   |    |           |
| 28  |     TABLE ACCESS FULL           |T_DEPT                     |30601| 896K|   56   (0)|
|*29  |  FILTER                         |                           |   |    |           |
|*30  |   CONNECT BY WITH FILTERING     |                           |   |    |           |
| 31  |    NESTED LOOPS                 |                           |   |    |           |
|*32  |     INDEX UNIQUE SCAN           |PK_T_COMPANY_ORGAN         |  1|   6|           |
| 33  |     TABLE ACCESS BY USER ROWID  |T_COMPANY_ORGAN            |   |    |           |
| 34  |    NESTED LOOPS                 |                           |   |    |           |
| 35  |     BUFFER SORT                 |                           |  7|  70|           |
| 36  |      CONNECT BY PUMP            |                           |   |    |           |
|*37  |     INDEX RANGE SCAN            |T_COMPANY_ORGAN_IDX_002    |  7|  70|    1   (0)|
--------------------------------------------------------------------------------

   2 - filter("SYS_ALIAS_1"."ITEM_ID"= (SELECT /*+ */ MAX("T_POLICY_PROBLEM"."ITEM_ID
") FROM
              "T_POLICY_PROBLEM" "T_POLICY_PROBLEM" WHERE "T_POLICY_PROBLEM"."NOTICE_
CODE"=:B1) AND
```

9.24 子查询非嵌套优化案例二

```
                    EXISTS (SELECT/*+ */ 0 FROM "T_DEPT" "T_DEPT" AND ("T_DEPT"."DEPT_ID"=
:B2)) AND    EXISTS
                    (SELECT /*+ */ 0 FROM"T_COMPANY_ORGAN" "T_COMPANY_ORGAN" WHERE
                    "T_COMPANY_ORGAN"."PARENT_ID"=NULL AND ("T_COMPANY_ORGAN"."ORGAN_ID"=:
B3)))
    8 - filter("SYS_ALIAS_1"."POLICY_ID" IS NOT NULL AND "SYS_ALIAS_1"."NOTICE_CODE" I
S NOT NULL AND
                    "SYS_ALIAS_1"."CHANGE_ID" IS NULL AND "SYS_ALIAS_1"."CASE_ID" IS NULL AND
                    "SYS_ALIAS_1"."GROUP_POLICY_ID" IS NULL AND TO_NUMBER("SYS_ALIAS_1"."OR
IGIN_TYPE")<>801 AND
                    TO_NUMBER("SYS_ALIAS_1"."ORIGIN_TYPE")<>802 AND "SYS_ALIAS_1"."PAY_ID"
IS NULL AND
                    "SYS_ALIAS_1"."CATEGORY_ID"<>130103 AND "SYS_ALIAS_1"."CATEGORY_ID"<>13
0104 AND
                    "SYS_ALIAS_1"."CATEGORY_ID"<>130102 AND "SYS_ALIAS_1"."CATEGORY_ID"<>14
0102 AND
                    "SYS_ALIAS_1"."CATEGORY_ID"<>140101 AND ("SYS_ALIAS_1"."FEE_ID" IS NULL
 OR
                    "SYS_ALIAS_1"."FEE_ID" IS NOT NULL AND TO_NUMBER("SYS_ALIAS_1"."ORIGIN_
TYPE")=701))
    9 - filter(TO_NUMBER("SYS_ALIAS_3"."POLICY_TYPE")=1)
   10 - access("SYS_ALIAS_1"."POLICY_ID"="SYS_ALIAS_3"."POLICY_ID")
   12 - access("SYS_ALIAS_1"."POLICY_ID"="CANCEL_APPOINT"."POLICY_ID"(+))
   13 - filter("PC"."NEED_PRITN"='Y')
   14 - access("SYS_ALIAS_1"."CATEGORY_ID"="PC"."CATEGORY_ID")
        filter("PC"."CATEGORY_ID"<>130103 AND "PC"."CATEGORY_ID"<>130104 AND "PC"."CAT
EGORY_ID"<>130102
                    AND "PC"."CATEGORY_ID"<>140102 AND "PC"."CATEGORY_ID"<>140101)
   16 - access("SYS_ALIAS_3"."AGENT_ID"="G"."AGENT_ID"(+))
   17 - access("LR"."MAIN_RECEIVER"='Y' AND "SYS_ALIAS_1"."ITEM_ID"="LR"."ITEM_ID")
   20 - access("T_POLICY_PROBLEM"."NOTICE_CODE"=:B1)
   21 - filter("T_DEPT"."DEPT_ID"=:B1)
   22 - filter("T_DEPT"."DEPT_ID"='1020200028')
   24 - access("T_DEPT"."DEPT_ID"='1020200028')
   29 - filter("T_COMPANY_ORGAN"."ORGAN_ID"=:B1)
   30 - filter("T_COMPANY_ORGAN"."ORGAN_ID"='10202')
   32 - access("T_COMPANY_ORGAN"."ORGAN_ID"='10202')
   37 - access("T_COMPANY_ORGAN"."PARENT_ID"=NULL)

77 rows selected.
```

从执行计划中 Id=2 看到，该 SQL 走了 Filter，Id=3、Id=18、Id=21、Id=29 都是 Id=2 的儿子。因为 Filter 类似嵌套循环，如果 Id=3 返回大量数据，会导致 Id=18、Id=21、Id=29 被多次扫描，正是因为 SQL 走的是 Filter，才导致 SQL 执行不出结果。

为什么会走 Filter 呢？我们注意查看 SQL 写法，SQL 语句中有两个 exists（子查询），子查询中有固化子查询关键字 start with，正是因为 SQL 写成了 exists，才导致走了 Filter。于是我们用 in 改写 exists。

```
select distinct decode(length(a.category_id),
                       5,
                       decode(a.origin_type, 801, 888888, 999999),
                       a.category_id) category_id,
                a.notice_code,
                a.treat_status,
                lr.real_name as receiver_name,
                f.send_code,
                f.policy_code,
                g.real_name agent_name,
```

```
                        f.organ_id,
                        f.dept_id,
                        a.policy_id,
                        a.change_id,
                        a.case_id,
                        a.group_policy_id,
                        a.fee_id,
                        a.auth_id,
                        a.pay_id,
                        cancel_appoint.appoint_time cancel_appoint_time,
                        a.insert_time,
                        a.send_time,
                        a.end_time,
                        f.agency_code,
                        a.REPLY_TIME,
                        a.REPLY_EMP_ID,
                        a.FIRST_DUTY,
                        a.NEED_SEND_PRINT,
                        11 source
    from t_policy_problem          a,
         t_policy                  f,
         t_agent                   g,
         t_letter_receiver         lr,
         t_problem_category        pc,
         t_policy_cancel_appoint cancel_appoint
  where f.agent_id = g.agent_id(+)
    and a.policy_id = f.policy_id(+)
    and lr.main_receiver = 'Y'
    and a.category_id = pc.category_id
    and a.item_id = lr.item_id
    and a.policy_id = cancel_appoint.policy_id(+)
    And a.Item_Id = (Select Max(item_id)
                       From t_Policy_Problem
                      Where notice_code = a.notice_code)
    and a.policy_id is not null
    and a.notice_code is not null
    and a.change_id is null
    and a.case_id is null
    and a.group_policy_id is null
    and a.origin_type not in (801, 802)
    and a.pay_id is null
    and a.category_id not in (130103, 130104, 130102, 140102, 140101)
    and f.policy_type = '1'
    and (a.fee_id is null or (a.fee_id is not null and a.origin_type = 701))
    and f.dept_id in (select dept_id
             from t_dept
            start with dept_id = '1020200028'
           connect by parent_id = prior dept_id))
    and f.organ_id in (select organ_id
          from T_COMPANY_ORGAN
         start with organ_id = '10202'
        connect by parent_id = prior organ_id)
    and pc.NEED_PRITN = 'Y';
```

改写后的执行计划如下。

```
---------------------------------------------------------------------------
| Id|Operation                     | Name        |Rows |Bytes|Cost(%CPU)|
---------------------------------------------------------------------------
|  0|SELECT STATEMENT              |             |   1|  259|  742   (1)|
|  1| SORT UNIQUE                  |             |   1|  259|  740   (0)|
|* 2|  FILTER                      |             |    |     |           |
```

9.24 子查询非嵌套优化案例二

```
|* 3|   HASH JOIN                         |                           |    1|  259|  725   (1)|
|  4|    NESTED LOOPS                     |                           |    1|  253|  723   (1)|
|  5|     NESTED LOOPS                    |                           |    1|  236|  722   (1)|
|  6|      NESTED LOOPS OUTER             |                           |    1|  229|  721   (1)|
|  7|       NESTED LOOPS OUTER            |                           |    1|  215|  720   (1)|
|* 8|        HASH JOIN                    |                           |    1|  199|  719   (1)|
|  9|         NESTED LOOPS                |                           |    1|  182|  662   (1)|
|*10|          TABLE ACCESS FULL          |T_POLICY_PROBLEM           |    1|  107|  660   (0)|
|*11|          TABLE ACCESS BY INDEX ROWID|T_POLICY                   |    1|   75|    2  (50)|
|*12|           INDEX UNIQUE SCAN         |PK_T_POLICY                |    1|     |    1   (0)|
| 13|         VIEW                        |VW_NSO_1                   |30601| 508K|           |
|*14|          CONNECT BY WITH FILTERING  |                           |     |     |           |
| 15|           NESTED LOOPS              |                           |     |     |           |
|*16|            INDEX UNIQUE SCAN        |PK_T_DEPT                  |    1|   17|    1   (0)|
| 17|            TABLE ACCESS BY USER ROWID|T_DEPT                    |     |     |           |
| 18|           HASH JOIN                 |                           |     |     |           |
| 19|            CONNECT BY PUMP          |                           |     |     |           |
| 20|            TABLE ACCESS FULL        |T_DEPT                     |30601| 896K|   56   (0)|
| 21|        TABLE ACCESS BY INDEX ROWID  |T_AGENT                    |    1|   16|    2  (50)|
|*22|         INDEX UNIQUE SCAN           |PK_T_AGENT                 |    1|     |           |
| 23|       TABLE ACCESS BY INDEX ROWID   |T_POLICY_CANCEL_APPOINT    |    1|   14|    2  (50)|
|*24|        INDEX UNIQUE SCAN            |UK1_POLICY_CANCEL_APPOINT  |    1|     |           |
|*25|      TABLE ACCESS BY INDEX ROWID    |T_PROBLEM_CATEGORY         |    1|    7|    2  (50)|
|*26|       INDEX UNIQUE SCAN             |PK_T_PROBLEM_CATEGORY      |    1|     |           |
|*27|     INDEX RANGE SCAN                |T_LETTER_RECEIVER_IDX_001  |    1|   17|    2   (0)|
| 28|    VIEW                             |VW_NSO_2                   |    7|   42|           |
|*29|     CONNECT BY WITH FILTERING       |                           |     |     |           |
| 30|      NESTED LOOPS                   |                           |     |     |           |
|*31|       INDEX UNIQUE SCAN             |PK_T_COMPANY_ORGAN         |    1|    6|           |
| 32|       TABLE ACCESS BY USER ROWID    |T_COMPANY_ORGAN            |     |     |           |
| 33|      NESTED LOOPS                   |                           |     |     |           |
| 34|       BUFFER SORT                   |                           |    7|   70|           |
| 35|        CONNECT BY PUMP              |                           |     |     |           |
|*36|       INDEX RANGE SCAN              |T_COMPANY_ORGAN_IDX_002    |    7|   70|    1   (0)|
| 37| SORT AGGREGATE                      |                           |    1|   21|           |
| 38|  TABLE ACCESS BY INDEX ROWID        |T_POLICY_PROBLEM           |    1|   21|    2  (50)|
|*39|   INDEX RANGE SCAN                  |IDX_POLICY_PROBLEM__N_CODE |    1|     |    3   (0)|
-----------------------------------------------------------------------------------------------

Predicate Information (identified by operation id):
---------------------------------------------------

   2 - filter("SYS_ALIAS_1"."ITEM_ID"= (SELECT /*+ */ MAX("T_POLICY_PROBLEM"."ITEM_ID
") FROM
              "T_POLICY_PROBLEM" "T_POLICY_PROBLEM" WHERE "T_POLICY_PROBLEM"."NOTICE_
CODE"=:B1))
   3 - access("F"."ORGAN_ID"="VW_NSO_2"."$nso_col_1")
   8 - access("F"."DEPT_ID"="VW_NSO_1"."$nso_col_1")
  10 - filter("SYS_ALIAS_1"."POLICY_ID" IS NOT NULL AND "SYS_ALIAS_1"."NOTICE_CODE" I
S NOT NULL AND
              "SYS_ALIAS_1"."CHANGE_ID" IS NULL AND "SYS_ALIAS_1"."CASE_ID" IS NULL AND
              "SYS_ALIAS_1"."GROUP_POLICY_ID"
       IS NULL AND TO_NUMBER("SYS_ALIAS_1"."ORIGIN_TYPE")<>801 AND
              TO_NUMBER("SYS_ALIAS_1"."ORIGIN_TYPE")<>802
              AND "SYS_ALIAS_1"."PAY_ID" IS NULL AND "SYS_ALIAS_1"."CATEGORY_ID"<>13
0103 AND
              "SYS_ALIAS_1"."CATEGORY_ID"<>130104 AND "SYS_ALIAS_1"."CATEGORY_ID"<>13
0102 AND
              "SYS_ALIAS_1"."CATEGORY_ID"<>140102 AND "SYS_ALIAS_1"."CATEGORY_ID"<>14
0101 AND
              ("SYS_ALIAS_1"."FEE_ID" IS NULL OR "SYS_ALIAS_1"."FEE_ID" IS NOT NULL AND
              TO_NUMBER("SYS_ALIAS_1"."ORIGIN_TYPE")=701))
```

```
11 - filter("F"."POLICY_TYPE"='1')
12 - access("SYS_ALIAS_1"."POLICY_ID"="F"."POLICY_ID")
14 - filter("T_DEPT"."DEPT_ID"='1020200028')
16 - access("T_DEPT"."DEPT_ID"='1020200028')
22 - access("F"."AGENT_ID"="G"."AGENT_ID"(+))
24 - access("SYS_ALIAS_1"."POLICY_ID"="CANCEL_APPOINT"."POLICY_ID"(+))
25 - filter("PC"."NEED_PRITN"='Y')
26 - access("SYS_ALIAS_1"."CATEGORY_ID"="PC"."CATEGORY_ID")
     filter("PC"."CATEGORY_ID"<>130103 AND "PC"."CATEGORY_ID"<>130104 AND "PC"."CAT
EGORY_ID"<>130102
            AND "PC"."CATEGORY_ID"<>140102 AND "PC"."CATEGORY_ID"<>140101)
27 - access("LR"."MAIN_RECEIVER"='Y' AND "SYS_ALIAS_1"."ITEM_ID"="LR"."ITEM_ID")
29 - filter("T_COMPANY_ORGAN"."ORGAN_ID"='10202')
31 - access("T_COMPANY_ORGAN"."ORGAN_ID"='10202')
36 - access("T_COMPANY_ORGAN"."PARENT_ID"=NULL)
39 - access("T_POLICY_PROBLEM"."NOTICE_CODE"=:B1)
```

SQL 改写之后，可以在 35 秒左右出结果，而之前是很久跑不出结果。用 in 代替 exists 之后，两个 in 子查询因为进行了 Subquery Unnesting，消除了 Filter。从执行计划中我们可以看到，两个子查询都走的是 HASH 连接，这样两个 in 子查询都只会被扫描一次。用 in 代替 exists 之后，执行计划中还有 Filter，这时的 Filter 来自于 t_Policy_Problem 自关联。

```
And a.Item_Id = (Select Max(item_id)
                   From t_Policy_Problem
                   Where notice_code = a.notice_code)
```

在第 8 章中讲到，可以利用分析函数改写自关联。因为当时朋友对 35 秒出结果已经很满意，所以我们没有进一步改写 SQL。本以为能逃过帮忙改写 SQL "一劫"，但是 2012 年刚过完春节，就被朋友骚扰了，朋友要求继续优化，有兴趣的读者可以查看博客：http://blog.csdn.net/robinson1988/article/details/7219958。

通过阅读本案例，相信大家应该纠正了 exists 效率比 in 高这种错误认识。如果 where 子查询中没有固化子查询关键字，不管写成 in 还是写成 exists，效率都是一样的，因为 CBO 始终能将子查询展开（unnest）。如果 where 子查询中有固化子查询关键字，这时我们最好用 in 而不是 exists，因为 in 可以控制子查询是否展开，而 exists 无法展开。至于 where 子查询是展开性能好还是不展开性能好，我们要具体情况具体分析。

9.25 烂用外连接导致无法谓词推入

2015 年，一位甲骨文公司的朋友请求协助优化。有个 SQL 单次执行需要 26.57 秒，一共要执行 226 次，如图 9-12 所示。

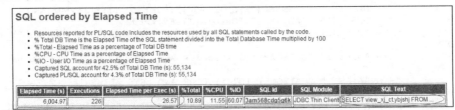

图 9-12

9.25 烂用外连接导致无法谓词推入

SQL 代码如下。

```sql
SELECT view_xj_ct.ybjshj  FROM view_xj_ct
WHERE view_xj_ct.ct_code = :1 AND view_xj_ct.pk_corp = :2
```

view_xj_ct 是一个视图，视图定义如下。

```sql
CREATE OR REPLACE FORCE VIEW "JXNC"."VIEW_XJ_CT" ("CT_CODE", "PK_CT_MANAGE", "YBJSHJ"
, "FKHJ", "KPJE", "JE", "PK_CORP") AS
  select a."CT_CODE",
         a."PK_CT_MANAGE",
         a."YBJSHJ",
         a."FKHJ",
         b.kpje,
         (case
           when b.kpje >= a.ybjshj then
             b.kpje
           else
             a.ybjshj
         end) je,
         pk_corp
    from (select cth.ct_code,
                 cth.pk_ct_manage,
                 sum(ctb.oritaxsummny) ybjshj,
                 sum(ctv.ljfk) fkhj,
                 ctb.pk_corp
            from ct_manage_b ctb
            left join ct_manage cth
              on ctb.pk_ct_manage = cth.pk_ct_manage
            left join view_xj_ct_fukuan ctv
              on ctv.pk_ct_manage_b = ctb.pk_ct_manage_b
             and ctv.pk_ct_manage = cth.pk_ct_manage
           where activeflag = 0
             and cth.dr = 0
             and ctb.dr = 0
           group by cth.ct_code, cth.pk_ct_manage, ctb.pk_corp) a
    left join (select cth.pk_ct_manage, sum(fp.noriginalsummny) kpje
                 from po_invoice_b fp
                 left join po_order_b dd
                   on fp.csourcebillrowid = dd.corder_bid
                 left join ct_manage_b ct
                   on ct.pk_ct_manage_b = dd.csourcerowid
                 left join ct_manage cth
                   on ct.pk_ct_manage = cth.pk_ct_manage
                where fp.dr = 0
                  and dd.cupsourcebilltype = 'Z2'
                group by cth.pk_ct_manage) b
      on b.pk_ct_manage = a.pk_ct_manage;
```

代码中：表 ct_manage_b 有数据 266 274（26 万条记录），表 ct_manage 有数据 88 563（8.8 万条记录），表 po_invoice_b 有数据 294 467（29 万条记录），表 po_order_b 有数据 143122（14 万条记录）。

上面视图 view_xj_ct 中又内嵌一个视图 view_xj_ct_fukuan，视图代码如下。

```sql
CREATE OR REPLACE FORCE VIEW "JXNC"."VIEW_XJ_CT_FUKUAN" ("DDHH", "PK_CORP", "PK_CT_MA
NAGE_B", "PK_CT_MANAGE", "LJFK", "CT_CODE") AS
  select ddhh,
         a.pk_corp,
         a.pk_ct_manage_b,
```

```
             ctb.pk_ct_manage,
             sum(a.ljfk) ljfk,
             cth.ct_code
  from (select a.ddhh,
               a.dwbm pk_corp,
               a.zyx5 pk_ct_manage_b,
               a.jfybje ljfk
          from arap_djfb a
          left join arap_djzb b on a.vouchid = b.vouchid
         where a.dr = 0
           and b.dr = 0
           and a.djlxbm = 'D3'
           and a.jsfsbm in ('Z2','Z5','D1')
           and b.djzt not in ('-99','1')) a
  left join ct_manage_b ctb on ctb.pk_ct_manage_b = a.pk_ct_manage_b
  left join ct_manage cth on cth.pk_ct_manage = ctb.pk_ct_manage
 group by ddhh, a.pk_corp, a.pk_ct_manage_b, ctb.pk_ct_manage, cth.ct_code
 order by a.pk_ct_manage_b;
```

其中：表 arap_djfb 有数据 1 175 707（117 万条记录），表 arap_djzb 有数据 149 157（15 万条记录），表 ct_manage_b 有数据 266 274（26 万条记录），表 ct_manage 有数据 88 563（8.8 万条记录）。

SQL 语句的执行计划如下。

```
SQL> explain plan for SELECT view_xj_ct.ybjshj  FROM view_xj_ct
  2  WHERE view_xj_ct.ct_code = :1 AND view_xj_ct.pk_corp = :2;

Explained.

SQL> select * from table(dbms_xplan.display);

PLAN_TABLE_OUTPUT
-----------------------------------------------------------------------------------------
Plan hash value: 3563589558
-----------------------------------------------------------------------------------------
| Id|Operation                         |Name              |Rows  |Bytes |TempSpc|Cost (%CPU)|Time     |
-----------------------------------------------------------------------------------------
|  0|SELECT STATEMENT                  |                  |    1 |   57 |       |49994   (1)|00:10:00 |
|* 1|  HASH JOIN OUTER                 |                  |    1 |   57 |       |49994   (1)|00:10:00 |
|  2|   VIEW                           |                  |    1 |   35 |       |32190   (1)|00:06:27 |
|  3|    HASH GROUP BY                 |                  |    1 |   74 |       |32190   (1)|00:06:27 |
|* 4|     HASH JOIN OUTER              |                  |    1 |   74 |       |32189   (1)|00:06:27 |
|  5|      VIEW                        |                  |    1 |   35 |       |    2   (0)|00:00:01 |
|  6|       NESTED LOOPS               |                  |      |      |       |           |         |
|  7|        NESTED LOOPS              |                  |    1 |   95 |       |    2   (0)|00:00:01 |
|* 8|         TABLE ACCESS BY INDEX ROWID|CT_MANAGE       |    1 |   40 |       |    1   (0)|00:00:01 |
|* 9|          INDEX RANGE SCAN        |I_CT_M_1          |    2 |      |       |    1   (0)|00:00:01 |
|*10|         INDEX RANGE SCAN         |I_CT_M_B_1        |    3 |      |       |    1   (0)|00:00:01 |
|*11|        TABLE ACCESS BY INDEX ROWID|CT_MANAGE_B     |    1 |   55 |       |    1   (0)|00:00:01 |
| 12|      VIEW                        |VIEW_XJ_CT_FUKUAN |39191 |1492K |       |32186   (1)|00:06:27 |
| 13|       HASH GROUP BY              |                  |39191 |6468K | 6976K |32186   (1)|00:06:27 |
|*14|        HASH JOIN RIGHT OUTER     |                  |39191 |6468K | 3976K |30726   (1)|00:06:09 |
| 15|         TABLE ACCESS FULL        |CT_MANAGE         |88505 |2938K |       | 1621   (2)|00:00:20 |
|*16|         HASH JOIN OUTER          |                  |39191 |5166K | 4024K |28636   (1)|00:05:44 |
|*17|          HASH JOIN               |                  |39191 |3559K | 2952K |23574   (1)|00:04:43 |
| 18|           INLIST ITERATOR        |                  |      |      |       |           |         |
|*19|            TABLE ACCESS BY INDEX ROWID|ARAP_DJFB    |39191 |2487K |       |20692   (1)|00:04:09 |
|*20|             INDEX RANGE SCAN     |I_ARAP_DJFB_JSZC02| 337K |      |       |  251   (2)|00:00:04 |
|*21|           TABLE ACCESS FULL      |ARAP_DJZB         | 127K |3476K |       | 2494   (2)|00:00:30 |
| 22|          TABLE ACCESS FULL       |CT_MANAGE_B       | 266K |  10M |       | 4179   (2)|00:00:51 |
```

9.25 烂用外连接导致无法谓词推入

```
|  23|     VIEW                        |                  |88480|1900K|       |17802   (1)|00:03:34|
|  24|      HASH GROUP BY              |                  |88480|  10M|   16M |17802   (1)|00:03:34|
|*25|        HASH JOIN                 |                  | 120K|  14M|  5024K|14906   (1)|00:02:59|
|*26|         TABLE ACCESS FULL        |PO_INVOICE_B      | 138K|3389K|       | 5263   (1)|00:01:04|
|*27|         HASH JOIN RIGHT OUTER    |                  |98165|   9M|  2856K| 8850   (2)|00:01:47|
|  28|          INDEX FAST FULL SCAN   |PK_CT_MANAGE      |88505|1815K|       |  107   (2)|00:00:02|
|*29|          HASH JOIN OUTER         |                  |98165|8052K|  5184K| 8154   (2)|00:01:38|
|*30|           TABLE ACCESS FULL      |PO_ORDER_B        |98165|4026K|       | 3035   (1)|00:00:37|
|  31|           TABLE ACCESS FULL     |CT_MANAGE_B       | 266K|  10M|       | 4179   (2)|00:00:51|
------------------------------------------------------------------------------------------------

Predicate Information (identified by operation id):
---------------------------------------------------

   1 - access("B"."PK_CT_MANAGE"(+)="A"."PK_CT_MANAGE")
   4 - access("CTV"."PK_CT_MANAGE"(+)="CTH"."PK_CT_MANAGE" AND
              "CTV"."PK_CT_MANAGE_B"(+)="CTB"."PK_CT_MANAGE_B")
   8 - filter("CTH"."DR"=0 AND "CTH"."ACTIVEFLAG"=0)
   9 - access("CTH"."CT_CODE"=:1)
  10 - access("CTB"."PK_CT_MANAGE"="CTH"."PK_CT_MANAGE")
  11 - filter("CTB"."PK_CORP"=:2 AND "CTB"."DR"=0)
  14 - access("CTH"."PK_CT_MANAGE"(+)="CTB"."PK_CT_MANAGE")
  16 - access("CTB"."PK_CT_MANAGE_B"(+)="A"."ZYX5")
  17 - access("A"."VOUCHID"="B"."VOUCHID")
  19 - filter("A"."DJLXBM"='D3' AND "A"."DR"=0)
  20 - access("A"."JSFSBM"='D1' OR "A"."JSFSBM"='Z2' OR "A"."JSFSBM"='Z5')
  21 - filter("B"."DR"=0 AND "B"."DJZT"<>1 AND "B"."DJZT"<>(-99))
  25 - access("FP"."CSOURCEBILLROWID"="DD"."CORDER_BID")
  26 - filter("FP"."CSOURCEBILLROWID" IS NOT NULL AND "FP"."DR"=0)
  27 - access("CT"."PK_CT_MANAGE"="CTB"."PK_CT_MANAGE"(+))
  29 - access("CT"."PK_CT_MANAGE_B"(+)="DD"."CSOURCEROWID")
  30 - filter("DD"."CUPSOURCEBILLTYPE"='Z2')

60 rows selected.
```

对于上述的执行计划,甲骨文公司的朋友创建了一个 index。

```
create index idx_jszc1026 on ARAP_djfb(jsfsbm,djlxbm,dr);
```

之前大约 26 秒出结果,创建新 index 后速度是 2.6 秒出结果,新建索引后的执行计划如下。

```
PLAN_TABLE_OUTPUT
-------------------------------------------------------------------------------------------
Plan hash value: 2820245905

| Id|Operation                         |Name             |Rows |Bytes|TempSpc|Cost(%CPU)|Time    |
-------------------------------------------------------------------------------------------
|  0|SELECT STATEMENT                  |                 |   1|   57|       |32043   (1)|00:06:25|
|* 1| HASH JOIN OUTER                  |                 |   1|   57|       |32043   (1)|00:06:25|
|  2|  VIEW                            |                 |   1|   35|       |14239   (2)|00:02:51|
|  3|   HASH GROUP BY                  |                 |   1|   74|       |14239   (2)|00:02:51|
|* 4|    HASH JOIN OUTER               |                 |   1|   74|       |14238   (2)|00:02:51|
|  5|     VIEW                         |                 |   1|   35|       |    2   (0)|00:00:01|
|  6|      NESTED LOOPS                |                 |    |     |       |           |        |
|  7|       NESTED LOOPS               |                 |   1|   95|       |    2   (0)|00:00:01|
|* 8|        TABLE ACCESS BY INDEX ROWID|CT_MANAGE       |   1|   40|       |    1   (0)|00:00:01|
|* 9|         INDEX RANGE SCAN         |I_CT_M_1         |   2|     |       |    1   (0)|00:00:01|
|*10|        INDEX RANGE SCAN          |I_CT_M_B_1       |   3|     |       |    1   (0)|00:00:01|
|*11|       TABLE ACCESS BY INDEX ROWID|CT_MANAGE_B      |   1|   55|       |    1   (0)|00:00:01|
| 12|     VIEW                         |VIEW_XJ_CT_FUKUAN|39191|1492K|       |14234   (2)|00:02:51|
| 13|      HASH GROUP BY               |                 |39191|6468K|  6976K|14234   (2)|00:02:51|
|*14|       HASH JOIN RIGHT OUTER      |                 |39191|6468K|  3976K|12775   (2)|00:02:34|
```

```
|  15|         TABLE ACCESS FULL           |CT_MANAGE     | 88505| 2938K|      | 1621  (2)|00:00:20|
|* 16|         HASH JOIN OUTER             |              | 39191| 5166K| 4024K|10685  (2)|00:02:09|
|* 17|          HASH JOIN                  |              | 39191| 3559K| 2952K| 5622  (1)|00:01:08|
|  18|           INLIST ITERATOR           |              |      |      |      |          |        |
|  19|            TABLE ACCESS BY INDEX ROWID|ARAP_DJFB    | 39191| 2487K|      | 2740  (1)|00:00:33|
|* 20|             INDEX RANGE SCAN        |IDX_JSZC1026  | 39212|      |      |   43  (3)|00:00:01|
|* 21|           TABLE ACCESS FULL         |ARAP_DJZB     |  127K| 3476K|      | 2494  (2)|00:00:30|
|  22|          TABLE ACCESS FULL          |CT_MANAGE_B   |  266K|   10M|      | 4179  (2)|00:00:51|
|  23|   VIEW                              |              | 88480| 1900K|      |17802  (1)|00:03:34|
|  24|    HASH GROUP BY                    |              | 88480|   10M|   16M|17802  (1)|00:03:34|
|* 25|     HASH JOIN                       |              |  120K|   14M| 5024K|14906  (1)|00:02:59|
|* 26|      TABLE ACCESS FULL              |PO_INVOICE_B  |  138K| 3389K|      | 5263  (1)|00:01:04|
|* 27|      HASH JOIN RIGHT OUTER          |              | 98165|    9M| 2856K| 8850  (2)|00:01:47|
|  28|       INDEX FAST FULL SCAN          |PK_CT_MANAGE  | 88505| 1815K|      |  107  (2)|00:00:02|
|* 29|       HASH JOIN OUTER               |              | 98165| 8052K| 5184K| 8154  (2)|00:01:38|
|* 30|        TABLE ACCESS FULL            |PO_ORDER_B    | 98165| 4026K|      | 3035  (1)|00:00:37|
|  31|        TABLE ACCESS FULL            |CT_MANAGE_B   |  266K|   10M|      | 4179  (2)|00:00:51|
--------------------------------------------------------------------------------

Predicate Information (identified by operation id):
---------------------------------------------------

   1 - access("B"."PK_CT_MANAGE"(+)="A"."PK_CT_MANAGE")
   4 - access("CTV"."PK_CT_MANAGE"(+)="CTH"."PK_CT_MANAGE" AND
              "CTV"."PK_CT_MANAGE_B"(+)="CTB"."PK_CT_MANAGE_B")
   8 - filter("CTH"."DR"=0 AND "CTH"."ACTIVEFLAG"=0)
   9 - access("CTH"."CT_CODE"=:1)
  10 - access("CTB"."PK_CT_MANAGE"="CTH"."PK_CT_MANAGE")
  11 - filter("CTB"."PK_CORP"=:2 AND "CTB"."DR"=0)
  14 - access("CTH"."PK_CT_MANAGE"(+)="CTB"."PK_CT_MANAGE")
  16 - access("CTB"."PK_CT_MANAGE_B"(+)="A"."ZYX5")
  17 - access("A"."VOUCHID"="B"."VOUCHID")
  20 - access(("A"."JSFSBM"='D1' OR "A"."JSFSBM"='Z2' OR "A"."JSFSBM"='Z5') AND "A"."
DJLXBM"='D3' AND
              "A"."DR"=0)
  21 - filter("B"."DR"=0 AND "B"."DJZT"<>1 AND "B"."DJZT"<>(-99))
  25 - access("FP"."CSOURCEBILLROWID"="DD"."CORDER_BID")
  26 - filter("FP"."CSOURCEBILLROWID" IS NOT NULL AND "FP"."DR"=0)
  27 - access("CT"."PK_CT_MANAGE"="CTH"."PK_CT_MANAGE"(+))
  29 - access("CT"."PK_CT_MANAGE_B"(+)="DD"."CSOURCEROWID")
  30 - filter("DD"."CUPSOURCEBILLTYPE"='Z2')

60 rows selected.
```

如图 9-13 所示,做一笔单据在后台要多次调用这个语句。

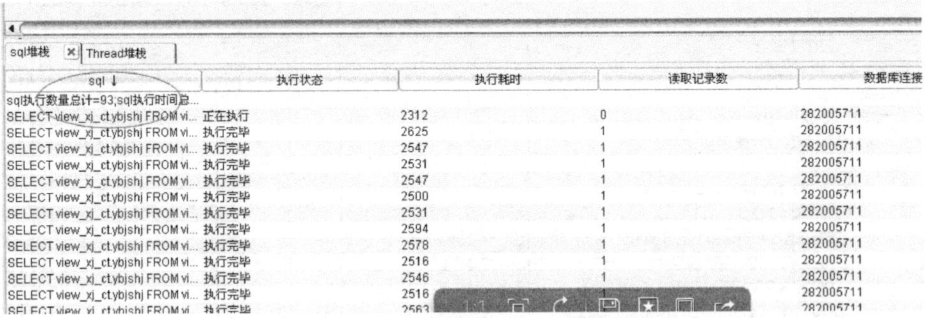

图 9-13

100 个 SQL 语句每个执行 2.6 秒，全部执行就要 260 秒，将近 4 分钟。到这里，甲骨文的朋友问能否进一步优化该 SQL。

下面是分析过程。

在尝试优化 SQL 之前，首先询问该 SQL 返回多少行数据，甲骨文的朋友回答返回 1 行数据。在进行 SQL 优化的时候，我们必须知道一个 SQL 最终应该返回多少行数据，因为知道了 SQL 最终返回数据，就能判断表连接究竟是采用嵌套循环还是采用 HASH 连接，这至关重要。因为 SQL 最终返回一行数据，所以判断 SQL 的执行计划应该走嵌套循环。但是本 SQL 执行计划中几乎全是 HASH 连接。根据 SQL 语句过滤条件入手，一步一步分析执行计划，看哪里出了问题。

SQL 语句的过滤条件是 WHERE view_xj_ct.ct_code = :1 AND view_xj_ct.pk_corp = :2。

这两个过滤条件已经在书中用阴影部分标注，为了方便读者查看现将其摘抄下来。

```
select cth.ct_code,
       cth.pk_ct_manage,
       sum(ctb.oritaxsummny) ybjshj,
       sum(ctv.ljfk) fkhj,
       ctb.pk_corp
  from ct_manage_b ctb
  left join ct_manage cth
    on ctb.pk_ct_manage = cth.pk_ct_manage
  left join view_xj_ct_fukuan ctv
    on ctv.pk_ct_manage_b = ctb.pk_ct_manage_b
   and ctv.pk_ct_manage = ctb.pk_ct_manage
 where activeflag = 0
   and cth.dr = 0
   and ctb.dr = 0
 group by cth.ct_code, cth.pk_ct_manage, ctb.pk_corp
```

过滤条件分别针对 cth 和 ctb 进行过滤，执行计划中 Id=9 走的是 cth.ct_code 的索引，这说明此处发生了常量谓词推入，将过滤条件(常量过滤条件)推入到视图中进行了过滤。Id=9 属于 cth，它与 id=10(ctb)走的是嵌套循环。cth 与 ctb 关联的结果集在执行计划中是 Id=5 这步，Id=5 与 Id=12(view_xj_ct_fukuan)进行的是 HASH 连接。Id=12 是一个视图。因为该 SQL 最终只返回 1 行数据，应该全走嵌套循环才对，但是关联到视图 view_xj_ct_fukuan 的时候居然走的是 HASH 连接，所以笔者判断 Id=5 与 Id=12 关联方式出错。SQL 语句中，视图 view_xj_ct_fukuan 的别名是 ctv，ctv 分别与 cth 和 ctb 进行了关联。

```
left join view_xj_ct_fukuan ctv
   on ctv.pk_ct_manage_b = ctb.pk_ct_manage_b
  and ctv.pk_ct_manage = cth.pk_ct_manage
```

如果能让 cth 与 ctb 关联之后得到的结果集通过 ctv 的连接列传值给 ctv，通过连接列将数据将数据推入到视图中，这样就可以让视图走嵌套循环了，这种方式就是连接列谓词推入，但是执行计划并没有这样做。

于是查看如下视图 view_xj_ct_fukuan 的源代码。

```
CREATE OR REPLACE FORCE VIEW "JXNC"."VIEW_XJ_CT_FUKUAN" ("DDHH", "PK_CORP", "PK_CT_MA
NAGE_B", "PK_CT_MANAGE", "LJFK", "CT_CODE") AS
  select ddhh,
         a.pk_corp,
         a.pk_ct_manage_b,
         ctb.pk_ct_manage,
         sum(a.ljfk) ljfk,
         cth.ct_code
    from (select a.ddhh,
                 a.dwbm pk_corp,
                 a.zyx5 pk_ct_manage_b,
                 a.jfybje ljfk
            from arap_djfb a
            left join arap_djzb b on a.vouchid = b.vouchid
           where a.dr = 0
             and b.dr = 0
             and a.djlxbm = 'D3'
             and a.jsfsbm in ('Z2','Z5','D1')
             and b.djzt not in ('-99', '1')) a
    left join ct_manage_b ctb on ctb.pk_ct_manage_b = a.pk_ct_manage_b
    left join ct_manage cth on cth.pk_ct_manage = ctb.pk_ct_manage
   group by ddhh, a.pk_corp, a.pk_ct_manage_b, ctb.pk_ct_manage, cth.ct_code
   order by a.pk_ct_manage_b;
```

视图 ctv.pk_ct_manage 字段来自于 ctb，而 ctb 与 a 是外连接，而且 ctb 是从表，并不是主表。

正是因为 ctb 是视图中外连接的从表，而且视图 ctv 也是外连接的从表，所以导致 cth 不能通过连接列 pk_ct_manage 将谓词推入到 ctv.pk_ct_manage 中，从而导致走了 HASH 连接。

```
left join view_xj_ct_fukuan ctv
   on ctv.pk_ct_manage_b = ctb.pk_ct_manage_b
  and ctv.pk_ct_manage = cth.pk_ct_manage
```

如果能将视图中的外连接改成内连接，就可以将谓词推入到 ctv 中，从而走嵌套循环。

通过反复分析 SQL 写法，我们确认可以将视图中的外连接改写为内连接。于是新建了一个视图，专门用于本 SQL，将外连接改写为内连接，而且将后面的子查询也改成了内连接。最终 SQL 能在 0.01 秒内执行完毕，执行 100 个 SQL 也仅需耗时 1 秒，从而将原本要执行 4 分钟的单据业务优化到 1 秒。

接下来，我们通过实验为大家模拟当时情况。

```
SQL> create table emp_new as select * from emp;

Table created.

SQL> create index idx_ename on emp(ename);

Index created.
```

视图（e）里面表关联是外连接，而且视图（e）作为外连接从表，视图（e）连接列来自从表。

```
select /*+ push_pred(e) */ *
  from emp_new a
  left join (select d.dname, e.ename, sum(e.sal) total_sal
               from dept d
               left join emp e on d.deptno = e.deptno
              group by dname, ename) e on a.ename = e.ename
```

9.25 烂用外连接导致无法谓词推入

```
where empno = 7900;
```

执行计划如下。

```
SQL> select /*+ push_pred(e) */ *
  2    from emp_new a
  3    left join (select d.dname, e.ename, sum(e.sal) total_sal
  4                 from dept d
  5                 left join emp e on d.deptno = e.deptno
  6                group by dname, ename) e on a.ename = e.ename
  7   where empno = 7900;

Execution Plan
----------------------------------------------------------
Plan hash value: 3023292314
---------------------------------------------------------------------------------
| Id|Operation                        |Name    | Rows | Bytes | Cost (%CPU)| Time     |
---------------------------------------------------------------------------------
|  0|SELECT STATEMENT                 |        |    1 |  116 |   10  (30)| 00:00:01 |
|* 1| HASH JOIN OUTER                 |        |    1 |  116 |   10  (30)| 00:00:01 |
|* 2|  TABLE ACCESS FULL              |EMP_NEW |    1 |   87 |    2   (0)| 00:00:01 |
|  3|  VIEW                           |        |   14 |  406 |    7  (29)| 00:00:01 |
|  4|   HASH GROUP BY                 |        |   14 |  364 |    7  (29)| 00:00:01 |
|  5|    MERGE JOIN OUTER             |        |   14 |  364 |    6  (17)| 00:00:01 |
|  6|     TABLE ACCESS BY INDEX ROWID |DEPT    |    4 |   52 |    2   (0)| 00:00:01 |
|  7|      INDEX FULL SCAN            |PK_DEPT |    4 |      |    1   (0)| 00:00:01 |
|* 8|     SORT JOIN                   |        |   14 |  182 |    4  (25)| 00:00:01 |
|  9|      TABLE ACCESS FULL          |EMP     |   14 |  182 |    3   (0)| 00:00:01 |
---------------------------------------------------------------------------------

Predicate Information (identified by operation id):
---------------------------------------------------

   1 - access("A"."ENAME"="E"."ENAME"(+))
   2 - filter("A"."EMPNO"=7900)
   8 - access("D"."DEPTNO"="E"."DEPTNO"(+))
       filter("D"."DEPTNO"="E"."DEPTNO"(+))
```

当视图里面表关联是外连接，而且视图与其他表关联作为外连接从表，视图连接列来自视图里面的从表，此时不能谓词推入。

我们将视图里面表关联改成内连接。

```
select /*+ push_pred(e) */ *
  from emp_new a
  left join (select d.dname, e.ename, sum(e.sal) total_sal
               from dept d
               join emp e on d.deptno = e.deptno
              group by dname, ename) e on a.ename = e.ename
 where empno = 7900;
```

执行计划如下。

```
SQL> select /*+ push_pred(e) */ *
  2    from emp_new a
  3    left join (select d.dname, e.ename, sum(e.sal) total_sal
  4                 from dept d
  5                 join emp e on d.deptno = e.deptno
  6                group by dname, ename) e on a.ename = e.ename
  7   where empno = 7900;
```

```
Execution Plan
----------------------------------------------------------
Plan hash value: 3258229530

--------------------------------------------------------------------------------
| Id|Operation                       |Name     | Rows | Bytes | Cost(%CPU)| Time     |
--------------------------------------------------------------------------------
|  0|SELECT STATEMENT                |         |   1|   111 |   6  (17)| 00:00:01 |
|  1| NESTED LOOPS OUTER             |         |   1|   111 |   6  (17)| 00:00:01 |
|* 2|  TABLE ACCESS FULL             |EMP_NEW  |   1|    87 |   2   (0)| 00:00:01 |
|  3|  VIEW PUSHED PREDICATE         |         |   1|    24 |   4  (25)| 00:00:01 |
|  4|   SORT GROUP BY                |         |   1|    26 |   4  (25)|          |
|  5|    NESTED LOOPS                |         |    |       |          |          |
|  6|     NESTED LOOPS               |         |   1|    26 |   3   (0)| 00:00:01 |
|  7|      TABLE ACCESS BY INDEX ROWID|EMP     |   1|    13 |   2   (0)| 00:00:01 |
|* 8|       INDEX RANGE SCAN         |IDX_ENAME|   1|       |   1   (0)| 00:00:01 |
|* 9|      INDEX UNIQUE SCAN         |PK_DEPT  |   1|       |   0   (0)| 00:00:01 |
| 10|     TABLE ACCESS BY INDEX ROWID|DEPT     |   1|    13 |   1   (0)| 00:00:01 |
--------------------------------------------------------------------------------

Predicate Information (identified by operation id):
---------------------------------------------------

   2 - filter("A"."EMPNO"=7900)
   8 - access("E"."ENAME"="A"."ENAME")
   9 - access("D"."DEPTNO"="E"."DEPTNO")
```

将视图里面的外连接改成内连接之后，我们就可以将谓词推入到视图中了。

如果不改视图中的外连接，将 SQL 语句中的外连接改成内连接也可以将谓词推入视图。

```
select /*+ push_pred(e) */ *
  from emp_new a
  join (select d.dname, e.ename, sum(e.sal) total_sal
          from dept d
          left join emp e on d.deptno = e.deptno
         group by dname, ename) e on a.ename = e.ename
 where empno = 7900;
```

执行计划如下。

```
SQL> select /*+ push_pred(e) */ *
  2    from emp_new a
  3    join (select d.dname, e.ename, sum(e.sal) total_sal
  4            from dept d
  5            left join emp e on d.deptno = e.deptno
  6           group by dname, ename) e on a.ename = e.ename
  7   where empno = 7900;

Execution Plan
----------------------------------------------------------
Plan hash value: 3747089680

--------------------------------------------------------------------------------
| Id|Operation                       |Name     | Rows | Bytes | Cost(%CPU)| Time     |
--------------------------------------------------------------------------------
|  0|SELECT STATEMENT                |         |   1 |  125 |   5  (20)| 00:00:01 |
|  1| HASH GROUP BY                  |         |   1 |  125 |   5  (20)| 00:00:01 |
|  2|  NESTED LOOPS                  |         |     |      |          |          |
|  3|   NESTED LOOPS                 |         |   1 |  125 |   4   (0)| 00:00:01 |
|  4|    NESTED LOOPS                |         |   1 |  112 |   3   (0)| 00:00:01 |
|* 5|     TABLE ACCESS FULL          |EMP_NEW  |   1 |   99 |   2   (0)| 00:00:01 |
|  6|     TABLE ACCESS BY INDEX ROWID|EMP      |   1 |   13 |   1   (0)| 00:00:01 |
```

```
|*  7|        INDEX RANGE SCAN      |IDX_ENAME|     1 |       |     0   (0)| 00:00:01 |
|*  8|        INDEX UNIQUE SCAN     |PK_DEPT  |     1 |       |     0   (0)| 00:00:01 |
|   9|     TABLE ACCESS BY INDEX ROWID|DEPT   |     1 |    13 |     1   (0)| 00:00:01 |
----------------------------------------------------------------------------------------

Predicate Information (identified by operation id):
---------------------------------------------------

   5 - filter("A"."EMPNO"=7900)
   7 - access("A"."ENAME"="E"."ENAME")
   8 - access("D"."DEPTNO"="E"."DEPTNO")
```

笔者当时究竟是怎么判断可以将 view_xj_ct_fukuan ctv 里面的视图改成内连接的呢？请大家注意观察原始 view_xj_ct 部分代码。

```
select cth.ct_code,
       cth.pk_ct_manage,
       sum(ctb.oritaxsummny) ybjshj,
       sum(ctv.ljfk) fkhj,
       ctb.pk_corp
  from ct_manage_b ctb
  left join ct_manage cth
    on ctb.pk_ct_manage = cth.pk_ct_manage
  left join view_xj_ct_fukuan ctv
    on ctv.pk_ct_manage_b = ctb.pk_ct_manage_b
   and ctv.pk_ct_manage = cth.pk_ct_manage
 where activeflag = 0
   and cth.dr = 0
   and ctb.dr = 0
 group by cth.ct_code, cth.pk_ct_manage, ctb.pk_corp
```

注意观察阴影部分连接条件，视图 ctv 中的连接列也是来自 cth 和 ctb。

```
CREATE OR REPLACE FORCE VIEW "JXNC"."VIEW_XJ_CT_FUKUAN" ("DDHH", "PK_CORP", "PK_CT_MA
NAGE_B", "PK_CT_MANAGE", "LJFK", "CT_CODE") AS
  select ddhh,
         a.pk_corp,
         a.pk_ct_manage_b,
         ctb.pk_ct_manage,
         sum(a.ljfk) ljfk,
         cth.ct_code
    from (select a.ddhh,
                 a.dwbm pk_corp,
                 a.zyx5 pk_ct_manage_b,
                 a.jfybje ljfk
            from arap_djfb a
            left join arap_djzb b on a.vouchid = b.vouchid
           where a.dr = 0
             and b.dr = 0
             and a.djlxbm = 'D3'
             and a.jsfsbm in ('Z2', 'Z5','D1')
             and b.djzt not in ('-99', '1')) a
    left join ct_manage_b ctb on ctb.pk_ct_manage_b = a.pk_ct_manage_b
    left join ct_manage cth on cth.pk_ct_manage = ctb.pk_ct_manage
   group by ddhh, a.pk_corp, a.pk_ct_manage_b, ctb.pk_ct_manage, cth.ct_code
   order by a.pk_ct_manage_b;
```

同时视图 ctv 中有对连接列进行汇总，这其实相当于如下 SQL。

```
select e.empno, sum(sum_sal)
  from emp e
```

```
       left join (select d.deptno, sum(sal) sum_sal
                    from dept d
                    left join emp e on d.deptno = e.deptno
                   group by d.deptno) d on e.deptno = d.deptno
 group by empno;
```

上面 SQL 可以安全地将 left join 改写为 inner join。

```
select e.empno, sum(sum_sal)
  from emp e
  left join (select d.deptno, sum(sal) sum_sal
                    from dept d
                    join emp e on d.deptno = e.deptno
                   group by d.deptno) d on e.deptno = d.deptno
 group by empno;
```

同理，原始 SQL 中后面的子查询也能改写为 inner join。

想要优化本案例中的 SQL，必须具备较强的 SQL 优化能力以及较强的 SQL 改写能力，这两种能力缺一不可。通过本案例，我们也要反思，为什么开发人员在 SQL 中一直写 left join？我们甚至怀疑是不是开发人员只会 left join，或者不管写什么 SQL，一直 left join，这太可怕了，由此可见，在系统上线之前，SQL 审核是多么重要！

9.26 谓词推入优化案例

2011 年，一位 ITPUB 的网友请求优化如下 SQL。

```
SELECT *
  FROM (SELECT  A.INVOICE_ID,
                A.VENDOR_ID,
                A.INVOICE_NUM,
                A.INVOICE_AMOUNT,
                A.GL_DATE,
                A.INVOICE_CURRENCY_CODE,
                SUM(NVL(B.PREPAY_AMOUNT_APPLIED, 0)) PAID_AMOUNT,
                A.INVOICE_AMOUNT - SUM(NVL(B.PREPAY_AMOUNT_APPLIED, 0)) REMAIN
           FROM ap.AP_INVOICES_ALL A, APPS.AP_UNAPPLY_PREPAYS_V B
          WHERE A.INVOICE_ID = B.INVOICE_ID(+)
            AND A.ORG_ID = 126 /*:B4*/
            AND A.SOURCE = 'OSM IMPORTED' /*:B3*/
            AND A.INVOICE_NUM BETWEEN NVL( /*:B2*/ null, A.INVOICE_NUM) AND
                 NVL( /*:B1*/ null, A.INVOICE_NUM)
          GROUP BY A.INVOICE_ID,
                   A.INVOICE_NUM,
                   A.INVOICE_AMOUNT,
                   A.VENDOR_ID,
                   A.GL_DATE,
                   A.INVOICE_CURRENCY_CODE)
 WHERE REMAIN > 0 ;
```

该 SQL 要执行 1 个多小时，AP_UNAPPLY_PREPAYS_V 是一个视图，代码如下。

```
CREATE OR REPLACE VIEW APPS.AP_UNAPPLY_PREPAYS_V AS
SELECT AID1.ROWID ROW_ID,
       AID1.INVOICE_ID INVOICE_ID,
       AID1.INVOICE_DISTRIBUTION_ID INVOICE_DISTRIBUTION_ID,
       AID1.PREPAY_DISTRIBUTION_ID PREPAY_DISTRIBUTION_ID,
```

9.26 谓词推入优化案例

```
            AID1.DISTRIBUTION_LINE_NUMBER PREPAY_DIST_NUMBER,
            (-1) * AID1.AMOUNT PREPAY_AMOUNT_APPLIED,
            nvl(AID2.PREPAY_AMOUNT_REMAINING, AID2.AMOUNT) PREPAY_AMOUNT_REMAINING,
            AID1.DIST_CODE_COMBINATION_ID DIST_CODE_COMBINATION_ID,
            AID1.ACCOUNTING_DATE ACCOUNTING_DATE,
            AID1.PERIOD_NAME PERIOD_NAME,
            AID1.SET_OF_BOOKS_ID SET_OF_BOOKS_ID,
            AID1.DESCRIPTION DESCRIPTION,
            AID1.PO_DISTRIBUTION_ID PO_DISTRIBUTION_ID,
            AID1.RCV_TRANSACTION_ID RCV_TRANSACTION_ID,
            AID1.ORG_ID ORG_ID,
            AI.INVOICE_NUM PREPAY_NUMBER,
            AI.VENDOR_ID VENDOR_ID,
            AI.VENDOR_SITE_ID VENDOR_SITE_ID,
            ATC.TAX_ID TAX_ID,
            ATC.NAME TAX_CODE,
            PH.SEGMENT1 PO_NUMBER,
            PV.VENDOR_NAME VENDOR_NAME,
            PV.SEGMENT1 VENDOR_NUMBER,
            PVS.VENDOR_SITE_CODE VENDOR_SITE_CODE,
            RSH.RECEIPT_NUM RECEIPT_NUMBER
       FROM AP_INVOICES              AI,
            AP_INVOICE_DISTRIBUTIONS AID1,
            AP_INVOICE_DISTRIBUTIONS AID2,
            AP_TAX_CODES             ATC,
            PO_VENDORS               PV,
            PO_VENDOR_SITES          PVS,
            PO_DISTRIBUTIONS         PD,
            PO_HEADERS               PH,
            PO_LINES                 PL,
            PO_LINE_LOCATIONS        PLL,
            RCV_TRANSACTIONS         RTXNS,
            RCV_SHIPMENT_HEADERS     RSH,
            RCV_SHIPMENT_LINES       RSL
      WHERE AID1.PREPAY_DISTRIBUTION_ID = AID2.INVOICE_DISTRIBUTION_ID
        AND AI.INVOICE_ID = AID2.INVOICE_ID
        AND AID1.AMOUNT < 0
        AND nvl(AID1.REVERSAL_FLAG, 'N') != 'Y'
        AND AID1.TAX_CODE_ID = ATC.TAX_ID(+)
        AND AID1.LINE_TYPE_LOOKUP_CODE = 'PREPAY'
        AND AI.VENDOR_ID = PV.VENDOR_ID
        AND AI.VENDOR_SITE_ID = PVS.VENDOR_SITE_ID
        AND AID1.PO_DISTRIBUTION_ID = PD.PO_DISTRIBUTION_ID(+)
        AND PD.PO_HEADER_ID = PH.PO_HEADER_ID(+)
        AND PD.LINE_LOCATION_ID = PLL.LINE_LOCATION_ID(+)
        AND PLL.PO_LINE_ID = PL.PO_LINE_ID(+)
        AND AID1.RCV_TRANSACTION_ID = RTXNS.TRANSACTION_ID(+)
        AND RTXNS.SHIPMENT_LINE_ID = RSL.SHIPMENT_LINE_ID(+)
        AND RSL.SHIPMENT_HEADER_ID = RSH.SHIPMENT_HEADER_ID(+);
```

执行计划如下。

```
-----------------------------------------------------------------------------------
| Id  |Operation                       | Name                |Rows |Bytes|Cost |
-----------------------------------------------------------------------------------
|  0  |SELECT STATEMENT                |                     |   1 |  69 | 722 |
|* 1  | FILTER                         |                     |     |     |     |
|  2  |  SORT GROUP BY                 |                     |   1 |  69 | 722 |
|  3  |   NESTED LOOPS OUTER           |                     |   3 | 207 | 697 |
|* 4  |    TABLE ACCESS FULL           |AP_INVOICES_ALL      |   3 | 153 | 694 |
|  5  |    VIEW PUSHED PREDICATE       |AP_UNAPPLY_PREPAYS_V |   1 |  18 |   1 |
|  6  |     NESTED LOOPS               |                     |   1 | 372 |   3 |
```

```
|   7  |              NESTED LOOPS                   |                              |  1|  368|   3|
|   8  |               NESTED LOOPS                  |                              |  1|  361|   2|
|   9  |                NESTED LOOPS                 |                              |  1|  347|   1|
|  10  |                 NESTED LOOPS OUTER          |                              |  1|  334|   1|
|  11  |                  NESTED LOOPS OUTER         |                              |  1|  321|   1|
|  12  |                   NESTED LOOPS OUTER        |                              |  1|  295|   1|
|  13  |                    NESTED LOOPS OUTER       |                              |  1|  269|   1|
|  14  |                     NESTED LOOPS OUTER      |                              |  1|  243|   1|
|  15  |                      NESTED LOOPS OUTER     |                              |  1|  197|   1|
|  16  |                       NESTED LOOPS OUTER    |                              |  1|  157|   1|
|  17  |                        NESTED LOOPS OUTER   |                              |  1|   98|   1|
|* 18  |                         TABLE ACCESS BY INDEX ROWID |AP_INVOICE_DISTRIBUTIONS_ALL|  1|   72|   1|
|* 19  |                          INDEX FULL SCAN    |AP_INVOICE_DISTRIBUTIONS_N20  |  1|     |    |
|* 20  |                         TABLE ACCESS BY INDEX ROWID |AP_TAX_CODES_ALL      |  1|   26|    |
|* 21  |                          INDEX UNIQUE SCAN  |AP_TAX_CODES_U1               |  1|     |    |
|* 22  |                        TABLE ACCESS BY INDEX ROWID |PO_DISTRIBUTIONS_ALL   |  1|   59|    |
|* 23  |                         INDEX UNIQUE SCAN   |PO_DISTRIBUTIONS_U1           |  1|     |    |
|* 24  |                       TABLE ACCESS BY INDEX ROWID |PO_HEADERS_ALL           |  1|   40|    |
|* 25  |                        INDEX UNIQUE SCAN    |PO_HEADERS_U1                 |  1|     |    |
|* 26  |                      TABLE ACCESS BY INDEX ROWID |PO_LINE_LOCATIONS_ALL     |  1|   46|    |
|* 27  |                       INDEX UNIQUE SCAN     |PO_LINE_LOCATIONS_U1          |  1|     |    |
|* 28  |                     TABLE ACCESS BY INDEX ROWID |PO_LINES_ALL               |  1|   26|    |
|* 29  |                      INDEX UNIQUE SCAN      |PO_LINES_U1                   |  1|     |    |
|  30  |                    TABLE ACCESS BY INDEX ROWID |RCV_TRANSACTIONS           |  1|   26|    |
|* 31  |                     INDEX UNIQUE SCAN       |RCV_TRANSACTIONS_U1           |  1|     |    |
|  32  |                   TABLE ACCESS BY INDEX ROWID |RCV_SHIPMENT_LINES          |  1|   26|    |
|* 33  |                    INDEX UNIQUE SCAN        |RCV_SHIPMENT_LINES_U1         |  1|     |    |
|* 34  |                   INDEX UNIQUE SCAN         |RCV_SHIPMENT_HEADERS_U1       |  1|   13|    |
|* 35  |                  TABLE ACCESS BY INDEX ROWID |AP_INVOICE_DISTRIBUTIONS_ALL |  1|   13|    |
|* 36  |                   INDEX UNIQUE SCAN         |AP_INVOICE_DISTRIBUTIONS_U2   |  1|     |    |
|* 37  |                 TABLE ACCESS BY INDEX ROWID |AP_INVOICES_ALL                |  1|   14|   1|
|* 38  |                  INDEX UNIQUE SCAN          |AP_INVOICES_U1                |  1|     |    |
|* 39  |                TABLE ACCESS BY INDEX ROWID  |PO_VENDOR_SITES_ALL           |  1|    7|   1|
|* 40  |                 INDEX UNIQUE SCAN           |PO_VENDOR_SITES_U1            |  1|     |    |
|* 41  |               INDEX UNIQUE SCAN             |PO_VENDORS_U1                 |  1|    4|    |
```

```
Predicate Information (identified by operation id):
---------------------------------------------------

   1 - filter("A"."INVOICE_AMOUNT"-SUM(NVL("B"."PREPAY_AMOUNT_APPLIED",0))>0)
   4 - filter("A"."ORG_ID"=126 AND "A"."SOURCE"='OSM IMPORTED' AND
              "A"."INVOICE_NUM">=NVL(NULL,"A"."INVOICE_NUM") AND "A"."INVOICE_NUM"<=N
VL(NULL,"A"."INVOICE_NUM"))
  18 - filter("A"."INVOICE_ID"="AP_INVOICE_DISTRIBUTIONS_ALL"."INVOICE_ID" AND
              "AP_INVOICE_DISTRIBUTIONS_ALL"."AMOUNT"<0 AND NVL("AP_INVOICE_DISTRIBUT
IONS_ALL"."REVERSAL_FLAG",'N')<>'Y'
              AND "AP_INVOICE_DISTRIBUTIONS_ALL"."LINE_TYPE_LOOKUP_CODE"='PREPAY' AND
              NVL("AP_INVOICE_DISTRIBUTIONS_ALL"."ORG_ID",NVL(TO_NUMBER(DECODE(SUBSTR
B(:B1,1,1),'
              ',NULL,SUBSTRB(:B2,1,10))),(-99)))=NVL(TO_NUMBER(DECODE(SUBSTRB(:B3,1,1
),' ',NULL,SUBSTRB(:B4,1,10))),(-99))
  19 - filter("AP_INVOICE_DISTRIBUTIONS_ALL"."PREPAY_DISTRIBUTION_ID" IS NOT NULL)
  20 - filter(NVL("AP_TAX_CODES_ALL"."ORG_ID"(+),NVL(TO_NUMBER(DECODE(SUBSTRB(:B1,1,1),'
              ',NULL,SUBSTRB(:B2,1,10))),(-99)))=NVL(TO_NUMBER(DECODE(SUBSTRB(:B3,1,1
),' ',NULL,SUBSTRB(:B4,1,10))),(-99))
  21 - access("AP_INVOICE_DISTRIBUTIONS_ALL"."TAX_CODE_ID"="AP_TAX_CODES_ALL"."TAX_ID
"(+))
  22 - filter(NVL("PO_DISTRIBUTIONS_ALL"."ORG_ID"(+),NVL(TO_NUMBER(DECODE(SUBSTRB(:B1
,1,1),'
              ',NULL,SUBSTRB(:B2,1,10))),(-99)))=NVL(TO_NUMBER(DECODE(SUBSTRB(:B3,1,1
),' ',NULL,SUBSTRB(:B4,1,10))),(-99))
```

9.26 谓词推入优化案例

```
  23 - access("AP_INVOICE_DISTRIBUTIONS_ALL"."PO_DISTRIBUTION_ID"="PO_DISTRIBUTIONS_A
LL"."PO_DISTRIBUTION_ID"
            (+))
  24 - filter(NVL("PO_HEADERS_ALL"."ORG_ID"(+),NVL(TO_NUMBER(DECODE(SUBSTRB(:B1,1,1),'
            ',NULL,SUBSTRB(:B2,1,10))),(-99)))=NVL(TO_NUMBER(DECODE(SUBSTRB(:B3,1,1
),' ',NULL,SUBSTRB(:B4,1,10))),(-99)))
  25 - access("PO_DISTRIBUTIONS_ALL"."PO_HEADER_ID"="PO_HEADERS_ALL"."PO_HEADER_ID"(+))
  26 - filter(NVL("PO_LINE_LOCATIONS_ALL"."ORG_ID"(+),NVL(TO_NUMBER(DECODE(SUBSTRB(:B
1,1,1),'
            ',NULL,SUBSTRB(:B2,1,10))),(-99)))=NVL(TO_NUMBER(DECODE(SUBSTRB(:B3,1,1
),' ',NULL,SUBSTRB(:B4,1,10))),(-99)))
  27 - access("PO_DISTRIBUTIONS_ALL"."LINE_LOCATION_ID"="PO_LINE_LOCATIONS_ALL"."LINE
_LOCATION_ID"(+))
  28 - filter(NVL("PO_LINES_ALL"."ORG_ID"(+),NVL(TO_NUMBER(DECODE(SUBSTRB(:B1,1,1),'
            ',NULL,SUBSTRB(:B2,1,10))),(-99)))=NVL(TO_NUMBER(DECODE(SUBSTRB(:B3,1,1
),' ',NULL,SUBSTRB(:B4,1,10))),(-99)))
  29 - access("PO_LINE_LOCATIONS_ALL"."PO_LINE_ID"="PO_LINES_ALL"."PO_LINE_ID"(+))
  31 - access("AP_INVOICE_DISTRIBUTIONS_ALL"."RCV_TRANSACTION_ID"="RTXNS"."TRANSACTIO
N_ID"(+))
  33 - access("RTXNS"."SHIPMENT_LINE_ID"="RSL"."SHIPMENT_LINE_ID"(+))
  34 - access("RSL"."SHIPMENT_HEADER_ID"="RSH"."SHIPMENT_HEADER_ID"(+))
  35 - filter(NVL("AP_INVOICE_DISTRIBUTIONS_ALL"."ORG_ID",NVL(TO_NUMBER(DECODE(SUBSTR
B(:B1,1,1),'
            ',NULL,SUBSTRB(:B2,1,10))),(-99)))=NVL(TO_NUMBER(DECODE(SUBSTRB(:B3,1,1
),' ',NULL,SUBSTRB(:B4,1,10))),(-99)))
  36 - access("AP_INVOICE_DISTRIBUTIONS_ALL"."PREPAY_DISTRIBUTION_ID"="AP_INVOICE_DIS
TRIBUTIONS_ALL"."INVOICE
            _DISTRIBUTION_ID")
  37 - filter(NVL("AP_INVOICES_ALL"."ORG_ID",NVL(TO_NUMBER(DECODE(SUBSTRB(:B1,1,1),'
            ',NULL,SUBSTRB(:B2,1,10))),(-99)))=NVL(TO_NUMBER(DECODE(SUBSTRB(:B3,1,1
),' ',NULL,SUBSTRB(:B4,1,10))),(-99)))
  38 - access("AP_INVOICES_ALL"."INVOICE_ID"="AP_INVOICE_DISTRIBUTIONS_ALL"."INVOICE_
ID")
  39 - filter(NVL("PO_VENDOR_SITES_ALL"."ORG_ID",NVL(TO_NUMBER(DECODE(SUBSTRB(:B1,1,1),'
            ',NULL,SUBSTRB(:B2,1,10))),(-99)))=NVL(TO_NUMBER(DECODE(SUBSTRB(:B3,1,1
),' ',NULL,SUBSTRB(:B4,1,10))),(-99)))
  40 - access("AP_INVOICES_ALL"."VENDOR_SITE_ID"="PO_VENDOR_SITES_ALL"."VENDOR_SITE_I
D")
  41 - access("AP_INVOICES_ALL"."VENDOR_ID"="PV"."VENDOR_ID")

Note: cpu costing is off
```

从执行计划中 Id=5 看到,该 SQL 发生了连接列谓词推入,视图 AP_UNAPPLY_PREPAYS_V 被当作了嵌套循环的被驱动表。原始 SQL 中,两表的关联条件如下。

```
WHERE A.INVOICE_ID = B.INVOICE_ID(+)
```

视图中 B.INVOICE_ID 来自于 AID1.INVOICE_ID INVOICE_ID,因此,我们应该检查执行计划中 AID1.INVOICE_ID INVOICE_ID 是否走了索引。我们从执行计划中 Id=18 发现如下。

```
18 - filter("A"."INVOICE_ID"="AP_INVOICE_DISTRIBUTIONS_ALL"."INVOICE_ID")
```

这里是将连接列谓词推入到执行计划中 Id=18 进行的过滤操作,并不是将连接列谓词推入视图让表 AP_INVOICE_DISTRIBUTIONS 走 INVOICE_ID 的索引。这显然大错特错了。

因为发生了谓词推入,视图 AP_UNAPPLY_PREPAYS_V 作为嵌套循环被驱动表会被多次扫描。这里的谓词推入的时候只是起的过滤作用,并没有走谓词连接列索引。因此,我们使用 HINT: USE_HASH(A,B),让两表走 HASH 连接,从而避免视图被多次反复扫描。添加 HINT

之后，SQL 能在 1 秒返回结果。

我们也可以调整隐含参数，关闭连接列谓词推入。

```
ALTER SESSION SET "_push_join_predicate" = FALSE;
```

禁止连接列谓词推入，也能达到效果。

我们还可以检查表 AP_INVOICE_DISTRIBUTIONS 表的 INVOICE_ID 列是否存在索引，如果没有索引，可以建立一个索引，从而实现真正的连接列谓词推入。但是因为当时使用 USE_HASH 已经优化了 SQL，所以没有继续检查。

最终的 SQL 如下。

```
SELECT *
  FROM (SELECT /*+ use_hash(a,b) */ A.INVOICE_ID,
               A.VENDOR_ID,
               A.INVOICE_NUM,
               A.INVOICE_AMOUNT,
               A.GL_DATE,
               A.INVOICE_CURRENCY_CODE,
               SUM(NVL(B.PREPAY_AMOUNT_APPLIED, 0)) PAID_AMOUNT,
               A.INVOICE_AMOUNT - SUM(NVL(B.PREPAY_AMOUNT_APPLIED, 0)) REMAIN
          FROM ap.AP_INVOICES_ALL A, APPS.AP_UNAPPLY_PREPAYS_V B
         WHERE A.INVOICE_ID = B.INVOICE_ID(+)
           AND A.ORG_ID = 126 /*:B4*/
           AND A.SOURCE = 'OSM IMPORTED' /*:B3*/
           AND A.INVOICE_NUM BETWEEN NVL( /*:B2*/ null, A.INVOICE_NUM) AND
               NVL( /*:B1*/ null, A.INVOICE_NUM)
         GROUP BY A.INVOICE_ID,
                  A.INVOICE_NUM,
                  A.INVOICE_AMOUNT,
                  A.VENDOR_ID,
                  A.GL_DATE,
                  A.INVOICE_CURRENCY_CODE)
 WHERE REMAIN > 0 ;
```

添加 HINT 后的执行计划如下。

```
---------------------------------------------------------------------------------
| Id  | Operation                           | Name                       |Rows|Bytes|Cost|
---------------------------------------------------------------------------------
|   0 | SELECT STATEMENT                    |                            |  1 |  69 | 723|
|*  1 |  FILTER                             |                            |    |     |    |
|   2 |   SORT GROUP BY                     |                            |  1 |  69 | 723|
|*  3 |    HASH JOIN OUTER                  |                            |  3 | 207 | 698|
|*  4 |     TABLE ACCESS FULL               | AP_INVOICES_ALL            |  3 | 153 | 694|
|   5 |     VIEW                            | AP_UNAPPLY_PREPAYS_V       |  1 |  18 |   3|
|   6 |      NESTED LOOPS                   |                            |  1 | 372 |   3|
|   7 |       NESTED LOOPS                  |                            |  1 | 368 |   3|
|   8 |        NESTED LOOPS                 |                            |  1 | 361 |   2|
|   9 |         NESTED LOOPS                |                            |  1 | 347 |   1|
|  10 |          NESTED LOOPS OUTER         |                            |  1 | 334 |   1|
|  11 |           NESTED LOOPS OUTER        |                            |  1 | 321 |   1|
|  12 |            NESTED LOOPS OUTER       |                            |  1 | 295 |   1|
|  13 |             NESTED LOOPS OUTER      |                            |  1 | 269 |   1|
|  14 |              NESTED LOOPS OUTER     |                            |  1 | 243 |   1|
|  15 |               NESTED LOOPS OUTER    |                            |  1 | 197 |   1|
|  16 |                NESTED LOOPS OUTER   |                            |  1 | 157 |   1|
|  17 |                 NESTED LOOPS OUTER  |                            |  1 |  98 |   1|
|* 18 |                  TABLE ACCESS BY INDEX ROWID| AP_INVOICE_DISTRIBUTIONS_ALL|  1 |  72 |   1|
```

9.26 谓词推入优化案例

```
|*19 |          INDEX FULL SCAN             |AP_INVOICE_DISTRIBUTIONS_N20|   1|     |    |
|*20 |        TABLE ACCESS BY INDEX ROWID   |AP_TAX_CODES_ALL            |   1|  26 |    |
|*21 |          INDEX UNIQUE SCAN           |AP_TAX_CODES_U1             |   1|     |    |
|*22 |       TABLE ACCESS BY INDEX ROWID    |PO_DISTRIBUTIONS_ALL        |   1|  59 |    |
|*23 |         INDEX UNIQUE SCAN            |PO_DISTRIBUTIONS_U1         |   1|     |    |
|*24 |        TABLE ACCESS BY INDEX ROWID   |PO_HEADERS_ALL              |   1|  40 |    |
|*25 |          INDEX UNIQUE SCAN           |PO_HEADERS_U1               |   1|     |    |
|*26 |       TABLE ACCESS BY INDEX ROWID    |PO_LINE_LOCATIONS_ALL       |   1|  46 |    |
|*27 |         INDEX UNIQUE SCAN            |PO_LINE_LOCATIONS_U1        |   1|     |    |
|*28 |      TABLE ACCESS BY INDEX ROWID     |PO_LINES_ALL                |   1|  26 |    |
|*29 |        INDEX UNIQUE SCAN             |PO_LINES_U1                 |   1|     |    |
| 30 |     TABLE ACCESS BY INDEX ROWID      |RCV_TRANSACTIONS            |   1|  26 |    |
|*31 |       INDEX UNIQUE SCAN              |RCV_TRANSACTIONS_U1         |   1|     |    |
| 32 |    TABLE ACCESS BY INDEX ROWID       |RCV_SHIPMENT_LINES          |   1|  26 |    |
|*33 |      INDEX UNIQUE SCAN               |RCV_SHIPMENT_LINES_U1       |   1|     |    |
|*34 |     INDEX UNIQUE SCAN                |RCV_SHIPMENT_HEADERS_U1     |   1|  13 |    |
|*35 |   TABLE ACCESS BY INDEX ROWID        |AP_INVOICE_DISTRIBUTIONS_ALL|   1|  13 |    |
|*36 |     INDEX UNIQUE SCAN                |AP_INVOICE_DISTRIBUTIONS_U2 |   1|     |    |
|*37 |  TABLE ACCESS BY INDEX ROWID         |AP_INVOICES_ALL             |   1|  14 |   1|
|*38 |    INDEX UNIQUE SCAN                 |AP_INVOICES_U1              |   1|     |    |
|*39 | TABLE ACCESS BY INDEX ROWID          |PO_VENDOR_SITES_ALL         |   1|   7 |   1|
|*40 |   INDEX UNIQUE SCAN                  |PO_VENDOR_SITES_U1          |   1|     |    |
|*41 |   INDEX UNIQUE SCAN                  |PO_VENDORS_U1               |   1|   4 |    |
----------------------------------------------------------------------------------------

Predicate Information (identified by operation id):
---------------------------------------------------

   1 - filter("A"."INVOICE_AMOUNT"-SUM(NVL("B"."PREPAY_AMOUNT_APPLIED",0))>0)
   3 - access("A"."INVOICE_ID"="B"."INVOICE_ID"(+))
   4 - filter("A"."ORG_ID"=126 AND "A"."SOURCE"='OSM IMPORTED' AND
              "A"."INVOICE_NUM">=NVL(NULL,"A"."INVOICE_NUM") AND "A"."INVOICE_NUM"<=N
VL(NULL,"A"."INVOICE_NUM"))
  18 - filter("AP_INVOICE_DISTRIBUTIONS_ALL"."AMOUNT"<0 AND
              NVL("AP_INVOICE_DISTRIBUTIONS_ALL"."REVERSAL_FLAG",'N')<>'Y' AND
              "AP_INVOICE_DISTRIBUTIONS_ALL"."LINE_TYPE_LOOKUP_CODE"='PREPAY' AND
              NVL("AP_INVOICE_DISTRIBUTIONS_ALL"."ORG_ID",NVL(TO_NUMBER(DECODE(SUBSTR
B(:B1,1,1),'
              ',NULL,SUBSTRB(:B2,1,10))),(-99)))=NVL(TO_NUMBER(DECODE(SUBSTRB(:B3,1,1
),' ',NULL,SUBSTRB(:B4,1,10))),(-99)))
  19 - filter("AP_INVOICE_DISTRIBUTIONS_ALL"."PREPAY_DISTRIBUTION_ID" IS NOT NULL)
  20 - filter(NVL("AP_TAX_CODES_ALL"."ORG_ID"(+),NVL(TO_NUMBER(DECODE(SUBSTRB(:B1,1,1),'
              ',NULL,SUBSTRB(:B2,1,10))),(-99)))=NVL(TO_NUMBER(DECODE(SUBSTRB(:B3,1,1
),' ',NULL,SUBSTRB(:B4,1,10))),(-99)))
  21 - access("AP_INVOICE_DISTRIBUTIONS_ALL"."TAX_CODE_ID"="AP_TAX_CODES_ALL"."TAX_ID
"(+))
  22 - filter(NVL("PO_DISTRIBUTIONS_ALL"."ORG_ID"(+),NVL(TO_NUMBER(DECODE(SUBSTRB(:B1
,1,1),'
              ',NULL,SUBSTRB(:B2,1,10))),(-99)))=NVL(TO_NUMBER(DECODE(SUBSTRB(:B3,1,1
),' ',NULL,SUBSTRB(:B4,1,10))),(-99)))
  23 - access("AP_INVOICE_DISTRIBUTIONS_ALL"."PO_DISTRIBUTION_ID"="PO_DISTRIBUTIONS_A
LL"."PO_DISTRIBUTION_ID"
              (+))
  24 - filter(NVL("PO_HEADERS_ALL"."ORG_ID"(+),NVL(TO_NUMBER(DECODE(SUBSTRB(:B1,1,1),'
              ',NULL,SUBSTRB(:B2,1,10))),(-99)))=NVL(TO_NUMBER(DECODE(SUBSTRB(:B3,1,1
),' ',NULL,SUBSTRB(:B4,1,10))),(-99)))
  25 - access("PO_DISTRIBUTIONS_ALL"."PO_HEADER_ID"="PO_HEADERS_ALL"."PO_HEADER_ID"(+))
  26 - filter(NVL("PO_LINE_LOCATIONS_ALL"."ORG_ID"(+),NVL(TO_NUMBER(DECODE(SUBSTRB(:B
1,1,1),'
              ',NULL,SUBSTRB(:B2,1,10))),(-99)))=NVL(TO_NUMBER(DECODE(SUBSTRB(:B3,1,1
),' ',NULL,SUBSTRB(:B4,1,10))),(-99)))
```

```
        27 - access("PO_DISTRIBUTIONS_ALL"."LINE_LOCATION_ID"="PO_LINE_LOCATIONS_ALL"."LINE
_LOCATION_ID"(+))
        28 - filter(NVL("PO_LINES_ALL"."ORG_ID"(+),NVL(TO_NUMBER(DECODE(SUBSTRB(:B1,1,1),'
              ',NULL,SUBSTRB(:B2,1,10))),(-99)))=NVL(TO_NUMBER(DECODE(SUBSTRB(:B3,1,1
),' ',NULL,SUBSTRB(:B4,1,10))),(-99)))
        29 - access("PO_LINE_LOCATIONS_ALL"."PO_LINE_ID"="PO_LINES_ALL"."PO_LINE_ID"(+))
        31 - access("AP_INVOICE_DISTRIBUTIONS_ALL"."RCV_TRANSACTION_ID"="RTXNS"."TRANSACTIO
N_ID"(+))
        33 - access("RTXNS"."SHIPMENT_LINE_ID"="RSL"."SHIPMENT_LINE_ID"(+))
        34 - access("RSL"."SHIPMENT_HEADER_ID"="RSH"."SHIPMENT_HEADER_ID"(+))
        35 - filter(NVL("AP_INVOICE_DISTRIBUTIONS_ALL"."ORG_ID",NVL(TO_NUMBER(DECODE(SUBSTR
B(:B1,1,1),'
              ',NULL,SUBSTRB(:B2,1,10))),(-99)))=NVL(TO_NUMBER(DECODE(SUBSTRB(:B3,1,1
),' ',NULL,SUBSTRB(:B4,1,10))),(-99)))
        36 - access("AP_INVOICE_DISTRIBUTIONS_ALL"."PREPAY_DISTRIBUTION_ID"="AP_INVOICE_DIS
TRIBUTIONS_ALL"."INVOICE
              _DISTRIBUTION_ID")
        37 - filter(NVL("AP_INVOICES_ALL"."ORG_ID",NVL(TO_NUMBER(DECODE(SUBSTRB(:B1,1,1),'
              ',NULL,SUBSTRB(:B2,1,10))),(-99)))=NVL(TO_NUMBER(DECODE(SUBSTRB(:B3,1,1
),' ',NULL,SUBSTRB(:B4,1,10))),(-99)))
        38 - access("AP_INVOICES_ALL"."INVOICE_ID"="AP_INVOICE_DISTRIBUTIONS_ALL"."INVOICE_
ID")
        39 - filter(NVL("PO_VENDOR_SITES_ALL"."ORG_ID",NVL(TO_NUMBER(DECODE(SUBSTRB(:B1,1,1),'
              ',NULL,SUBSTRB(:B2,1,10))),(-99)))=NVL(TO_NUMBER(DECODE(SUBSTRB(:B3,1,1
),' ',NULL,SUBSTRB(:B4,1,10))),(-99)))
        40 - access("AP_INVOICES_ALL"."VENDOR_SITE_ID"="PO_VENDOR_SITES_ALL"."VENDOR_SITE_I
D")
        41 - access("AP_INVOICES_ALL"."VENDOR_ID"="PV"."VENDOR_ID")
```

水平高的读者或许有疑问，执行计划 Id=19 是 INDEX FULL SCAN，然后再回表过滤，这里也有性能问题，全表扫描效率应该也比 INDEX FULL SCAN 再回表效率高！是的，我们也发现了这个地方有性能问题，但是既然 SQL 都执行到 1 秒了，也就没继续优化了，千万别得了优化强迫症。

最后，我们再次强调，如果发生了连接列谓词推入，一定要检查执行计划中是否走了谓词被推入的表的连接列索引。

9.27 使用 CARDINALITY 优化 SQL

2011 年，一位 ITPUB 的网友请求优化如下 SQL，该 SQL 执行不出结果。

```
SQL>  explain plan for    select ((v.yvalue * 300) / (u.xvalue * 50)), u.xtime
  2      from (select x.index_value xvalue, substr(x.update_time, 1, 14) xtime
  3              from tb_indexs x
  4             where x.id in (select  min(a.id)
  5                              from tb_indexs a
  6                             where a.code = 'HSI'
  7                               and a.update_time > 20110701000000
  8                               and a.update_time < 20110722000000
  9                             group by a.update_time) u,
 10          (select  y.index_value yvalue, substr(y.update_time, 1, 14) ytime
 11             from tb_indexs y
 12            where y.id in (select  min(b.id)
 13                             from tb_indexs b
 14                            where b.code = '000300'
 15                              and b.update_time > 20110701000000
 16                              and b.update_time < 20110722000000
```

```
17                          group by b.update_time)) v
18   where u.xtime = v.ytime
19   order by u.xtime;
Explained.

SQL> select * from table(dbms_xplan.display);

PLAN_TABLE_OUTPUT
-------------------------------------------------------------------------------------

Plan hash value: 573554298

-------------------------------------------------------------------------------------
| Id  | Operation                         | Name           | Rows | Bytes | Cost(%CPU)|
-------------------------------------------------------------------------------------
|   0 | SELECT STATEMENT                  |                |   1  |  54   |  13   (8) |
|   1 |  SORT ORDER BY                    |                |   1  |  54   |  13   (8) |
|   2 |   NESTED LOOPS                    |                |   1  |  54   |  12   (0) |
|   3 |    MERGE JOIN CARTESIAN           |                |   1  |  33   |  10   (0) |
|   4 |     NESTED LOOPS                  |                |   1  |  27   |   6   (0) |
|   5 |      VIEW                         | VW_NSO_2       |   1  |   6   |   4   (0) |
|   6 |       HASH GROUP BY               |                |   1  |  41   |   4   (0) |
|   7 |        TABLE ACCESS BY INDEX ROWID| TB_INDEXS      |   1  |  41   |   4   (0) |
|*  8 |         INDEX RANGE SCAN          | IDX_UPDATE_TIME|   1  |       |   3   (0) |
|   9 |      TABLE ACCESS BY INDEX ROWID  | TB_INDEXS      |   1  |  21   |   2   (0) |
|* 10 |       INDEX UNIQUE SCAN           | PK_INDEXS      |   1  |       |   1   (0) |
|  11 |     BUFFER SORT                   |                |   1  |   6   |   8   (0) |
|  12 |      VIEW                         | VW_NSO_1       |   1  |   6   |   4   (0) |
|  13 |       HASH GROUP BY               |                |   1  |  41   |   4   (0) |
|  14 |        TABLE ACCESS BY INDEX ROWID| TB_INDEXS      |   1  |  41   |   4   (0) |
|* 15 |         INDEX RANGE SCAN          | IDX_UPDATE_TIME|   1  |       |   3   (0) |
|* 16 |    TABLE ACCESS BY INDEX ROWID    | TB_INDEXS      |   1  |  21   |   2   (0) |
|* 17 |     INDEX UNIQUE SCAN             | PK_INDEXS      |   1  |       |   1   (0) |
-------------------------------------------------------------------------------------

Predicate Information (identified by operation id):
---------------------------------------------------

   8 - access("A"."UPDATE_TIME">20110701000000 AND "A"."CODE"='HSI' AND
              "A"."UPDATE_TIME"<20110722000000)
       filter("A"."CODE"='HSI')
  10 - access("X"."ID"="$nso_col_1")
  15 - access("B"."UPDATE_TIME">20110701000000 AND "B"."CODE"='000300' AND
              "B"."UPDATE_TIME"<20110722000000)
       filter("B"."CODE"='000300')
  16 - filter(SUBSTR(TO_CHAR("X"."UPDATE_TIME"),1,14)=SUBSTR(TO_CHAR("Y"."UPDATE_TIME
              "),1,14)
              )
  17 - access("Y"."ID"="$nso_col_1")

38 rows selected.
```

大家请仔细观察 SQL 语句，该 SQL 访问的都是同一个表 TB_INDEXS，表在 SQL 语句中被访问了 4 次，我们可以对 SQL 进行等价改写，让 SQL 只访问一次，从而就达到了优化目的。

但是，网友希望在不改写 SQL 的前提下优化该 SQL 语句，因此只能从执行计划入手优化 SQL。执行计划中，Id=3 是笛卡儿积，这就是为什么该 SQL 执行不出结果。为什么会产生笛卡儿积呢？因为执行计划中所有的步骤 Rows 都估算返回为 1 行数据，所以优化器选择了笛卡几积连接（在5.4节中我们讲过，离笛卡儿积关键字最近的"表"被错误地估算为1行的时候，优化器很容易选择走笛卡儿积连接）。

执行计划的入口是Id=8，也就是SQL语句中的in子查询，优化器评估Id=8返回1行数据，但是实际上Id=8要返回2万行数据。笔者曾经尝试对表TB_INDEXS重新收集统计信息，但是收集完统计信息之后，优化器还是评估Id=8返回1行数据。

为什么优化器会评估Id=8返回1行数据呢？这是因为字段UPDATE_TIME被设计为了NUMBER类型，而实际上UPDATE_TIME应该是DATE类型，同时where条件中还有一个选择性较低的过滤条件，优化器估算返回的行数等于表的总行数与UPDATE_TIME的选择性、CODE的选择性的乘积。UPDATE_TIME因为字段类型设计错误，本来应该估算返回21天的数据，但是因为UPDATE_TIME设计为了NUMBER类型，导致优化器在估算返回行数的时候不是利用DATE类型估算返回行数，而是利用NUMBER类型估算返回行数。大家请注意观察UPDATE_TIME的过滤条件，将年月日存储为NUMBER类型是一个天文数字，然后where条件只是取出一个天文数字中极小一部分数据，因此估算返回的行数始终会被估算为1行。

因为执行计划入口的Rows估算错误，所以后面的执行计划不用看，全是错误的。因为UPDATE_TIME已经被设计为NUMBER类型了，想要通过修改UPDATE_TIME为DATE类型来纠正优化器估算返回的Rows是不可行的，因为需要申请停机时间。

怎么才可以让优化器知道真实Rows呢？我们可以使用HINT：CARDINALITY。

/*+ cardinality(a 10000) */ 表示指定a表有1万行数据。

/*+ cardinality(@a 10000) */ 表示指定query block a有1万行数据。

添加完HINT后的执行计划如下。

```
SQL> set autot trace
SQL> select /*+ cardinality(@a 20000) cardinality(@b 20000) */((v.yvalue * 300)/(u.xv
alue * 50)), u.xtime
  2    from (select x.index_value xvalue, substr(x.update_time, 1, 14) xtime
  3            from tb_indexs x
  4           where x.id in (select /*+ QB_NAME(a) */ min(a.id)
  5                            from tb_indexs a
  6                           where a.code = 'HSI'
  7                             and a.update_time > 20110701000000
  8                             and a.update_time < 20110722000000
  9                           group by a.update_time)) u,
 10         (select y.index_value yvalue, substr(y.update_time, 1, 14) ytime
 11            from tb_indexs y
 12           where y.id in (select /*+ QB_NAME(b) */ min(b.id)
 13                            from tb_indexs b
 14                           where b.code = '000300'
 15                             and b.update_time > 20110701000000
 16                             and b.update_time < 20110722000000
 17                           group by b.update_time)) v
 18   where u.xtime = v.ytime
 19   order by u.xtime;

3032 rows selected.

Elapsed: 00:00:15.07

Execution Plan
```

9.27 使用 CARDINALITY 优化 SQL

```
--------------------------------------------------------------------
Plan hash value: 2679503093

--------------------------------------------------------------------
| Id  | Operation                       | Name            | Rows  | Bytes  |Cost(%CPU)|
--------------------------------------------------------------------
|   0 | SELECT STATEMENT                |                 |   935 | 50490  | 1393   (7)|
|   1 |  SORT ORDER BY                  |                 |   935 | 50490  | 1393   (7)|
|*  2 |   HASH JOIN                     |                 |   935 | 50490  | 1392   (7)|
|   3 |    VIEW                         | VW_NSO_1        | 20000 |  117K  |    4   (0)|
|   4 |     HASH GROUP BY               |                 | 20000 |  800K  |    4   (0)|
|   5 |      TABLE ACCESS BY INDEX ROWID| TB_INDEXS       |     1 |   41   |    4   (0)|
|*  6 |       INDEX RANGE SCAN          | IDX_UPDATE_TIME |     1 |        |    3   (0)|
|*  7 |    HASH JOIN                    |                 | 31729 | 1487K  | 1386   (7)|
|*  8 |     HASH JOIN                   |                 | 20000 |  527K  |  695   (7)|
|   9 |      VIEW                       | VW_NSO_2        | 20000 |  117K  |    4   (0)|
|  10 |       HASH GROUP BY             |                 | 20000 |  800K  |    4   (0)|
|  11 |        TABLE ACCESS BY INDEX ROWID| TB_INDEXS     |     1 |   41   |    4   (0)|
|* 12 |         INDEX RANGE SCAN        | IDX_UPDATE_TIME |     1 |        |    3   (0)|
|  13 |      TABLE ACCESS FULL          | TB_INDEXS       |  678K |   13M  |  678   (5)|
|  14 |     TABLE ACCESS FULL           | TB_INDEXS       |  678K |   13M  |  678   (5)|
--------------------------------------------------------------------

Predicate Information (identified by operation id):
---------------------------------------------------

   2 - access("Y"."ID"="$nso_col_1")
   6 - access("B"."UPDATE_TIME">20110701000000 AND "B"."CODE"='000300' AND
              "B"."UPDATE_TIME"<20110722000000)
       filter("B"."CODE"='000300')
   7 - access(SUBSTR(TO_CHAR("X"."UPDATE_TIME"),1,14)=SUBSTR(TO_CHAR("Y"."UPDATE_TIME
"),1,14)
              )
   8 - access("X"."ID"="$nso_col_1")
  12 - access("A"."UPDATE_TIME">20110701000000 AND "A"."CODE"='HSI' AND
              "A"."UPDATE_TIME"<20110722000000)
       filter("A"."CODE"='HSI')


Statistics
----------------------------------------------------------
         29  recursive calls
          0  db block gets
       8351  consistent gets
       4977  physical reads
         72  redo size
     141975  bytes sent via SQL*Net to client
       2622  bytes received via SQL*Net from client
        204  SQL*Net roundtrips to/from client
          1  sorts (memory)
          0  sorts (disk)
       3032  rows processed
```

通过指定执行计划入口（子查询）返回 2 万行数据，纠正了之前错误的执行计划，SQL 最终执行了 15 秒就返回了所有的结果。

如果不知道有 CARDINALITY 这个 HINT，怎么优化 SQL 呢？我们可以启用动态采样 Level 4 及以上（最好别超过 6），让优化器能较为准确地评估出子查询返回的 Rows，这样也能达到优化目的。如果不知道动态采样怎么优化 SQL 呢？我们可以直接使用 HINT，比如

USE_HASH 等，让 SQL 走我们认为正确的执行计划也能达到优化目的。当然了，最佳的优化方法应该是直接从业务上入手，从表设计上入手，从 SQL 写法上入手，而不是退而求其次从执行计划入手，但是很多时候我们往往只能从执行计划上入手优化 SQL，这或许是绝大多数 DBA 的无奈。

本案例博客地址：http://blog.csdn.net/robinson1988/article/details/6626384。

9.28 利用等待事件优化 SQL

本案例发生在 2010 年，当时作者罗老师在惠普担任开发 DBA，支撑宝洁公司的数据仓库项目。ETL 开发人员需要帮助调查一个 long running 的 JOB，该 JOB 执行了 7 个小时还没执行完。

数据库环境为 11.1.0.7（RAC，4 节点）。

```
SQL> select * from v$version;

BANNER
--------------------------------------------------------------------------------
Oracle Database 11g Enterprise Edition Release 11.1.0.7.0 - 64bit Production
```

数据块大小为 16k。

```
SQL> show parameter db_block_size

NAME                                 TYPE                             VALUE
------------------------------------ -------------------------------- ------
db_block_size                        integer                          16384
```

执行得慢的 JOB 是一个 insert into ...select ...语句。一般情况下，如果 select 语句跑得快，那么整个 JOB 也就跑得快，因此我们应该把主要精力放在 select 语句上面。select 部分的 SQL 语句如下，这是一个接近 400 行的 SQL（因为 SQL 实在太长，所以没有对 SQL 格式化）。

```
SELECT   ACTVY_SKID,FUND_SKID,PRMTN_SKID,PROD_SKID,DATE_SKID,
ACCT_SKID,BUS_UNIT_SKID,FY_DATE_SKID,ESTMT_VAR_COST_AMT,ESTMT_FIXED_COST_AMT,
REVSD_ESTMT_VAR_COST_AMT,ACTL_VAR_COST_AMT,ACTL_FIXED_COST_AMT,COST_PLAN_AMT,
COST_CMMT_AMT,COST_BOOK_AMT,ESTMT_COST_OVRRD_AMT,LA_TOT_BOOK_AMT,
MANUL_COST_OVRRD_AMT,ACTL_COST_AMT
FROM   (SELECT ACTVY_SKID,FUND_SKID,PROD_SKID,PRMTN_SKID,DATE_SKID,ACCT_SKID,
BUS_UNIT_SKID,FY_DATE_SKID,ESTMT_VAR_COST_AMT,ESTMT_FIXED_COST_AMT,
REVSD_ESTMT_VAR_COST_AMT,0 as ACTL_COST_AMT,ACTL_VAR_COST_AMT,ACTL_FIXED_COST_AMT,
MANUL_COST_OVRRD_AMT,ESTMT_COST_OVRRD_AMT,COST_BOOK_AMT,
-- Updated by Luke for QC3369
-- If the committed amount on Activity level <0 then return 0
(CASE WHEN SUM(ESTMT_COST_OVRRD_AMT - ACTL_VAR_COST_AMT -
ACTL_FIXED_COST_AMT) OVER(PARTITION BY ACTVY_SKID) < 0 THEN 0
ELSE COST_CMMT_AMT END) AS COST_CMMT_AMT,
-- Updated by Luke for QC3369
(CASE WHEN SUM(ESTMT_COST_OVRRD_AMT - ACTL_VAR_COST_AMT -
ACTL_FIXED_COST_AMT) OVER(PARTITION BY ACTVY_SKID) < 0 THEN 0
ELSE COST_PLAN_AMT END) AS COST_PLAN_AMT,LA_TOT_BOOK_AMT
FROM (SELECT ACTVY_SKID,FUND_SKID,PROD_SKID,PRMTN_SKID,
DATE_SKID,ACCT_SKID,BUS_UNIT_SKID,FY_DATE_SKID,ESTMT_VAR_COST_AMT,
```

9.28 利用等待事件优化 SQL

```
ESTMT_FIXED_COST_AMT,REVSD_ESTMT_VAR_COST_AMT,ACTL_VAR_COST_AMT,
ACTL_FIXED_COST_AMT,MANUL_COST_OVRRD_AMT,
(CASE WHEN SUBSTR(ESTMT_COST_IND, 1, 1) = 'E' THEN
ESTMT_FIXED_COST_AMT + ESTMT_VAR_COST_AMT WHEN SUBSTR(ESTMT_COST_IND, 1, 1) = 'R' THE
N ESTMT_FIXED_COST_AMT + DECODE(REVSD_BPT_COST_AMT,0,REVSD_ESTMT_VAR_COST_AMT,
--Ax Revised Estimated Variable Cost REVSD_BPT_COST_AMT) --BPT Revised Cost
WHEN SUBSTR(ESTMT_COST_IND, 1, 1) = 'M' THEN MANUL_COST_OVRRD_AMT
WHEN ESTMT_COST_IND IS NULL THEN DECODE(CORP_PRMTN_TYPE_CODE,
'Annual Agreement',ESTMT_FIXED_COST_AMT + DECODE(REVSD_BPT_COST_AMT,0,
REVSD_ESTMT_VAR_COST_AMT, --Ax Revised Estimated Variable Cost
REVSD_BPT_COST_AMT), --BPT Revised Cost
ESTMT_FIXED_COST_AMT + ESTMT_VAR_COST_AMT) END) AS ESTMT_COST_OVRRD_AMT,
(ACTL_VAR_COST_AMT + ACTL_FIXED_COST_AMT) AS COST_BOOK_AMT,
DECODE(PRMTN_STTUS_CODE,'Confirmed',
--Estimate Total Cost - Actual Cost
--Add the logic of Activity Stop date and Pyment allow IND
--For Defect 2913 Luke 2010-5-5
(CASE WHEN (ACTVY_STOP_DATE IS NULL OR ACTVY_STOP_DATE > SYSDATE OR
NVL(PYMT_ALLWD_STOP_IND, 'N') = 'Y') THEN (CASE WHEN SUBSTR(ESTMT_COST_IND, 1, 1) = '
E' THEN ESTMT_FIXED_COST_AMT + ESTMT_VAR_COST_AMT WHEN SUBSTR(ESTMT_COST_IND, 1, 1) =
 'R' THEN ESTMT_FIXED_COST_AMT + DECODE(REVSD_BPT_COST_AMT,0,REVSD_ESTMT_VAR_COST_AMT,
--Ax Revised Estimated Variable Cost
REVSD_BPT_COST_AMT) --BPT Revised Cost
WHEN SUBSTR(ESTMT_COST_IND, 1, 1) = 'M' THEN MANUL_COST_OVRRD_AMT
WHEN ESTMT_COST_IND IS NULL THEN DECODE(CORP_PRMTN_TYPE_CODE,'Annual Agreement',
ESTMT_FIXED_COST_AMT + DECODE(REVSD_BPT_COST_AMT,0,REVSD_ESTMT_VAR_COST_AMT,
--Ax Revised Estimated Variable Cost
REVSD_BPT_COST_AMT), --BPT Revised Cost
ESTMT_FIXED_COST_AMT + ESTMT_VAR_COST_AMT) END) - (ACTL_VAR_COST_AMT + ACTL_FIXED_COS
T_AMT)
ELSE 0 END), 0) AS COST_CMMT_AMT,(CASE WHEN (PRMTN_STTUS_CODE IN ('Planned', 'Revised
') AND NVL(APPRV_STTUS_CODE, 'Nothing') <> 'Rejected' AND
--Add the logic of Activity Stop date and Pyment allow IND
--For Defect 2913 Luke 2010-5-5
(ACTVY_STOP_DATE IS NULL OR ACTVY_STOP_DATE > SYSDATE OR NVL(PYMT_ALLWD_STOP_IND, 'N'
) = 'Y'))
THEN (CASE WHEN SUBSTR(ESTMT_COST_IND, 1, 1) = 'E' THEN ESTMT_FIXED_COST_AMT + ESTMT_
VAR_COST_AMT
WHEN SUBSTR(ESTMT_COST_IND, 1, 1) = 'R' THEN ESTMT_FIXED_COST_AMT + DECODE(REVSD_BPT_
COST_AMT,0, REVSD_ESTMT_VAR_COST_AMT, --Ax Revised Estimated Variable Cost REVSD_BPT_
COST_AMT) --BPT Revised Cost
WHEN SUBSTR(ESTMT_COST_IND, 1, 1) = 'M' THEN MANUL_COST_OVRRD_AMT WHEN ESTMT_COST_IND
 IS NULL THEN DECODE(CORP_PRMTN_TYPE_CODE,'Annual Agreement',ESTMT_FIXED_COST_AMT +DE
CODE(REVSD_BPT_COST_AMT,0,
REVSD_ESTMT_VAR_COST_AMT, --Ax Revised Estimated Variable Cost
REVSD_BPT_COST_AMT), --BPT Revised Cost
ESTMT_FIXED_COST_AMT + ESTMT_VAR_COST_AMT) END) - (ACTL_VAR_COST_AMT + ACTL_FIXED_COS
T_AMT)  ELSE 0 END) AS COST_PLAN_AMT,(CASE WHEN MTH_START_DATE > TRUNC(SYSDATE, 'MM')
 AND PRMTN_STTUS_CODE IN ('Planned', 'Confirmed', 'Revised') THEN (CASE WHEN SUBSTR(E
STMT_COST_IND, 1, 1)= 'E' THEN ESTMT_FIXED_COST_AMT + ESTMT_VAR_COST_AMT WHEN SUBSTR(
ESTMT_COST_IND, 1, 1) = 'R' THEN ESTMT_FIXED_COST_AMT + DECODE(REVSD_BPT_COST_AMT,0,R
EVSD_ESTMT_VAR_COST_AMT,
--Ax Revised Estimated Variable Cost
REVSD_BPT_COST_AMT) --BPT Revised Cost
WHEN SUBSTR(ESTMT_COST_IND, 1, 1) = 'M' THEN MANUL_COST_OVRRD_AMT WHEN ESTMT_COST_IND
 IS NULL THEN DECODE(CORP_PRMTN_TYPE_CODE,'Annual Agreement',ESTMT_FIXED_COST_AMT +DE
CODE(REVSD_BPT_COST_AMT,0,
REVSD_ESTMT_VAR_COST_AMT, --Ax Revised Estimated Variable Cost
REVSD_BPT_COST_AMT), --BPT Revised Cost
ESTMT_FIXED_COST_AMT + ESTMT_VAR_COST_AMT) END)
```

```sql
     WHEN MTH_START_DATE <= TRUNC(SYSDATE, 'MM') THEN (ACTL_VAR_COST_AMT + ACTL_FIXED_COST
     _AMT) ELSE 0 END) AS LA_TOT_BOOK_AMT FROM (SELECT ACTVY_MTH_GTIN.ACTVY_SKID,ACTVY_MT
     H_GTIN.FUND_SKID,ACTVY_MTH_GTIN.PROD_SKID,
     ACTVY_MTH_GTIN.PRMTN_SKID,ACTVY_MTH_GTIN.MTH_SKID AS DATE_SKID,ACTVY_MTH_GTIN.ACCT_SK
     ID,ACTVY_MTH_GTIN.BUS_UNIT_SKID,
     ACTVY_MTH_GTIN.FY_DATE_SKID,PRMTN.PRMTN_STTUS_CODE,
     PRMTN.APPRV_STTUS_CODE,ACTVY.ESTMT_COST_IND,ACTVY.CORP_PRMTN_TYPE_CODE,ACTVY.ACTVY_ST
     OP_DATE,
     ACTVY.PYMT_ALLWD_STOP_IND,CAL.MTH_START_DATE,ROUND(NVL(DECODE(ACTVY.COST_TYPE_CODE,'%
      Fund',(ACTVY_MTH_GTIN.ESTMT_VAR_COST * -- added by Rita for defect 3105 in R10
     ACTVY_MTH_GTIN.ACTVY_GTIN_ESTMT_WGHT_RATE),DECODE(ACTVY.CORP_PRMTN_TYPE_CODE,
     'AnnualAgreement',AA.ESTMT_VAR_COST_AMT,ESTMT_VAR_COST.ESTMT_VAR_COST_AMT)),0),7) AS
     ESTMT_VAR_COST_AMT,
     -- Modified by Simon For CR389 in R10 on 2010-3-18
     ROUND(NVL(DECODE(ACTVY.COST_TYPE_CODE,
     -- % Fund
     '% Fund',ACTVY_MTH_GTIN.ESTMT_FIX_COST * ACTVY_MTH_GTIN.ACTVY_GTIN_ESTMT_WGHT_RATE,
     -- Fixed
     'Fixed',ACTVY_MTH_GTIN.ESTMT_FIX_COST * ACTVY_MTH_GTIN.ACTVY_GTIN_ESTMT_WGHT_RATE,
     -- Not % Fund or Fixed
     DECODE(DECODE(ACTVY.CORP_PRMTN_TYPE_CODE,'Annual Agreement',
     SUM(NVL(AA.ESTMT_VAR_COST_AMT,0))OVER(PARTITION BY ACTVY_MTH_GTIN.ACTVY_SKID),
     SUM(NVL(ESTMT_VAR_COST.ESTMT_VAR_COST_AMT,0))OVER(PARTITION BY ACTVY_MTH_GTIN.ACTVY_S
     KID)),
     0,ACTVY_MTH_GTIN.ESTMT_FIX_COST * BRAND_MTH_RATE,ACTVY_MTH_GTIN.ESTMT_FIX_COST * NVL(
     DECODE(ACTVY.CORP_PRMTN_TYPE_CODE,'AnnualAgreement',AA.ESTMT_VAR_COST_AMT,ESTMT_VAR_C
     OST.ESTMT_VAR_COST_AMT),0) / DECODE(ACTVY.CORP_PRMTN_TYPE_CODE,'Annual Agreement',SUM
     (NVL(AA.ESTMT_VAR_COST_AMT,0))
     OVER(PARTITION BY ACTVY_MTH_GTIN.ACTVY_SKID),SUM(NVL(ESTMT_VAR_COST.ESTMT_VAR_COST_AM
     T,0))
     OVER(PARTITION BY ACTVY_MTH_GTIN.ACTVY_SKID)))),0),7) AS ESTMT_FIXED_COST_AMT,
     -- Change in R10 for Revised Cost logic
     ROUND(NVL(DECODE(ACTVY.CORP_PRMTN_TYPE_CODE,'Annual Agreement',AA.REVSD_ESTMT_VAR_COS
     T_AMT,
     REVSD_VAR_COST.REVSD_ESTMT_VAR_COST_AMT),0),7) AS REVSD_ESTMT_VAR_COST_AMT,
     ROUND(NVL(ESTMT_VAR_COST.REVSD_BPT_COST_AMT, 0), 7) AS REVSD_BPT_COST_AMT,
     ROUND(NVL((ACTVY_MTH_GTIN.ACTL_VAR_COST * ACTVY_MTH_GTIN.ACTVY_GTIN_ACTL_WGHT_RATE),0
     ),7)
     AS ACTL_VAR_COST_AMT,ROUND(NVL((ACTVY_MTH_GTIN.ACTL_FIX_COST * ACTVY_MTH_GTIN.ACTVY_G
     TIN_ACTL_WGHT_RATE),0),7) AS ACTL_FIXED_COST_AMT,ROUND(NVL(DECODE(ACTVY.COST_TYPE_COD
     E,'% Fund',
     ACTVY_MTH_GTIN.MANUL_COST_OVRRD_AMT * ACTVY_MTH_GTIN.ACTVY_GTIN_ESTMT_WGHT_RATE,
     'Fixed',ACTVY_MTH_GTIN.MANUL_COST_OVRRD_AMT * ACTVY_MTH_GTIN.ACTVY_GTIN_ESTMT_WGHT_RA
     TE,
     DECODE(DECODE(ACTVY.CORP_PRMTN_TYPE_CODE,'Annual Agreement',SUM(NVL(AA.ESTMT_VAR_COST
     _AMT,0))
     OVER(PARTITION BY ACTVY_MTH_GTIN.ACTVY_SKID),SUM(NVL(ESTMT_VAR_COST.ESTMT_VAR_COST_AM
     T,0))
     OVER(PARTITION BY ACTVY_MTH_GTIN.ACTVY_SKID)),0,ACTVY_MTH_GTIN.MANUL_COST_OVRRD_AMT *
     BRAND_MTH_RATE,ACTVY_MTH_GTIN.MANUL_COST_OVRRD_AMT *
     NVL(DECODE(ACTVY.CORP_PRMTN_TYPE_CODE,'Annual Agreement',AA.ESTMT_VAR_COST_AMT,
     ESTMT_VAR_COST.ESTMT_VAR_COST_AMT),0) /DECODE(ACTVY.CORP_PRMTN_TYPE_CODE,
     'Annual Agreement',SUM(NVL(AA.ESTMT_VAR_COST_AMT,0))
     OVER(PARTITION BY ACTVY_MTH_GTIN.ACTVY_SKID),SUM(NVL(ESTMT_VAR_COST.ESTMT_VAR_COST_AM
     T,0))
     OVER(PARTITION BY ACTVY_MTH_GTIN.ACTVY_SKID)))),0),7) AS MANUL_COST_OVRRD_AMT
     FROM OPT_ACTVY_DIM ACTVY,OPT_PRMTN_DIM PRMTN,OPT_CAL_MASTR_DIM CAL,
     (SELECT ACTVY.ACTVY_SKID,ACTVY_GTIN_BRAND.ACTVY_ID,ACTVY.FUND_SKID,
     ACTVY.ACCT_PRMTN_SKID AS ACCT_SKID,ACTVY_GTIN_BRAND.PROD_SKID,ACTVY_GTIN_BRAND.PROD_ID,
     ACTVY_GTIN_BRAND.PRMTN_SKID,ACTVY.BUS_UNIT_SKID,ACTVY_GTIN_BRAND.MTH_SKID,
     ACTVY_GTIN_BRAND.FY_DATE_SKID,ACTVY.VAR_COST_ESTMT_AMT AS ESTMT_VAR_COST,
     ACTVY.PRDCT_FIXED_COST_AMT AS ESTMT_FIX_COST,ACTVY.CALC_INDEX_NUM AS ACTL_FIX_COST,
```

9.28 利用等待事件优化 SQL

```
ACTVY.ACTL_VAR_COST_NUM AS ACTL_VAR_COST,ACTVY.ESTMT_COST_OVRRD_AMT,ACTVY.MANUL_COST_
OVRRD_AMT,
ACTVY_GTIN_BRAND.ACTVY_GTIN_ACTL_WGHT_RATE,ACTVY_GTIN_BRAND.ACTVY_GTIN_ESTMT_WGHT_RATE,
ACTVY_GTIN_BRAND.BRAND_MTH_RATE FROM OPT_ACTVY_FCT ACTVY,
OPT_ACTVY_GTIN_BRAND_SFCT ACTVY_GTIN_BRAND, OPT_ACCT_DIM  ACCT
WHERE ACTVY.ACTVY_SKID = ACTVY_GTIN_BRAND.ACTVY_SKID AND ACCT.ACCT_SKID = ACTVY.ACCT_
PRMTN_SKID
-- Optimal1, B018, 9-Oct-2010, Kingham, filter out TSP account
AND ACCT.FUND_FRCST_MODEL_DESC not like 'TSP%') ACTVY_MTH_GTIN,
--Estamate variable cost aggregated to brand level
(SELECT  ESTMT.ACTVY_ID AS ACTVY_ID,BRAND_HIER.BRAND_ID AS PROD_ID,
ESTMT.DATE_SKID AS DATE_SKID,ESTMT.BUS_UNIT_SKID AS BUS_UNIT_SKID,
SUM(ESTMT.ESTMT_VAR_COST_AMT) AS ESTMT_VAR_COST_AMT,
SUM(ESTMT.REVSD_BPT_COST_AMT) AS REVSD_BPT_COST_AMT
FROM OPT_ACTVY_GTIN_ESTMT_SFCT ESTMT, -- add by rita
OPT_PROD_BRAND_ASSOC_DIM  BRAND_HIER,CAL_MASTR_DIM    CAL
WHERE ESTMT.PROD_ID = BRAND_HIER.PROD_ID AND ESTMT.DATE_SKID = CAL.CAL_MASTR_SKID
AND CAL.FISC_YR_SKID = BRAND_HIER.FY_DATE_SKID GROUP BY ESTMT.ACTVY_ID,
BRAND_HIER.BRAND_ID,ESTMT.DATE_SKID,ESTMT.BUS_UNIT_SKID) ESTMT_VAR_COST,
--Revised variable cost aggregated to brand level
(SELECT REVSD.ACTVY_ID AS ACTVY_ID,BRAND_HIER.BRAND_ID AS PROD_ID,
REVSD.DATE_SKID AS DATE_SKID,REVSD.BUS_UNIT_SKID AS BUS_UNIT_SKID,
SUM(REVSD.REVSD_ESTMT_VAR_COST_AMT) AS REVSD_ESTMT_VAR_COST_AMT
FROM OPT_ACTVY_GTIN_REVSD_SFCT REVSD,OPT_PROD_BRAND_ASSOC_DIM   BRAND_HIER,
CAL_MASTR_DIM   CAL WHERE REVSD.PROD_ID = BRAND_HIER.PROD_ID
AND REVSD.DATE_SKID = CAL.CAL_MASTR_SKID AND CAL.FISC_YR_SKID = BRAND_HIER.FY_DATE_SKID
GROUP BY REVSD.ACTVY_ID,
BRAND_HIER.BRAND_ID,REVSD.DATE_SKID,REVSD.BUS_UNIT_SKID) REVSD_VAR_COST,
--AA Variable Cost aggregated to Brand Level
(SELECT   AA.ACTVY_ID AS ACTVY_ID,BRAND_HIER.BRAND_ID AS PROD_ID,AA.MTH_SKID AS DATE_S
KID,
AA.BUS_UNIT_SKID AS BUS_UNIT_SKID,SUM(AA.ESTMT_VAR_COST_AMT) AS ESTMT_VAR_COST_AMT,
SUM(AA.REVSD_VAR_ESTMT_COST_AMT) AS REVSD_ESTMT_VAR_COST_AMT FROM OPT_ACTVY_BUOM_GTIN
_COST_TFADS AA,
OPT_PROD_BRAND_ASSOC_DIM BRAND_HIER WHERE AA.BUOM_GTIN_PROD_SKID = BRAND_HIER.PROD_SKID
AND BRAND_HIER.FY_DATE_SKID = AA.FY_DATE_SKID GROUP BY AA.ACTVY_ID,
BRAND_HIER.BRAND_ID,AA.MTH_SKID,AA.BUS_UNIT_SKID) AA
WHERE ACTVY_MTH_GTIN.ACTVY_ID = ESTMT_VAR_COST.ACTVY_ID(+)
AND ACTVY_MTH_GTIN.MTH_SKID = ESTMT_VAR_COST.DATE_SKID(+)
AND ACTVY_MTH_GTIN.PROD_ID = ESTMT_VAR_COST.PROD_ID(+)
AND ACTVY_MTH_GTIN.ACTVY_ID = REVSD_VAR_COST.ACTVY_ID(+)
AND ACTVY_MTH_GTIN.MTH_SKID = REVSD_VAR_COST.DATE_SKID(+)
AND ACTVY_MTH_GTIN.PROD_ID = REVSD_VAR_COST.PROD_ID(+)
AND ACTVY_MTH_GTIN.ACTVY_ID = AA.ACTVY_ID(+)
AND ACTVY_MTH_GTIN.MTH_SKID = AA.DATE_SKID(+)
AND ACTVY_MTH_GTIN.PROD_ID = AA.PROD_ID(+)
AND ACTVY_MTH_GTIN.ACTVY_SKID = ACTVY.ACTVY_SKID
AND ACTVY_MTH_GTIN.PRMTN_SKID = PRMTN.PRMTN_SKID
AND ACTVY_MTH_GTIN.MTH_SKID = CAL.CAL_MASTR_SKID))
```

SQL 的执行计划如下（为了方便排版，我们删除了部分无关紧要的信息）。

```
SQL> select * from table(dbms_xplan.display);

PLAN_TABLE_OUTPUT
--------------------------------------------------------------------------------
Plan hash value: 2005223222
--------------------------------------------------------------------------------
| Id |Operation                                        |Name        |Rows |
--------------------------------------------------------------------------------
|  0 |SELECT STATEMENT                                 |            |   1 |
|  1 | VIEW                                            |            |   1 |
```

```
|  2 |   WINDOW BUFFER                                 |                                    |     1|
|  3 |    VIEW                                         |                                    |     1|
|  4 |     WINDOW SORT                                 |                                    |     1|
|  5 |      NESTED LOOPS                               |                                    |      |
|  6 |       NESTED LOOPS                              |                                    |     1|
|  7 |        NESTED LOOPS                             |                                    |     1|
|* 8 |         HASH JOIN OUTER                         |                                    |     1|
|* 9 |          HASH JOIN OUTER                        |                                    |     1|
|*10 |           HASH JOIN OUTER                       |                                    |     1|
|*11 |            HASH JOIN                            |                                    |     1|
| 12 |             NESTED LOOPS                        |                                    |      |
| 13 |              NESTED LOOPS                       |                                    |     1|
|*14 |               HASH JOIN                         |                                    |     1|
| 15 |                PARTITION LIST ALL               |                                    |     1|
|*16 |                 TABLE ACCESS FULL               |OPT_ACCT_DIM                        |     1|
| 17 |                PARTITION LIST ALL               |                                    |  114K|
| 18 |                 TABLE ACCESS FULL               |OPT_ACTVY_FCT                       |  114K|
|*19 |               INDEX RANGE SCAN                  |OPT_ACTVY_DIM_PK                    |     1|
| 20 |              TABLE ACCESS BY GLOBAL INDEX ROWID |OPT_ACTVY_DIM                       |     1|
| 21 |             PARTITION LIST ALL                  |                                    |   19M|
| 22 |              TABLE ACCESS FULL                  |OPT_ACTVY_GTIN_BRAND_SFCT           |   19M|
| 23 |            VIEW                                 |                                    |     1|
| 24 |             HASH GROUP BY                       |                                    |     1|
| 25 |              NESTED LOOPS                       |                                    |      |
| 26 |               NESTED LOOPS                      |                                    |     1|
| 27 |                TABLE ACCESS FULL                |OPT_ACTVY_BUOM_GTIN_COST_TFADS      |     1|
|*28 |                INDEX RANGE SCAN                 |OPT_PROD_BRAND_ASSOC_DIM_PK         |     1|
| 29 |               TABLE ACCESS BY GLOBAL INDEX ROWID|OPT_PROD_BRAND_ASSOC_DIM            |     1|
| 30 |           VIEW                                  |                                    |   718|
| 31 |            HASH GROUP BY                        |                                    |   718|
|*32 |             HASH JOIN                           |                                    |   718|
|*33 |              HASH JOIN                          |                                    |   872|
| 34 |               PARTITION LIST ALL                |                                    |   872|
| 35 |                TABLE ACCESS FULL                |OPT_ACTVY_GTIN_REVSD_SFCT           |   872|
| 36 |               TABLE ACCESS FULL                 |OPT_CAL_MASTR_DIM                   | 36826|
| 37 |              PARTITION LIST ALL                 |                                    |  671K|
| 38 |               TABLE ACCESS FULL                 |OPT_PROD_BRAND_ASSOC_DIM            |  671K|
| 39 |          VIEW                                   |                                    |  6174|
| 40 |           HASH GROUP BY                         |                                    |  6174|
|*41 |            HASH JOIN                            |                                    |  6174|
|*42 |             HASH JOIN                           |                                    |  8998|
| 43 |              PARTITION LIST ALL                 |                                    |  8998|
| 44 |               TABLE ACCESS FULL                 |OPT_ACTVY_GTIN_ESTMT_SFCT           |  8998|
| 45 |              TABLE ACCESS FULL                  |OPT_CAL_MASTR_DIM                   | 36826|
| 46 |             PARTITION LIST ALL                  |                                    |  671K|
| 47 |              TABLE ACCESS FULL                  |OPT_PROD_BRAND_ASSOC_DIM            |  671K|
| 48 |        TABLE ACCESS BY INDEX ROWID              |OPT_CAL_MASTR_DIM                   |     1|
|*49 |         INDEX UNIQUE SCAN                       |OPT_CAL_MASTR_DIM_PK                |     1|
|*50 |       INDEX RANGE SCAN                          |OPT_PRMTN_DIM_PK                    |     1|
| 51 |      TABLE ACCESS BY GLOBAL INDEX ROWID         |OPT_PRMTN_DIM                       |     1|
```

```
Predicate Information (identified by operation id):
---------------------------------------------------

   8 - access("ACTVY_GTIN_BRAND"."ACTVY_ID"="ESTMT_VAR_COST"."ACTVY_ID"(+) AND
              "ACTVY_GTIN_BRAND"."MTH_SKID"="ESTMT_VAR_COST"."DATE_SKID"(+) AND
              "ACTVY_GTIN_BRAND"."PROD_ID"="ESTMT_VAR_COST"."PROD_ID"(+))
   9 - access("ACTVY_GTIN_BRAND"."ACTVY_ID"="REVSD_VAR_COST"."ACTVY_ID"(+) AND
              "ACTVY_GTIN_BRAND"."MTH_SKID"="REVSD_VAR_COST"."DATE_SKID"(+) AND
              "ACTVY_GTIN_BRAND"."PROD_ID"="REVSD_VAR_COST"."PROD_ID"(+))
  10 - access("ACTVY_GTIN_BRAND"."ACTVY_ID"="AA"."ACTVY_ID"(+) AND
```

9.28 利用等待事件优化 SQL

```
                  "ACTVY_GTIN_BRAND"."MTH_SKID"="AA"."DATE_SKID"(+) AND
                  "ACTVY_GTIN_BRAND"."PROD_ID"="AA"."PROD_ID"(+))
  11 - access("ACTVY"."ACTVY_SKID"="ACTVY_GTIN_BRAND"."ACTVY_SKID")
  14 - access("ACCT"."ACCT_SKID"="ACTVY"."ACCT_PRMTN_SKID")
  16 - filter("ACCT"."FUND_FRCST_MODEL_DESC" NOT LIKE 'TSP%')
  19 - access("ACTVY"."ACTVY_SKID"="ACTVY"."ACTVY_SKID")
  28 - access("AA"."BUOM_GTIN_PROD_SKID"="BRAND_HIER"."PROD_SKID" AND
              "BRAND_HIER"."FY_DATE_SKID"="AA"."FY_DATE_SKID")
  32 - access("REVSD"."PROD_ID"="BRAND_HIER"."PROD_ID" AND
              "CAL"."FISC_YR_SKID"="BRAND_HIER"."FY_DATE_SKID")
  33 - access("REVSD"."DATE_SKID"="CAL"."CAL_MASTR_SKID")
  41 - access("ESTMT"."PROD_ID"="BRAND_HIER"."PROD_ID" AND
              "CAL"."FISC_YR_SKID"="BRAND_HIER"."FY_DATE_SKID")
  42 - access("ESTMT"."DATE_SKID"="CAL"."CAL_MASTR_SKID")
  49 - access("ACTVY_GTIN_BRAND"."MTH_SKID"="CAL"."CAL_MASTR_SKID")
  50 - access("ACTVY_GTIN_BRAND"."PRMTN_SKID"="PRMTN"."PRMTN_SKID")
```

该 SQL 是用来做数据清洗的（ETL），需要处理大量数据。处理大量数据应该走 HASH 连接，因此该执行计划是错误的，因为执行计划中有大量的嵌套循环。

注意观察执行计划，执行计划中 Id=16 和 Id=27 优化器评估只返回 1 行数据，因此怀疑 OPT_ACCT_DIM 和 OPT_ACTVY_BUOM_GTIN_COST_TFADS 这两个表统计信息有问题。对这两个表收集完统计信息之后，我们再来看一下执行计划。

```
SQL> select * from table(dbms_xplan.display);

PLAN_TABLE_OUTPUT
-------------------------------------------------------------------------------
Plan hash value: 183294992
-------------------------------------------------------------------------------
| Id  |Operation                       |Name                      |Rows  |
-------------------------------------------------------------------------------
|   0 |SELECT STATEMENT                |                          |  19M |
|   1 | VIEW                           |                          |  19M |
|   2 |  WINDOW BUFFER                 |                          |  19M |
|   3 |   VIEW                         |                          |  19M |
|   4 |    WINDOW SORT                 |                          |  19M |
|*  5 |     HASH JOIN                  |                          |  19M |
|   6 |      PARTITION LIST ALL        |                          |37880 |
|   7 |       TABLE ACCESS FULL        |OPT_PRMTN_DIM             |37880 |
|*  8 |      HASH JOIN                 |                          |  19M |
|   9 |       TABLE ACCESS FULL        |OPT_CAL_MASTR_DIM         |36826 |
|* 10 |       HASH JOIN RIGHT OUTER    |                          |  19M |
|  11 |        VIEW                    |                          | 6174 |
|  12 |         HASH GROUP BY          |                          | 6174 |
|* 13 |          HASH JOIN             |                          | 6174 |
|* 14 |           HASH JOIN            |                          | 8998 |
|  15 |            PARTITION LIST ALL  |                          | 8998 |
|  16 |             TABLE ACCESS FULL  |OPT_ACTVY_GTIN_ESTMT_SFCT | 8998 |
|  17 |            TABLE ACCESS FULL   |OPT_CAL_MASTR_DIM         |36826 |
|  18 |           PARTITION LIST ALL   |                          | 671K |
|  19 |            TABLE ACCESS FULL   |OPT_PROD_BRAND_ASSOC_DIM  | 671K |
|* 20 |        HASH JOIN RIGHT OUTER   |                          |  19M |
|  21 |         VIEW                   |                          |  718 |
|  22 |          HASH GROUP BY         |                          |  718 |
|* 23 |           HASH JOIN            |                          |  718 |
|* 24 |            HASH JOIN           |                          |  872 |
|  25 |             PARTITION LIST ALL |                          |  872 |
|  26 |              TABLE ACCESS FULL |OPT_ACTVY_GTIN_REVSD_SFCT |  872 |
|  27 |             TABLE ACCESS FULL  |OPT_CAL_MASTR_DIM         |36826 |
```

```
| 28 |              PARTITION LIST ALL          |                              |  671K|
| 29 |              TABLE ACCESS FULL           |OPT_PROD_BRAND_ASSOC_DIM      |  671K|
|*30 |         HASH JOIN RIGHT OUTER            |                              |   19M|
| 31 |          VIEW                            |                              |    1|
| 32 |           HASH GROUP BY                  |                              |    1|
| 33 |            NESTED LOOPS                  |                              |     |
| 34 |             NESTED LOOPS                 |                              |    1|
| 35 |              TABLE ACCESS FULL           |OPT_ACTVY_BUOM_GTIN_COST_TFADS|    1|
|*36 |              INDEX RANGE SCAN            |OPT_PROD_BRAND_ASSOC_DIM_PK   |    1|
| 37 |             TABLE ACCESS BY GLOBAL INDEX ROWID|OPT_PROD_BRAND_ASSOC_DIM |    1|
|*38 |          HASH JOIN                       |                              |   19M|
|*39 |           HASH JOIN                      |                              |  114K|
| 40 |            PARTITION LIST ALL            |                              |  115K|
| 41 |             TABLE ACCESS FULL            |OPT_ACTVY_DIM                 |  115K|
|*42 |            HASH JOIN                     |                              |  114K|
| 43 |             PARTITION LIST ALL           |                              | 94478|
|*44 |              TABLE ACCESS FULL           |OPT_ACCT_DIM                  | 94478|
| 45 |             PARTITION LIST ALL           |                              |  114K|
| 46 |              TABLE ACCESS FULL           |OPT_ACTVY_FCT                 |  114K|
| 47 |           PARTITION LIST ALL             |                              |   19M|
| 48 |            TABLE ACCESS FULL             |OPT_ACTVY_GTIN_BRAND_SFCT     |   19M|
---------------------------------------------------------------------------------------

Predicate Information (identified by operation id):
---------------------------------------------------

   5 - access("ACTVY_GTIN_BRAND"."PRMTN_SKID"="PRMTN"."PRMTN_SKID")
   8 - access("ACTVY_GTIN_BRAND"."MTH_SKID"="CAL"."CAL_MASTR_SKID")
  10 - access("ACTVY_GTIN_BRAND"."ACTVY_ID"="ESTMT_VAR_COST"."ACTVY_ID"(+)
              AND "ACTVY_GTIN_BRAND"."MTH_SKID"="ESTMT_VAR_COST"."DATE_SKID"(+) AND
              "ACTVY_GTIN_BRAND"."PROD_ID"="ESTMT_VAR_COST"."PROD_ID"(+))
  13 - access("ESTMT"."PROD_ID"="BRAND_HIER"."PROD_ID" AND
              "CAL"."FISC_YR_SKID"="BRAND_HIER"."FY_DATE_SKID")
  14 - access("ESTMT"."DATE_SKID"="CAL"."CAL_MASTR_SKID")
  20 - access("ACTVY_GTIN_BRAND"."ACTVY_ID"="REVSD_VAR_COST"."ACTVY_ID"(+)
              AND "ACTVY_GTIN_BRAND"."MTH_SKID"="REVSD_VAR_COST"."DATE_SKID"(+) AND
              "ACTVY_GTIN_BRAND"."PROD_ID"="REVSD_VAR_COST"."PROD_ID"(+))
  23 - access("REVSD"."PROD_ID"="BRAND_HIER"."PROD_ID" AND
              "CAL"."FISC_YR_SKID"="BRAND_HIER"."FY_DATE_SKID")
  24 - access("REVSD"."DATE_SKID"="CAL"."CAL_MASTR_SKID")
  30 - access("ACTVY_GTIN_BRAND"."ACTVY_ID"="AA"."ACTVY_ID"(+) AND
              "ACTVY_GTIN_BRAND"."MTH_SKID"="AA"."DATE_SKID"(+) AND
              "ACTVY_GTIN_BRAND"."PROD_ID"="AA"."PROD_ID"(+))
  36 - access("AA"."BUOM_GTIN_PROD_SKID"="BRAND_HIER"."PROD_SKID" AND
              "BRAND_HIER"."FY_DATE_SKID"="AA"."FY_DATE_SKID")
  38 - access("ACTVY"."ACTVY_SKID"="ACTVY_GTIN_BRAND"."ACTVY_SKID")
  39 - access("ACTVY"."ACTVY_SKID"="ACTVY"."ACTVY_SKID")
  42 - access("ACCT"."ACCT_SKID"="ACTVY"."ACCT_PRMTN_SKID")
  44 - filter("ACCT"."FUND_FRCST_MODEL_DESC" NOT LIKE 'TSP%')
```

执行计划中，除了 Id=35 和 Id=37 两个表没有走 HASH 连接之外，其余表都走了 HASH 连接。Id=35 的表 OPT_ACTVY_BUOM_GTIN_COST_TFADS 之前已经收集过统计信息，因此 Id=35 和 Id=37 的表走嵌套循环没有问题，那么整个 SQL 的执行计划现在是正确的。纠正完执行计划之后，笔者将 SQL 放在后台运行了大概两小时，发现 SQL 还没执行完毕。起初，笔者认为 SQL 执行 7 个小时还没跑完是因为 SQL 执行计划错误导致的，但是现在纠正了 SQL 的执行计划，SQL 执行了两小时还是没有跑完，于是监控 SQL 的等待事件，看 SQL 究竟在等什么。

```
SQL> select inst_id,sid,serial#,event,p1,p2,p3
  2  from gv$session where osuser='luobi';

   INST_ID        SID    SERIAL# EVENT                                  P1         P2         P3
---------- ---------- ---------- ------------------------------ ---------- ---------- ----------
         2       4754      10050 direct path write temp              20025     857328          7

SQL> /

   INST_ID        SID    SERIAL# EVENT                                  P1         P2         P3
---------- ---------- ---------- ------------------------------ ---------- ---------- ----------
         2       4754      10050 direct path write temp              20025     406768          7

SQL> /

   INST_ID        SID    SERIAL# EVENT                                  P1         P2         P3
---------- ---------- ---------- ------------------------------ ---------- ---------- ----------
         2       4754      10050 direct path write temp              20007    2849264          7

SQL> /

   INST_ID        SID    SERIAL# EVENT                                  P1         P2         P3
---------- ---------- ---------- ------------------------------ ---------- ---------- ----------
         2       4754      10050 direct path write temp              20007     115341          7

SQL> /

   INST_ID        SID    SERIAL# EVENT                                  P1         P2         P3
---------- ---------- ---------- ------------------------------ ---------- ---------- ----------
         2       4754      10050 direct path write temp              20007      81029          7
```

我们监控到该 SQL 的等待事件为 direct path write temp，该等待事件表示当前 SQL 正在进行排序或者正在进行 HASH 连接，但是因为 PGA 不够大，不能完全容纳需要排序或者需要 HASH 的数据，导致有部分数据被写入 temp 表空间。

为了追查究竟是因为排序还是因为 HASH 而引发的 direct path write temp 等待，使用以下脚本查看临时段数据类型。

```
SQL> select a.inst_id, a.sid, a.serial#, a.sql_id, b.tablespace, b.blocks*
  2  (select value from v$parameter where name='db_block_size')/1024/1024 "Size(M)",b
.segtype
  3  from gv$session a, gv$tempseg_usage b where a.inst_id=b.inst_id and a.saddr = b.
session_addr
  4  and a.inst_id=2 and a.sid=4754;

INST_ID       SID    SERIAL#       SQL_ID         TABLESPACE       Size(M) SEGTYPE
------- --------- ---------- -------------- --------------- ------------- --------
      2      4754      10050 6qsuc8mafy20m  TEMP                        1 DATA
      2      4754      10050 6qsuc8mafy20m  TEMP                        1 LOB_DATA
      2      4754      10050 6qsuc8mafy20m  TEMP                        1 INDEX
      2      4754      10050 6qsuc8mafy20m  TEMP                        1 LOB_DATA
      2      4754      10050 6qsuc8mafy20m  TEMP                     3304 HASH
```

从 SQL 查询中我们看到，临时段数据类型为 HASH，耗费了 3 304MB 的 temp 表空间，这表示 SQL 是因为 HASH 连接引发的 direct path write temp 等待。

大家请仔细观察等待事件 P3，它的值一直为 7，这表示 Oracle 一次只写入 7 个块到 temp 表空间，而且是一直只写入 7 个块到 temp 表空间。笔者在第 4 章中讲到，绝大多数的操作系统，一次 I/O 最多只能读取或者写入 1MB 数据。这里的数据块大小为 16KB，正常情况下应该是每次 I/O 写入 64 个块到 temp 表空间，但是每次 I/O 只写了 7 个块。于是怀疑是 PGA 中 work area 不够导致出现了该问题。

PGA 在自动管理的情况下，单个 PGA 进程的 work area 不能超过 1GB（想要超过 1GB 需要修改隐含参数，但是本书主题是 SQL 优化，因此不想太多涉及到 Oracle 内部原理），如果 PGA 是手动管理，单个 PGA 进程的 work area 可以接近 2GB，但是不能超过 2GB。

```
SQL> alter session set workarea_size_policy = manual;

Session altered.

SQL> alter session set hash_area_size = 2147483648;   ---2GB
alter session set hash_area_size = 2147483648
                                   *
ERROR at line 1:
ORA-02017: integer value required

SQL> alter session set hash_area_size = 2147483647;

Session altered.
```

将 PGA 的 work area 设置为接近 2GB 之后，重新运行了 SQL 并且监控等待事件。

```
SQL> select inst_id,sid,serial#,event,p1,p2,p3
  2  from gv$session where osuser='luobi';

  INST_ID        SID    SERIAL# EVENT                          P1         P2         P3
---------- ---------- ---------- -------------------- ---------- ---------- ----------
        2       4885      11759 direct path write temp     20012      71053         64
```

将 PGA 的 work area 设置为接近 2GB 之后，笔者发现 P3 可以达到 64，相比之前一次只能写入 7 个块速度提升了 9 倍。

有 direct path write temp 等待必然会出现 direct path read temp 等待，在没修改 PGA 的 work area 之前，不仅仅是单次 I/O 只能写入 7 个块，单次 I/O 读取也是只能读取 7 个块，因此，将 PGA 的 work area 设置为接近 2GB 之后，整个 SQL 的性能应该提升了 18 倍。

最后，经过对比测试，手动设置 work area 的 SQL 只需要 56 分钟左右就能执行完毕。

```
6889440 rows selected.

Elapsed: 00:56:36.08
```

而自动 work area 管理的 SQL 还在一直等待 direct path write temp，估计该 SQL 如果不手动设置 work area 可能跑一天一夜都跑不完。

优化完上述 SQL 之后，我们发现当时整个平台已经瘫痪，整个平台都出现了 P3=7 的问题，最后经过与 Oracle 确认，发现该问题是 11.1.0.7 版本在 HPUX 平台下的一个 bug。Oracle 开发补丁需要一定的时间，在此期间，使用本书给出的方法临时解决了项目中遇到的问题，确保项目不会因此延期。

第 10 章　全自动 SQL 审核

本章，我们为大家分享一些常用的全自动 SQL 审核脚本，在实际工作中，我们可以对脚本进行适当修改，以便适应自己的数据库环境，从而提升工作效率。因为本书的主题是 SQL 优化，所以本章不会涉及常用的数据库监控脚本和常用的 DBA 运维脚本。

10.1　抓出外键没创建索引的表

此脚本不依赖统计信息。

建议在外键列上创建索引，外键列不创建索引容易导致死锁。级联删除的时候，外键列没有索引会导致表被全表扫描。以下脚本抓出 Scott 账户下外键没创建索引的表。

```sql
with cons as (select /*+ materialize */ owner, table_name, constraint_name
                from dba_constraints
               where owner = 'SCOTT'
                 AND constraint_type = 'R'),
     idx as (
     select /*+ materialize */ table_owner,table_name, column_name
         from dba_ind_columns
        where table_owner = 'SCOTT')
select owner,table_name,constraint_name,column_name
  from dba_cons_columns
 where (owner,table_name, constraint_name) in
       (select * from cons)
   and (owner,table_name, column_name) not in
       (select * from idx);
```

在 Scott 账户中，EMP 表的 deptno 列引用了 DEPT 表的 deptno 列，但是没有创建索引，因此我们通过脚本可以将其抓出。

```
SQL> with cons as (select /*+ materialize */ owner, table_name, constraint_name
  2              from dba_constraints
  3             where owner = 'SCOTT'
  4               AND constraint_type = 'R'),
  5       idx as (
  6       select /*+ materialize */ table_owner,table_name, column_name
  7           from dba_ind_columns
  8          where table_owner = 'SCOTT')
  9  select owner,table_name,constraint_name,column_name
 10    from dba_cons_columns
 11   where (owner,table_name, constraint_name) in
 12         (select * from cons)
 13     and (owner,table_name, column_name) not in
 14         (select * from idx);

OWNER      TABLE_NAME      CONSTRAINT_NAME       COLUMN_NAME
---------- --------------- --------------------- --------------------
SCOTT      EMP             FK_DEPTNO             DEPTNO
```

10.2 抓出需要收集直方图的列

此脚本依赖统计信息。

当一个表比较大，列选择性低于 5%，而且列出现在 where 条件中，为了防止优化器估算 Rows 出现较大偏差，我们需要对这种列收集直方图。以下脚本抓出 Scott 账户下，表总行数大于 5 万行、列选择性低于 5%并且列出现在 where 条件中的表以及列信息。

```
select a.owner,
       a.table_name,
       a.column_name,
       b.num_rows,
       a.num_distinct Cardinality,
       round(a.num_distinct / b.num_rows * 100, 2) selectivity
  from dba_tab_col_statistics a, dba_tables b
 where a.owner = b.owner
   and a.table_name = b.table_name
   and a.owner = 'SCOTT'
   and round(a.num_distinct / b.num_rows * 100, 2) < 5
   and num_rows > 50000
   and (a.table_name, a.column_name) in
       (select o.name, c.name
          from sys.col_usage$ u, sys.obj$ o, sys.col$ c, sys.user$ r
         where o.obj# = u.obj#
           and c.obj# = u.obj#
           and c.col# = u.intcol#
           and r.name = 'SCOTT');
```

在 Scott 账户中，test 表总行数大于 5 万行，owner 列选择性小于 5%，而且出现在 where 条件中，通过以上脚本我们可以将其抓出。

```
SQL> select a.owner,
  2         a.table_name,
  3         a.column_name,
  4         b.num_rows,
  5         a.num_distinct Cardinality,
  6         round(a.num_distinct / b.num_rows * 100, 2) selectivity
  7    from dba_tab_col_statistics a, dba_tables b
  8   where a.owner = b.owner
  9     and a.table_name = b.table_name
 10     and a.owner = 'SCOTT'
 11     and round(a.num_distinct / b.num_rows * 100, 2) < 5
 12     and num_rows > 50000
 13     and (a.table_name, a.column_name) in
 14         (select o.name, c.name
 15            from sys.col_usage$ u, sys.obj$ o, sys.col$ c, sys.user$ r
 16           where o.obj# = u.obj#
 17             and c.obj# = u.obj#
 18             and c.col# = u.intcol#
 19             and r.name = 'SCOTT');

OWNER    TABLE_NAME    COLUMN_NAME      NUM_ROWS  CARDINALITY  SELECTIVITY
-------- ------------- ------------- ----------- ------------ ------------
SCOTT    TEST          OWNER               73020           29          .04
```

10.3 抓出必须创建索引的列

此脚本依赖统计信息。

当一个表比较大，列选择性超过 20%，列出现在 where 条件中并且没有创建索引，我们可以对该列创建索引从而提升 SQL 查询性能。以下脚本抓出 Scott 账户下表总行数大于 5 万行、列选择性超过 20%、列出现在 where 条件中并且没有创建索引。

```
select owner,
       table_name,
       column_name,
       num_rows,
       Cardinality,
       selectivity
  from (select a.owner,
               a.table_name,
               a.column_name,
               b.num_rows,
               a.num_distinct Cardinality,
               round(a.num_distinct / b.num_rows * 100, 2) selectivity
          from dba_tab_col_statistics a, dba_tables b
         where a.owner = b.owner
           and a.table_name = b.table_name
           and a.owner = 'SCOTT')
 where selectivity >= 20
   and num_rows > 50000
   and (table_name, column_name) not in
       (select table_name, column_name
          from dba_ind_columns
         where table_owner = 'SCOTT' and column_position=1)
   and (table_name, column_name) in
       (select o.name, c.name
          from sys.col_usage$ u, sys.obj$ o, sys.col$ c, sys.user$ r
         where o.obj# = u.obj#
           and c.obj# = u.obj#
           and c.col# = u.intcol#
           and r.name = 'SCOTT');
```

在 Scott 账户中，test 表总行数大于 5 万行，有两个列出现在 where 条件中，选择性大于 20%，而且没有创建索引，我们通过以上脚本将其抓出。

```
SQL> select owner,
  2         table_name,
  3         column_name,
  4         num_rows,
  5         Cardinality,
  6         selectivity
  7    from (select a.owner,
  8                 a.table_name,
  9                 a.column_name,
 10                 b.num_rows,
 11                 a.num_distinct Cardinality,
 12                 round(a.num_distinct / b.num_rows * 100, 2) selectivity
 13            from dba_tab_col_statistics a, dba_tables b
 14           where a.owner = b.owner
 15             and a.table_name = b.table_name
 16             and a.owner = 'SCOTT')
```

```
 17    where selectivity >= 20
 18      and num_rows > 50000
 19      and (table_name, column_name) not in
 20          (select table_name, column_name
 21             from dba_ind_columns
 22            where table_owner = 'SCOTT' and column_position=1)
 23      and (table_name, column_name) in
 24          (select o.name, c.name
 25             from sys.col_usage$ u, sys.obj$ o, sys.col$ c, sys.user$ r
 26            where o.obj# = u.obj#
 27              and c.obj# = u.obj#
 28              and c.col# = u.intcol#
 29              and r.name = 'SCOTT');

OWNER      TABLE_NAME      COLUMN_NAME        NUM_ROWS  CARDINALITY  SELECTIVITY
---------- --------------- --------------- ----------- ------------ ------------
SCOTT      TEST            OBJECT_ID             73020        73020          100
SCOTT      TEST            OBJECT_NAME           73020        41002        56.15
```

10.4 抓出 SELECT * 的 SQL

此脚本不依赖统计信息。

在开发过程中,我们应该尽量避免编写 SELECT * 这种 SQL。SELECT * 这种 SQL,走索引无法避免回表,走 HASH 连接的时候会将驱动表所有的列放入 PGA 中,浪费 PGA 内存。执行计划中(V$SQL_PLAN/PLAN_TABLE),projection 字段表示访问了哪些字段,如果 projection 字段中字段个数等于表的字段总个数,那么我们就可以判断 SQL 语句使用了 SELECT *。以下脚本抓出 SELECT * 的 SQL。

```
select a.sql_id, a.sql_text, c.owner, d.table_name, d.column_cnt, c.size_mb
  from v$sql a,
       v$sql_plan b,
       (select owner, segment_name, sum(bytes / 1024 / 1024) size_mb
          from dba_segments
         group by owner, segment_name) c,
       (select owner, table_name, count(*) column_cnt
          from dba_tab_cols
         group by owner, table_name) d
 where a.sql_id = b.sql_id
   and a.child_number = b.child_number
   and b.object_owner = c.owner
   and b.object_name = c.segment_name
   and b.object_owner = d.owner
   and b.object_name = d.table_name
   and REGEXP_COUNT(b.projection, ']') = d.column_cnt
   and c.owner = 'SCOTT'
 order by 6 desc;
```

我们在 Scott 账户中运行如下 SQL。

```
select * from t where object_id<1000;
```

我们使用脚本将其抓出。

```
SQL> select a.sql_id, a.sql_text, c.owner, d.table_name, d.column_cnt, c.size_mb
  2    from v$sql a,
  3         v$sql_plan b,
```

```
  4        (select owner, segment_name, sum(bytes / 1024 / 1024) size_mb
  5           from dba_segments
  6          group by owner, segment_name) c,
  7        (select owner, table_name, count(*) column_cnt
  8           from dba_tab_cols
  9          group by owner, table_name) d
 10   where a.sql_id = b.sql_id
 11     and a.child_number = b.child_number
 12     and b.object_owner = c.owner
 13     and b.object_name = c.segment_name
 14     and b.object_owner = d.owner
 15     and b.object_name = d.table_name
 16     and REGEXP_COUNT(b.projection, ']') = d.column_cnt
 17     and c.owner = 'SCOTT'
 18   order by 6 desc;

SQL_ID        SQL_TEXT                                OWNER   TABLE_NAME COLUMN_CNT  SIZE_MB
-----------   --------------------------------------- ------- ---------- ---------- ---------
ga64bhp5fxhtn select * from t where object_id<1000    SCOTT   T                  15         9
```

10.5 抓出有标量子查询的 SQL

此脚本不依赖统计信息。

在开发过程中,我们应该尽量避免编写标量子查询。我们可以通过分析执行计划,抓出标量子查询语句。同一个 SQL 语句,执行计划中如果有两个或者两个以上的 depth=1 的执行计划就表示 SQL 中出现了标量子查询。以下脚本抓出 Scott 账户下在 SQL*Plus 中运行过的标量子查询语句。

```
select sql_id, sql_text, module
  from v$sql
 where parsing_schema_name = 'SCOTT'
   and module = 'SQL*Plus'
   AND sql_id in
       (select sql_id
          from (select sql_id,
                       count(*) over(partition by sql_id, child_number, depth) cnt
                  from V$SQL_PLAN
                 where depth = 1
                   and (object_owner = 'SCOTT' or object_owner is null))
         where cnt >= 2);
```

我们在 SQL*Plus 中运行如下标量子查询语句。

```
SQL> select dname,
  2    (select max(sal) from emp where deptno = d.deptno) max_sal
  3   from dept d;

DNAME           MAX_SAL
-------------- ----------
ACCOUNTING        5000
RESEARCH          3000
SALES             2850
OPERATIONS
```

我们利用以上脚本将刚运行过的标量子查询抓出。

```
SQL> select sql_id, sql_text, module
```

```
  2    from v$sql
  3   where parsing_schema_name = 'SCOTT'
  4     and module = 'SQL*Plus'
  5     AND sql_id in
  6         (select sql_id
  7            from (select sql_id,
  8                         count(*) over(partition by sql_id, child_number, depth) cnt
  9                    from V$SQL_PLAN
 10                   where depth = 1
 11                     and (object_owner = 'SCOTT' or object_owner is null))
 12          where cnt >= 2);

SQL_ID          SQL_TEXT                                                         MODULE
--------------- ---------------------------------------------------------------- ----------------------
739fhcu0pbz28   select dname,  (select max(sal) from emp where                   SQL*Plus
                deptno = d.deptno) max_sal from dept d
```

10.6 抓出带有自定义函数的 SQL

此脚本不依赖统计信息。

在开发过程中，我们应该避免在 SQL 语句中调用自定义函数。我们可以通过以下 SQL 语句抓出 SQL 语句中调用了自定义函数的 SQL。

```
select distinct sql_id, sql_text, module
  from V$SQL,
       (select object_name
          from DBA_OBJECTS O
         where owner = 'SCOTT'
           and object_type in ('FUNCTION', 'PACKAGE'))
 where (instr(upper(sql_text), object_name) > 0)
   and plsql_exec_time > 0
   and regexp_like(upper(sql_fulltext), '^[SELECT]')
   and parsing_schema_name = 'SCOTT';
```

我们在 Scott 账户中创建如下函数。

```
create or replace function f_getdname(v_deptno in number) return varchar2 as
  v_dname dept.dname%type;
begin
  select dname into v_dname from dept where deptno = v_deptno;
  return v_dname;
end f_getdname;
/
```

然后我们在 Scott 账户中运行如下 SQL。

```
SQL> select empno,sal,f_getdname(deptno) dname from emp;

     EMPNO        SAL DNAME
---------- ---------- --------------------
      7369        800 RESEARCH
      7499       1600 SALES
      7521       1250 SALES
      7566       2975 RESEARCH
      7654       1250 SALES
      7698       2850 SALES
      7782       2450 ACCOUNTING
      7788       3000 RESEARCH
```

```
                7839         5000 ACCOUNTING
                7844         1500 SALES
                7876         1100 RESEARCH
                7900          950 SALES
                7902         3000 RESEARCH
                7934         1300 ACCOUNTING
```

我们通过脚本抓出刚执行过的 SQL 语句。

```
SQL> select distinct sql_id, sql_text, module
  2    from V$SQL,
  3         (select object_name
  4            from DBA_OBJECTS O
  5           where owner = 'SCOTT'
  6             and object_type in ('FUNCTION', 'PACKAGE'))
  7   where (instr(upper(sql_text), object_name) > 0
  8     and plsql_exec_time > 0
  9     and regexp_like(upper(sql_fulltext), '^[SELECT]')
 10     and parsing_schema_name = 'SCOTT';

SQL_ID          SQL_TEXT                                                         MODULE
--------------- ---------------------------------------------------------------- ---------
2ck71xc69j49u   select empno,sal,f_getdname(deptno) dname from emp               SQL*Plus
```

10.7 抓出表被多次反复调用 SQL

此脚本不依赖统计信息。

在开发过程中，我们应该避免在同一个 SQL 语句中对同一个表多次访问。我们可以通过下面 SQL 抓出同一个 SQL 语句中对某个表进行多次扫描的 SQL。

```
select a.parsing_schema_name schema,
       a.sql_id,
       a.sql_text,
       b.object_name,
       b.cnt
  from v$sql a,
       (select *
          from (select sql_id,
                       child_number,
                       object_owner,
                       object_name,
                       object_type,
                       count(*) cnt
                  from v$sql_plan
                 where object_owner = 'SCOTT'
                 group by sql_id,
                          child_number,
                          object_owner,
                          object_name,
                          object_type)
         where cnt >= 2) b
 where a.sql_id = b.sql_id
   and a.child_number = b.child_number;
```

我们在 Scott 账户中运行如下 SQL。

```
select ename,job,deptno from emp where sal>(select avg(sal) from emp);
```

以上 SQL 访问了 emp 表两次，我们可以通过脚本将其抓出。

```
SQL> select a.parsing_schema_name schema,
  2         a.sql_id,
  3         a.sql_text,
  4         b.object_name,
  5         b.cnt
  6    from v$sql a,
  7         (select *
  8            from (select sql_id,
  9                         child_number,
 10                         object_owner,
 11                         object_name,
 12                         object_type,
 13                         count(*) cnt
 14                    from v$sql_plan
 15                   where object_owner = 'SCOTT'
 16                   group by sql_id,
 17                            child_number,
 18                            object_owner,
 19                            object_name,
 20                            object_type)
 21           where cnt >= 2) b
 22   where a.sql_id = b.sql_id
 23     and a.child_number = b.child_number;

SCHEMA          SQL_ID          SQL_TEXT                              OBJECT_NAME          CNT
--------------- --------------- ------------------------------------- ------------- ----------
SCOTT           fdt0z70z43vgv   select ename,job,deptno from          EMP                    2
                                emp where sal>(select avg(sal)
                                  from emp)
```

10.8 抓出走了 FILTER 的 SQL

此脚本不依赖统计信息。

当 where 子查询没能 unnest，执行计划中就会出现 FILTER，对于此类 SQL，我们应该在上线之前对其进行改写，避免执行计划中出现 FILTER，以下脚本可以抓出 where 子查询没能 unnest 的 SQL。

```
select parsing_schema_name schema, sql_id, sql_text
  from v$sql
 where parsing_schema_name = 'SCOTT'
   and (sql_id, child_number) in
       (select sql_id, child_number
          from v$sql_plan
         where operation = 'FILTER'
           and filter_predicates like '%IS NOT NULL%'
        minus
        select sql_id, child_number
          from v$sql_plan
         where object_owner = 'SYS');
```

我们在 Scott 账户中运行如下 SQL 并且查看执行计划。

```
SQL> select *
  2    from dept
  3   where exists (select null
```

10.8 抓出走了 FILTER 的 SQL

```
  4            from emp
  5            where dept.deptno = emp.deptno
  6            start with empno = 7698
  7            connect by prior empno = mgr);

    DEPTNO DNAME      LOC
---------- ---------- --------------------------------
        30 SALES      CHICAGO

Elapsed: 00:00:00.00

Execution Plan
----------------------------------------------------------
Plan hash value: 4210865686

--------------------------------------------------------------------------------
| Id|Operation                              |Name|Rows|Bytes|Cost(%CPU)|Time    |
--------------------------------------------------------------------------------
|  0|SELECT STATEMENT                       |    |  1 |  20 |   9  (0) |00:00:01|
|* 1| FILTER                                |    |    |     |          |        |
|  2|  TABLE ACCESS FULL                    |DEPT|  4 |  80 |   3  (0) |00:00:01|
|* 3|  FILTER                               |    |    |     |          |        |
|* 4|   CONNECT BY NO FILTERING WITH SW (UNIQUE)| |    |     |          |        |
|  5|    TABLE ACCESS FULL                  |EMP | 14 | 154 |   3  (0) |00:00:01|
--------------------------------------------------------------------------------

Predicate Information (identified by operation id):
---------------------------------------------------

   1 - filter( EXISTS (SELECT 0 FROM "EMP" "EMP" WHERE "EMP"."DEPTNO"=:B1 START WITH
              "EMPNO"=7698 CONNECT BY "MGR"=PRIOR "EMPNO"))
   3 - filter("EMP"."DEPTNO"=:B1)
   4 - access("MGR"=PRIOR "EMPNO")
       filter("EMPNO"=7698)


Statistics
----------------------------------------------------------
          0  recursive calls
          0  db block gets
         36  consistent gets
          0  physical reads
          0  redo size
        550  bytes sent via SQL*Net to client
        419  bytes received via SQL*Net from client
          2  SQL*Net roundtrips to/from client
          8  sorts (memory)
          0  sorts (disk)
          1  rows processed
```

以上 SQL 执行计划中出现了 FILTER，我们通过脚本抓出走了 FILTER 的 SQL。

```
SQL> select parsing_schema_name schema, sql_id, sql_text
  2    from v$sql
  3   where parsing_schema_name = 'SCOTT'
  4     and (sql_id, child_number) in
  5         (select sql_id, child_number
  6            from v$sql_plan
  7           where operation = 'FILTER'
  8             and filter_predicates like '%IS NOT NULL%'
  9          minus
 10          select sql_id, child_number
```

```
      11                  from v$sql_plan
      12                 where object_owner = 'SYS');

SCHEMA      SQL_ID           SQL_TEXT
----------  ---------------  --------------------------------------------
SCOTT       8rmn2fn149y2z    select * from dept   where exists (select null from emp
                             where dept.deptno = emp.deptno   start with em
                             pno = 7698   connect by prior empno = mgr)
```

10.9 抓出返回行数较多的嵌套循环 SQL

此脚本不依赖统计信息。

两表关联返回少量数据应该走嵌套循环,如果返回大量数据,应该走 HASH 连接,或者是排序合并连接。如果一个 SQL 语句返回行数较多(大于 1 万行),SQL 的执行计划在最后几步(Id<=5)走了嵌套循环,我们可以判定该执行计划中的嵌套循环是有问题的,应该走 HASH 连接。以下脚本抓出返回行数较多的嵌套循环 SQL。

```
select *
  from (select parsing_schema_name schema,
               sql_id,
               sql_text,
               rows_processed / executions rows_processed
          from v$sql
         where parsing_schema_name = 'SCOTT'
           and executions > 0
           and rows_processed / executions > 10000
         order by 4 desc) a
 where a.sql_id in (select sql_id
                      from v$sql_plan
                     where operation like '%NESTED LOOPS%'
                       and id <= 5);
```

在 scott 账户中分别创建 a 表和 b 表以及一个索引。

```
SQL> create table a as select * from dba_objects;

Table created.

SQL> create table b as select * from dba_objects;

Table created.

SQL> create index idx_b on b(object_id);

Index created.
```

运行如下 SQL 并且查看执行计划。

```
SQL> select /*+ use_nl(a,b) */ * from a,b where a.object_id=b.object_id;

72695 rows selected.


Execution Plan
----------------------------------------------------------
Plan hash value: 2104163270
```

```
---------------------------------------------------------------------------------
| Id  | Operation                      | Name  | Rows  | Bytes | Cost (%CPU)| Time     |
---------------------------------------------------------------------------------
|  0  | SELECT STATEMENT               |       | 60140 |  23M  | 120K   (1)| 00:24:07 |
|  1  |  NESTED LOOPS                  |       |       |       |           |          |
|  2  |   NESTED LOOPS                 |       | 60140 |  23M  | 120K   (1)| 00:24:07 |
|  3  |    TABLE ACCESS FULL           | A     | 60140 |  11M  |  187   (2)| 00:00:03 |
|* 4  |    INDEX RANGE SCAN            | IDX_B |   1   |       |    1   (0)| 00:00:01 |
|  5  |   TABLE ACCESS BY INDEX ROWID  | B     |   1   |  207  |    2   (0)| 00:00:01 |
---------------------------------------------------------------------------------

Predicate Information (identified by operation id):
---------------------------------------------------

   4 - access("A"."OBJECT_ID"="B"."OBJECT_ID")

Note
-----
   - dynamic sampling used for this statement (level=2)

Statistics
----------------------------------------------------------
        632  recursive calls
          0  db block gets
      22985  consistent gets
       1196  physical reads
          0  redo size
    6085032  bytes sent via SQL*Net to client
      53725  bytes received via SQL*Net from client
       4848  SQL*Net roundtrips to/from client
          0  sorts (memory)
          0  sorts (disk)
      72695  rows processed
```

我们可以使用脚本将错误的嵌套循环抓出。

```
SQL> select *
  2    from (select parsing_schema_name schema,
  3                 sql_id,
  4                 sql_text,
  5                 rows_processed / executions rows_processed
  6            from v$sql
  7           where parsing_schema_name = 'SCOTT'
  8             and executions > 0
  9             and rows_processed / executions > 10000
 10           order by 4 desc) a
 11   where a.sql_id in (select sql_id
 12                        from v$sql_plan
 13                       where operation like '%NESTED LOOPS%'
 14                         and id <= 5);

SCHEMA          SQL_ID           SQL_TEXT                                      ROWS_PROCESSED
--------------  ---------------  --------------------------------------------  --------------
SCOTT           4dwp5u34yv7mj    select /*+ use_nl(a,b) */ *                           72695
                                 from a,b where a.object_id=b.object_id
```

10.10 抓出 NL 被驱动表走了全表扫描的 SQL

此脚本不依赖统计信息。

嵌套循环的被驱动表应该走索引，以下脚本抓出嵌套循环被驱动表走了全表扫描的 SQL，同时根据表大小降序显示。

```sql
select c.sql_text, a.sql_id, b.object_name, d.mb
  from v$sql_plan a,
       (select *
          from (select sql_id,
                       child_number,
                       object_owner,
                       object_name,
                       parent_id,
                       operation,
                       options,
                       row_number() over(partition by sql_id, child_number, parent_id order by id) rn
                  from v$sql_plan)
         where rn = 2) b,
       v$sql c,
       (select owner, segment_name, sum(bytes / 1024 / 1024) mb
          from dba_segments
         group by owner, segment_name) d
 where b.sql_id = c.sql_id
   and b.child_number = c.child_number
   and b.object_owner = 'SCOTT'
   and a.sql_id = b.sql_id
   and a.child_number = b.child_number
   and a.operation like '%NESTED LOOPS%'
   and a.id = b.parent_id
   and b.operation = 'TABLE ACCESS'
   and b.options = 'FULL'
   and b.object_owner = d.owner
   and b.object_name = d.segment_name
 order by 4 desc;
```

我们在 Scott 账户中运行如下 SQL，强制两表走嵌套循环，强制两表走全表扫描。

```sql
select /*+ use_nl(a,b) full(a) full(b) */ *
  from a, b
 where a.object_id = b.object_id;
```

我们通过以上脚本将其抓出。

```sql
SQL> select c.sql_text, a.sql_id, b.object_name, d.mb
  2    from v$sql_plan a,
  3         (select *
  4            from (select sql_id,
  5                         child_number,
  6                         object_owner,
  7                         object_name,
  8                         parent_id,
  9                         operation,
 10                         options,
 11                         row_number() over(partition by sql_id, child_number, parent_id order by id) rn
```

```
 12                    from v$sql_plan)
 13            where rn = 2) b,
 14        v$sql c,
 15        (select owner, segment_name, sum(bytes / 1024 / 1024) mb
 16           from dba_segments
 17          group by owner, segment_name) d
 18   where b.sql_id = c.sql_id
 19     and b.child_number = c.child_number
 20     and b.object_owner = 'SCOTT'
 21     and a.sql_id = b.sql_id
 22     and a.child_number = b.child_number
 23     and a.operation like '%NESTED LOOPS%'
 24     and a.id = b.parent_id
 25     and b.operation = 'TABLE ACCESS'
 26     and b.options = 'FULL'
 27     and b.object_owner = d.owner
 28     and b.object_name = d.segment_name
 29   order by 4 desc;

SQL_TEXT                                              SQL_ID           OBJECT_NAME           MB
----------------------------------------------------  ---------------  --------------  ----------
select /*+ use_nl(a,b) full(a) full(b) */ *           6prgcr0qcj3qr    B                        9
 from a, b  where a.object_id = b.object_id
```

10.11 抓出走了 TABLE ACCESS FULL 的 SQL

此脚本不依赖统计信息。

如果一个大表走了全表扫描，会严重影响 SQL 性能。这时我们可以查看大表与谁进行关联。如果大表与小表（小结果集）关联，我们可以考虑让大表作为嵌套循环被驱动表，大表走连接列索引。如果大表与大表（大结果集）关联，我们可以检查大表过滤条件是否可以走索引，也要检查大表被访问了多少个字段。假设大表有 50 个字段，但是只访问了其中 5 个字段，这时我们可以建立一个组合索引，将 where 过滤字段、表连接字段以及 select 访问的字段组合在一起，这样就可以直接从索引中获取数据，避免大表全表扫描，从而提升性能。下面脚本抓出走了全表扫描的 SQL，同时显示访问了表多少个字段，表一共有多少个字段以及表段大小。

```
select a.sql_id,
       a.sql_text,
       d.table_name,
       REGEXP_COUNT(b.projection, ']') ||'/'|| d.column_cnt  column_cnt,
       c.size_mb,
       b.FILTER_PREDICATES filter
  from v$sql a,
       v$sql_plan b,
       (select owner, segment_name, sum(bytes / 1024 / 1024) size_mb
          from dba_segments
         group by owner, segment_name) c,
       (select owner, table_name, count(*) column_cnt
          from dba_tab_cols
         group by owner, table_name) d
 where a.sql_id = b.sql_id
   and a.child_number = b.child_number
   and b.object_owner = c.owner
   and b.object_name = c.segment_name
   and b.object_owner = d.owner
   and b.object_name = d.table_name
```

```
        and c.owner = 'SCOTT'
        and b.operation = 'TABLE ACCESS'
        and b.options = 'FULL'
  order by 5 desc;
```

在 Scott 账户中运行如下 SQL。

```
select owner,object_name from t where object_id>100;
```

使用脚本将其抓出。

```
SQL> select a.sql_id,
  2         a.sql_text,
  3         d.table_name,
  4         REGEXP_COUNT(b.projection, ']') || '/' || d.column_cnt column_cnt,
  5         c.size_mb,
  6         b.FILTER_PREDICATES filter
  7    from v$sql a,
  8         v$sql_plan b,
  9         (select owner, segment_name, sum(bytes / 1024 / 1024) size_mb
 10            from dba_segments
 11           group by owner, segment_name) c,
 12         (select owner, table_name, count(*) column_cnt
 13            from dba_tab_cols
 14           group by owner, table_name) d
 15   where a.sql_id = b.sql_id
 16     and a.child_number = b.child_number
 17     and b.object_owner = c.owner
 18     and b.object_name = c.segment_name
 19     and b.object_owner = d.owner
 20     and b.object_name = d.table_name
 21     and c.owner = 'SCOTT'
 22     and b.operation = 'TABLE ACCESS'
 23     and b.options = 'FULL'
 24   order by 5 desc;

SQL_ID         SQL_TEXT                              TABLE_NAME  COLUMN_CNT  SIZE_MB  FILTER
-------------  ------------------------------------  ----------  ----------  -------  ------
51mu5j3aydw94  select owner,object_name from t       T           2/15              9
```

在实际工作中,我们可以对脚本适当修改,比如过滤出大于 1GB 的表、过滤出表总字段数大于 20 的表、过滤出访问了超过 10 个字段的表等。

10.12 抓出走了 INDEX FULL SCAN 的 SQL

此脚本不依赖统计信息。

我们在第 4 章中提到,INDEX FULL SCAN 会扫描索引中所有的叶子块,单块读。如果索引很大,执行计划中出现了 INDEX FULL SCAN,这时 SQL 会出现严重的性能问题,因此我们需要抓出走了 INDEX FULL SCAN 的 SQL。以下脚本抓出走了 INDEX FULL SCAN 的 SQL 并且根据索引段大小降序显示。

```
select c.sql_text, c.sql_id, b.object_name, d.mb
  from v$sql_plan b,
       v$sql c,
       (select owner, segment_name, sum(bytes / 1024 / 1024) mb
```

```
            from dba_segments
          group by owner, segment_name) d
 where b.sql_id = c.sql_id
   and b.child_number = c.child_number
   and b.object_owner = 'SCOTT'
   and b.operation = 'INDEX'
   and b.options = 'FULL SCAN'
   and b.object_owner = d.owner
   and b.object_name = d.segment_name
 order by 4 desc;
```

我们在 Scott 账户中运行如下 SQL。

```
select * from t where object_id is not null order by object_id;
```

在 object_id 列创建索引之后，执行上面 SQL 会自动走 INDEX FULL SCAN，使用脚本将其抓出。

```
SQL> select c.sql_text, c.sql_id, b.object_name, d.mb
  2    from v$sql_plan b,
  3         v$sql c,
  4         (select owner, segment_name, sum(bytes / 1024 / 1024) mb
  5            from dba_segments
  6           group by owner, segment_name) d
  7   where b.sql_id = c.sql_id
  8     and b.child_number = c.child_number
  9     and b.object_owner = 'SCOTT'
 10     and b.operation = 'INDEX'
 11     and b.options = 'FULL SCAN'
 12     and b.object_owner = d.owner
 13     and b.object_name = d.segment_name
 14   order by 4 desc;

SQL_TEXT                          SQL_ID           OBJECT_NAME            MB
-------------------------------   --------------   ---------------   -------
select * from t where object_id   fkan9h6frsn90    IDX_ID                  2
is not null order by object_id
```

在实际工作中，我们可以对脚本作适当修改，例如过滤出大于 10GB 的索引。

10.13 抓出走了 INDEX SKIP SCAN 的 SQL

此脚本不依赖统计信息。

当执行计划中出现了 INDEX SKIP SCAN，通常说明需要额外添加一个索引。以下脚本抓出走了 INDEX SKIP SCAN 的 SQL。

```
select c.sql_text, c.sql_id, b.object_name, d.mb
  from v$sql_plan b,
       v$sql c,
       (select owner, segment_name, sum(bytes / 1024 / 1024) mb
          from dba_segments
         group by owner, segment_name) d
 where b.sql_id = c.sql_id
   and b.child_number = c.child_number
   and b.object_owner = 'SCOTT'
   and b.operation = 'INDEX'
   and b.options = 'SKIP SCAN'
```

```
         and b.object_owner = d.owner
         and b.object_name = d.segment_name
  order by 4 desc;
```

在 Scott 账户中创建如下测试表。

```
SQL> create table t_skip as select * from dba_objects;

Table created.
```

在 owner 字段上创建一个索引。

```
SQL> create index idx_owner_id on t_skip(owner,object_id);

Index created.
```

对表收集统计信息。

```
SQL> BEGIN
  2      DBMS_STATS.GATHER_TABLE_STATS(ownname          => 'SCOTT',
  3                                    tabname          => 'T_SKIP',
  4                                    estimate_percent => 100,
  5                          method_opt       => 'for all columns size skewonly',
  6                                    no_invalidate    => FALSE,
  7                                    degree           => 1,
  8                                    cascade          => TRUE);
  9  END;
 10  /

PL/SQL procedure successfully completed.
```

执行如下 SQL 并且查看执行计划。

```
SQL> select * from t_skip where object_id < 100;

98 rows selected.


Execution Plan
----------------------------------------------------------
Plan hash value: 979686564

---------------------------------------------------------------------------------------
| Id  | Operation                   | Name         | Rows | Bytes | Cost (%CPU)| Time     |
---------------------------------------------------------------------------------------
|   0 | SELECT STATEMENT            |              |   91 |  8827 |    95   (0)| 00:00:02 |
|   1 |  TABLE ACCESS BY INDEX ROWID| T_SKIP       |   91 |  8827 |    95   (0)| 00:00:02 |
|*  2 |   INDEX SKIP SCAN           | IDX_OWNER_ID |   91 |       |    92   (0)| 00:00:02 |
---------------------------------------------------------------------------------------

Predicate Information (identified by operation id):
---------------------------------------------------

   2 - access("OBJECT_ID"<100)
       filter("OBJECT_ID"<100)
```

通过脚本抓出走了 INDEX SKIP SCAN 的 SQL。

```
SQL> select c.sql_text, c.sql_id, b.object_name, d.mb
  2    from v$sql_plan b,
  3         v$sql c,
  4         (select owner, segment_name, sum(bytes / 1024 / 1024) mb
```

```
  5              from dba_segments
  6             group by owner, segment_name) d
  7   where b.sql_id = c.sql_id
  8     and b.child_number = c.child_number
  9     and b.object_owner = 'SCOTT'
 10     and b.operation = 'INDEX'
 11     and b.options = 'SKIP SCAN'
 12     and b.object_owner = d.owner
 13     and b.object_name = d.segment_name
 14   order by 4 desc;

SQL_TEXT                                         SQL_ID          OBJECT_NAME             MB
------------------------------------------------ --------------- ----------------- --------
select * from t_skip where object_id < 100       0837hu8zxha2y   IDX_OWNER_ID             2
```

10.14 抓出索引被哪些 SQL 引用

此脚本不依赖统计信息。

有时开发人员可能会胡乱建立一些索引，但是这些索引在数据库中可能并不会被任何一个 SQL 使用。这样的索引会增加维护成本，我们可以将其删掉。下面脚本查询 SQL 使用哪些索引。

```
select a.sql_text, a.sql_id, b.object_owner, b.object_name, b.object_type
  from v$sql a, v$sql_plan b
 where a.sql_id = b.sql_id
   and a.child_number = b.child_number
   and object_owner = 'SCOTT'
   and object_type like '%INDEX%'
order by 3,4,5;
```

我们在 Scott 账户中运行下面 SQL 并且查看执行计划。

```
SQL> select * from t where object_id<100;

98 rows selected.

Execution Plan
----------------------------------------------------------
Plan hash value: 827754323

--------------------------------------------------------------------------------------
| Id  | Operation                   | Name   | Rows  | Bytes | Cost (%CPU)| Time     |
--------------------------------------------------------------------------------------
|   0 | SELECT STATEMENT            |        |    91 |  8827 |     4   (0)| 00:00:01 |
|   1 |  TABLE ACCESS BY INDEX ROWID| T      |    91 |  8827 |     4   (0)| 00:00:01 |
|*  2 |   INDEX RANGE SCAN          | IDX_ID |    91 |       |     2   (0)| 00:00:01 |
--------------------------------------------------------------------------------------

Predicate Information (identified by operation id):
---------------------------------------------------

   2 - access("OBJECT_ID"<100)
```

我们通过脚本将它抓出。

```
SQL> select a.sql_text, a.sql_id, b.object_owner, b.object_name, b.object_type
```

```
   2    from v$sql a, v$sql_plan b
   3   where a.sql_id = b.sql_id
   4     and a.child_number = b.child_number
   5     and object_owner = 'SCOTT'
   6     and object_type like '%INDEX%'
   7  order by 3,4,5;

SQL_TEXT                              SQL_ID          OBJECT_OWNER  OBJECT_NAME   OBJECT_TYPE
------------------------------------  --------------  ------------  ------------  -----------
select * from t where object_id<100   0nvp2p03p06k4   SCOTT         IDX_ID        INDEX
```

10.15 抓出走了笛卡儿积的 SQL

此脚本不依赖统计信息。

我们在第 5 章中提到过笛卡儿积连接。当两表没有关联条件的时候就会走笛卡儿积，当 Rows 被估算为 1 的时候，也可能走笛卡儿积连接。下面脚本抓出走了笛卡儿积的 SQL。

```
select c.sql_text,
       a.sql_id,
       b.object_name,
       a.filter_predicates filter,
       a.access_predicates predicate,
       d.mb
  from v$sql_plan a,
       (select *
          from (select sql_id,
                       child_number,
                       object_owner,
                       object_name,
                       parent_id,
                       operation,
                       options,
                       row_number() over(partition by sql_id, child_number, parent_id order by id) rn
                  from v$sql_plan)
         where rn = 1) b,
       v$sql c,
       (select owner, segment_name, sum(bytes / 1024 / 1024) mb
          from dba_segments
         group by owner, segment_name) d
 where b.sql_id = c.sql_id
   and b.child_number = c.child_number
   and b.object_owner = 'SCOTT'
   and a.sql_id = b.sql_id
   and a.child_number = b.child_number
   and a.operation = 'MERGE JOIN'
   and a.id = b.parent_id
   and a.options = 'CARTESIAN'
   and b.object_owner = d.owner
   and b.object_name = d.segment_name
 order by 4 desc;
```

在 Scott 账户中运行如下 SQL。

```
select * from a,b;
```

利用脚本将其抓出。

```
SQL> select c.sql_text,
  2         a.sql_id,
  3         b.object_name,
  4         a.filter_predicates filter,
  5         a.access_predicates predicate,
  6         d.mb
  7    from v$sql_plan a,
  8         (select *
  9            from (select sql_id,
 10                         child_number,
 11                         object_owner,
 12                         object_name,
 13                         parent_id,
 14                         operation,
 15                         options,
 16                  row_number() over(partition by sql_id, child_number, parent_id order by id) rn
 17                    from v$sql_plan)
 18           where rn = 1) b,
 19         v$sql c,
 20         (select owner, segment_name, sum(bytes / 1024 / 1024) mb
 21            from dba_segments
 22           group by owner, segment_name) d
 23   where b.sql_id = c.sql_id
 24     and b.child_number = c.child_number
 25     and b.object_owner = 'SCOTT'
 26     and a.sql_id = b.sql_id
 27     and a.child_number = b.child_number
 28     and a.operation = 'MERGE JOIN'
 29     and a.id = b.parent_id
 30     and a.options = 'CARTESIAN'
 31     and b.object_owner = d.owner
 32     and b.object_name = d.segment_name
 33   order by 4 desc;

SQL_TEXT            SQL_ID          OBJECT_NAME   FILTER     PREDICATE         MB
-----------------   -------------   -----------   --------   ----------   -------
select * from a,b   9kwdjbbs50kcu   A                                           9
```

10.16 抓出走了错误的排序合并连接的 SQL

此脚本不依赖统计信息。

排序合并连接一般用于非等值关联，如果两表是等值关联，我们建议使用 HASH 连接代替排序合并连接，因为 HASH 连接只需要将驱动表放入 PGA 中，而排序合并连接要么是将两个表放入 PGA 中，要么是将一个表放入 PGA 中、另外一个表走 INDEX FULL SCAN，然后回表。如果两表是等值关联并且两表比较大，这时应该走 HASH 连接而不是排序合并连接。下面脚本抓出两表等值关联但是走了排序合并连接的 SQL，同时显示离 MERGE JOIN 关键字较远的表的段大小（太大 PGA 放不下）。

```
select c.sql_id, c.sql_text, d.owner, d.segment_name, d.mb
  from v$sql_plan a,
       v$sql_plan b,
       v$sql c,
       (select owner, segment_name, sum(bytes / 1024 / 1024) mb
          from dba_segments
```

```
          group by owner, segment_name) d
 where a.sql_id = b.sql_id
   and a.child_number = b.child_number
   and b.operation = 'SORT'
   and b.options = 'JOIN'
   and b.access_predicates like '%"="%'
   and a.parent_id = b.id
   and a.object_owner = 'SCOTT'
   and a.sql_id = c.sql_id
   and b.child_number = c.child_number
   and a.object_owner = d.owner
   and a.object_name = d.segment_name
 order by 4 desc;
```

我们在 Scott 账户中运行下面 SQL 并且查看执行计划。

```
SQL> select /*+ use_merge(e,d) */ *
  2    from emp e, dept d
  3   where e.deptno = d.deptno;

14 rows selected.

Execution Plan
----------------------------------------------------------
Plan hash value: 844388907

-------------------------------------------------------------------------------------
| Id  | Operation                    | Name    | Rows  | Bytes | Cost (%CPU)| Time     |
-------------------------------------------------------------------------------------
|   0 | SELECT STATEMENT             |         |    14 |   812 |     6  (17)| 00:00:01 |
|   1 |  MERGE JOIN                  |         |    14 |   812 |     6  (17)| 00:00:01 |
|   2 |   TABLE ACCESS BY INDEX ROWID| DEPT    |     4 |    80 |     2   (0)| 00:00:01 |
|   3 |    INDEX FULL SCAN           | PK_DEPT |     4 |       |     1   (0)| 00:00:01 |
|*  4 |   SORT JOIN                  |         |    14 |   532 |     4  (25)| 00:00:01 |
|   5 |    TABLE ACCESS FULL         | EMP     |    14 |   532 |     3   (0)| 00:00:01 |
-------------------------------------------------------------------------------------

Predicate Information (identified by operation id):
---------------------------------------------------

   4 - access("E"."DEPTNO"="D"."DEPTNO")
       filter("E"."DEPTNO"="D"."DEPTNO")
```

我们使用脚本将走了排序合并连接的 SQL 抓出,同时显示离 MERGE JOIN 关键字较远的表的段大小。

```
SQL> select c.sql_id, c.sql_text, d.owner, d.segment_name, d.mb
  2    from v$sql_plan a,
  3         v$sql_plan b,
  4         v$sql c,
  5         (select owner, segment_name, sum(bytes / 1024 / 1024) mb
  6            from dba_segments
  7           group by owner, segment_name) d
  8   where a.sql_id = b.sql_id
  9     and a.child_number = b.child_number
 10     and b.operation = 'SORT'
 11     and b.options = 'JOIN'
 12     and b.access_predicates like '%"="%'
 13     and a.parent_id = b.id
 14     and a.object_owner = 'SCOTT'
 15     and b.sql_id = c.sql_id
```

```
 16       and b.child_number = c.child_number
 17       and a.object_owner = d.owner
 18       and a.object_name = d.segment_name
 19   order by 4 desc;

SQL_ID          SQL_TEXT                                        OWNER   SEGMENT_NAME        MB
-------------   ----------------------------------------        -------  -------------  --------
c7gd7wn0gx4vq   select /*+ use_merge(e,d) */ * from emp e,      SCOTT    EMP             .0625
                dept d  where e.deptno = d.deptno
```

10.17 抓出 LOOP 套 LOOP 的 PSQL

此脚本不依赖统计信息。

在编写 PLSQL 的时候，我们应该尽量避免 LOOP 套 LOOP，因为双层循环，最内层循环类似笛卡儿积。假设外层循环返回 1 000 行数据，内层循环返回 1 000 行数据，那么内层循环里面的代码就会执行 1000*1000 次。以下脚本可以抓出 LOOP 套 LOOP 的 PLSQL。

```
with x as
(select /*+ materialize */ owner,name,type,line,text,rownum rn from dba_source where
(upper(text) like '%END%LOOP%' or upper(text) like '%FOR%LOOP%'))
select a.owner,a.name,a.type from x a,x b
where ((upper(a.text) like '%END%LOOP%'
and upper(b.text) like '%END%LOOP%'
and a.rn+1=b.rn)
or (upper(a.text) like '%FOR%LOOP%'
and upper(b.text) like '%FOR%LOOP%'
and a.rn+1=b.rn))
and a.owner=b.owner
and a.name=b.name
and a.type=b.type
and a.owner='SCOTT';
```

我们在 Scott 账户中创建 LOOP 套 LOOP 的存储过程。

```
create or replace procedure p_99 is
begin
  for i in 1 .. 9 loop
    dbms_output.put_line('');
    for x in 1 .. 9 loop
      if (i >= x) then
        dbms_output.put(' ' || i || ' x ' || x || ' = ' || i * x);
      end if;
    end loop;
    dbms_output.put_line('');
  end loop;
end;
```

我们通过脚本将以上的存储过程抓出。

```
SQL> with x as
  2  (select /*+ materialize */ owner,name,type,line,text,rownum rn from   dba_source
  3  where (upper(text) like '%END%LOOP%' or upper(text) like '%FOR%LOOP%'))
  4  select distinct a.owner,a.name,a.type from x a,x b
  5  where ((upper(a.text) like '%END%LOOP%'
  6  and upper(b.text) like '%END%LOOP%'
  7  and a.rn+1=b.rn)
  8  or (upper(a.text) like '%FOR%LOOP%'
```

```
  9    and upper(b.text) like '%FOR%LOOP%'
 10    and a.rn+1=b.rn))
 11    and a.owner=b.owner
 12    and a.name=b.name
 13    and a.type=b.type
 14    and a.owner='SCOTT';

OWNER              NAME              TYPE
---------------    ---------------   ---------------
SCOTT              P_99              PROCEDURE
```

10.18 抓出走了低选择性索引的 SQL

此脚本依赖统计信息。

如果一个索引选择性很低，说明列数据分布不均衡。当 SQL 走了数据分布不均衡列的索引，很容易走错执行计划，此时我们应该检查 SQL 语句中是否有其他过滤条件，如果有其他过滤条件，可以考虑建立组合索引，将选择性高的列作为引导列；如果没有其他过滤条件，应该检查列是否有收集直方图。以下脚本抓出走了低选择性索引的 SQL。

```
select c.sql_id,
       c.sql_text,
       b.index_name,
       e.table_name,
       trunc(d.num_distinct / e.num_rows * 100, 2) selectivity,
       d.num_distinct,
       e.num_rows
  from v$sql_plan a,
       (select *
          from (select index_owner,
                       index_name,
                       table_owner,
                       table_name,
                       column_name,
                       count(*) over(partition by index_owner, index_name, table_owne
r, table_name) cnt
                  from dba_ind_columns)
         where cnt = 1) b,
       v$sql c,
       dba_tab_col_statistics d,
       dba_tables e
 where a.object_owner = b.index_owner
   and a.object_name = b.index_name
   and b.index_owner = 'SCOTT'
   and a.access_predicates is not null
   and a.sql_id = c.sql_id
   and a.child_number = c.child_number
   and d.owner = e.owner
   and d.table_name = e.table_name
   and b.table_owner = e.owner
   and b.table_name = e.table_name
   and d.column_name = b.column_name
   and d.table_name = b.table_name
   and d.num_distinct / e.num_rows < 0.1;
```

我们在 Scott 账户中执行如下 SQL 并且查看执行计划。

10.18 抓出走了低选择性索引的 SQL

```
SQL> select * from t where owner='SYS';

23654 rows selected.

Execution Plan
----------------------------------------------------------
Plan hash value: 2480948561

--------------------------------------------------------------------------------
| Id  | Operation                   | Name      | Rows  | Bytes | Cost(%CPU)|Time     |
--------------------------------------------------------------------------------
|   0 | SELECT STATEMENT            |           |  2346 |  222K |    68   (0)|00:00:01|
|   1 |  TABLE ACCESS BY INDEX ROWID| T         |  2346 |  222K |    68   (0)|00:00:01|
|*  2 |   INDEX RANGE SCAN          | IDX_OWNER |  2346 |       |     6   (0)|00:00:01|
--------------------------------------------------------------------------------

Predicate Information (identified by operation id):
---------------------------------------------------

   2 - access("OWNER"='SYS')


Statistics
----------------------------------------------------------
          1  recursive calls
          0  db block gets
       3819  consistent gets
          0  physical reads
          0  redo size
    2680901  bytes sent via SQL*Net to client
      17756  bytes received via SQL*Net from client
       1578  SQL*Net roundtrips to/from client
          0  sorts (memory)
          0  sorts (disk)
      23654  rows processed
```

我们使用脚本将以上 SQL 抓出。

```
SQL> select c.sql_id,
  2         c.sql_text,
  3         b.index_name,
  4         e.table_name,
  5         trunc(d.num_distinct / e.num_rows * 100, 2) selectivity,
  6         d.num_distinct,
  7         e.num_rows
  8    from v$sql_plan a,
  9         (select *
 10            from (select index_owner,
 11                         index_name,
 12                         table_owner,
 13                         table_name,
 14                         column_name,
 15                         count(*) over(partition by index_owner, index_name, table_owner, table_name) cnt
 16                    from dba_ind_columns)
 17            where cnt = 1) b,
 18         v$sql c,
 19         dba_tab_col_statistics d,
 20         dba_tables e
 21   where a.object_owner = b.index_owner
 22     and a.object_name = b.index_name
 23     and b.index_owner = 'SCOTT'
```

```
 24       and a.access_predicates is not null
 25       and a.sql_id = c.sql_id
 26       and a.child_number = c.child_number
 27       and d.owner = e.owner
 28       and d.table_name = e.table_name
 29       and b.table_owner = e.owner
 30       and b.table_name = e.table_name
 31       and d.column_name = b.column_name
 32       and d.table_name = b.table_name
 33       and d.num_distinct / e.num_rows < 0.1;

SQL_ID        SQL_TEXT                    INDEX_NAME   TABLE_NAME  SELECTIVITY  NUM_DISTINCT  NUM_ROWS
-----------   --------------------------  -----------  ----------  -----------  ------------  --------
6gzd8z5vm5k0t select * from t where owner='SYS'  IDX_OWNER     T              .04            31     72734
```

10.19 抓出可以创建组合索引的 SQL（回表再过滤选择性高的列）

此脚本依赖统计信息。

我们在第 1 章中讲到，回表次数太多会严重影响 SQL 性能。当执行计划中发生了回表再过滤并且过滤字段的选择性比较高，我们可以将过滤字段包含在索引中避免回表再过滤，从而减少回表次数，提升查询性能。以下脚本抓出回表再过滤选择性较高的列。

```
select a.sql_id,
       a.sql_text,
       f.table_name,
       c.size_mb,
       e.column_name,
       round(e.num_distinct / f.num_rows * 100, 2) selectivity
  from v$sql a,
       v$sql_plan b,
       (select owner, segment_name, sum(bytes / 1024 / 1024) size_mb
          from dba_segments
         group by owner, segment_name) c,
       dba_tab_col_statistics e,
       dba_tables f
 where a.sql_id = b.sql_id
   and a.child_number = b.child_number
   and b.object_owner = c.owner
   and b.object_name = c.segment_name
   and e.owner = f.owner
   and e.table_name = f.table_name
   and b.object_owner = f.owner
   and b.object_name = f.table_name
   and instr(b.filter_predicates, e.column_name) > 0
   and (e.num_distinct / f.num_rows) > 0.1
   and c.owner = 'SCOTT'
   and b.operation = 'TABLE ACCESS'
   and b.options = 'BY INDEX ROWID'
   and e.owner = 'SCOTT'
 order by 4 desc;
```

我们在 Scott 账户中运行如下 SQL。

```
SQL> select * from t2 where object_id<1000 and object_name like 'T%';

26 rows selected.
```

10.19 抓出可以创建组合索引的 SQL（回表再过滤选择性高的列）

```
Execution Plan
----------------------------------------------------------
Plan hash value: 921640168

--------------------------------------------------------------------------------
| Id  | Operation                    | Name     | Rows  | Bytes | Cost(%CPU)|Time     |
--------------------------------------------------------------------------------
|   0 | SELECT STATEMENT             |          |    12 |  1164 |    19   (0)|00:00:01|
|*  1 |  TABLE ACCESS BY INDEX ROWID | T2       |    12 |  1164 |    19   (0)|00:00:01|
|*  2 |   INDEX RANGE SCAN           | IDX_T2_ID|   917 |       |     4   (0)|00:00:01|
--------------------------------------------------------------------------------

Predicate Information (identified by operation id):
---------------------------------------------------

   1 - filter("OBJECT_NAME" LIKE 'T%')
   2 - access("OBJECT_ID"<1000)


Statistics
----------------------------------------------------------
          1  recursive calls
          0  db block gets
         19  consistent gets
          0  physical reads
          0  redo size
       2479  bytes sent via SQL*Net to client
        430  bytes received via SQL*Net from client
          3  SQL*Net roundtrips to/from client
          0  sorts (memory)
          0  sorts (disk)
         26  rows processed
```

执行计划中发生了回表再过滤，过滤字段的选择性较高，我们利用脚本将以上 SQL 抓出。

```
SQL> select a.sql_id,
  2         a.sql_text,
  3         f.table_name,
  4         c.size_mb,
  5         e.column_name,
  6         round(e.num_distinct / f.num_rows * 100, 2) selectivity
  7    from v$sql a,
  8         v$sql_plan b,
  9         (select owner, segment_name, sum(bytes / 1024 / 1024) size_mb
 10            from dba_segments
 11           group by owner, segment_name) c,
 12         dba_tab_col_statistics e,
 13         dba_tables f
 14   where a.sql_id = b.sql_id
 15     and a.child_number = b.child_number
 16     and b.object_owner = c.owner
 17     and b.object_name = c.segment_name
 18     and e.owner = f.owner
 19     and e.table_name = f.table_name
 20     and b.object_owner = f.owner
 21     and b.object_name = f.table_name
 22     and instr(b.filter_predicates, e.column_name) > 0
 23     and (e.num_distinct / f.num_rows) > 0.1
 24     and c.owner = 'SCOTT'
 25     and b.operation = 'TABLE ACCESS'
```

```
26       and b.options = 'BY INDEX ROWID'
27       and e.owner = 'SCOTT'
28   order by 4 desc;

SQL_ID         SQL_TEXT                          TABLE_NAME  SIZE_MB  COLUMN_NAME   SELECTIVITY
------------   --------------------------------  ----------  -------  -----------   -----------
faqathsuy5w3d  select * from t2 where object_id  T2          9        OBJECT_NAME   0.94
               <1000 and object_name like 'T%'
```

10.20 抓出可以创建组合索引的 SQL（回表只访问少数字段）

此脚本不依赖统计信息。

我们在第 1 章中讲到，回表次数太多会严重影响 SQL 性能。当 SQL 走索引回表只访问表中少部分字段，我们可以将这些字段与过滤条件组合起来建立为一个组合索引，这样就能避免回表，从而提升查询性能。下面脚本抓出回表只访问少数字段的 SQL。

```
select a.sql_id,
       a.sql_text,
       d.table_name,
       REGEXP_COUNT(b.projection, ']') ||'/'|| d.column_cnt  column_cnt,
       c.size_mb,
       b.FILTER_PREDICATES filter
  from v$sql a,
       v$sql_plan b,
       (select owner, segment_name, sum(bytes / 1024 / 1024) size_mb
          from dba_segments
         group by owner, segment_name) c,
       (select owner, table_name, count(*) column_cnt
          from dba_tab_cols
         group by owner, table_name) d
 where a.sql_id = b.sql_id
   and a.child_number = b.child_number
   and b.object_owner = c.owner
   and b.object_name = c.segment_name
   and b.object_owner = d.owner
   and b.object_name = d.table_name
   and c.owner = 'SCOTT'
   and b.operation = 'TABLE ACCESS'
   and b.options = 'BY INDEX ROWID'
   and  REGEXP_COUNT(b.projection, ']')/d.column_cnt<0.25
 order by 5 desc;
```

我们在 Scott 账户中运行如下 SQL。

```
SQL> select object_name from t2 where object_id<1000;

942 rows selected.

Execution Plan
----------------------------------------------------------
Plan hash value: 921640168

--------------------------------------------------------------------------------
| Id  | Operation          | Name | Rows  | Bytes | Cost (%CPU)| Time     |
--------------------------------------------------------------------------------
|   0 | SELECT STATEMENT   |      |   917 | 27510 |    19   (0)| 00:00:01 |
```

10.20 抓出可以创建组合索引的 SQL（回表只访问少数字段）

```
|   1 |   TABLE ACCESS BY INDEX ROWID| T2       |   917 | 27510 |    19   (0)|00:00:01|
|*  2 |    INDEX RANGE SCAN          | IDX_T2_ID|   917 |       |     4   (0)|00:00:01|
----------------------------------------------------------------------------------------

Predicate Information (identified by operation id):
---------------------------------------------------

   2 - access("OBJECT_ID"<1000)


Statistics
----------------------------------------------------------
          0  recursive calls
          0  db block gets
        141  consistent gets
          0  physical reads
          0  redo size
      24334  bytes sent via SQL*Net to client
       1102  bytes received via SQL*Net from client
         64  SQL*Net roundtrips to/from client
          0  sorts (memory)
          0  sorts (disk)
        942  rows processed
```

因为上面 SQL 回表只访问了 1 个字段，我们可以利用脚本将上面 SQL 抓出。

```
SQL> select a.sql_id,
  2         a.sql_text,
  3         d.table_name,
  4         REGEXP_COUNT(b.projection, ']') ||'/'|| d.column_cnt  column_cnt,
  5         c.size_mb,
  6         b.FILTER_PREDICATES filter
  7    from v$sql a,
  8         v$sql_plan b,
  9         (select owner, segment_name, sum(bytes / 1024 / 1024) size_mb
 10            from dba_segments
 11           group by owner, segment_name) c,
 12         (select owner, table_name, count(*) column_cnt
 13            from dba_tab_cols
 14           group by owner, table_name) d
 15   where a.sql_id = b.sql_id
 16     and a.child_number = b.child_number
 17     and b.object_owner = c.owner
 18     and b.object_name = c.segment_name
 19     and b.object_owner = d.owner
 20     and b.object_name = d.table_name
 21     and c.owner = 'SCOTT'
 22     and b.operation = 'TABLE ACCESS'
 23     and b.options = 'BY INDEX ROWID'
 24     and  REGEXP_COUNT(b.projection, ']')/d.column_cnt<0.25
 25   order by 5 desc;

SQL_ID         SQL_TEXT                          TABLE_NAME COLUMN_CNT  SIZE_MB  FILTER
-------------  --------------------------------  ---------- ----------  -------  ------
bzyprvnc41ak8  select object_name from t2        T2         1/15              9
               where object_id<1000
```

欢迎来到异步社区！

异步社区的来历

异步社区（www.epubit.com.cn）是人民邮电出版社旗下 IT 专业图书旗舰社区，于 2015 年 8 月上线运营。

异步社区依托于人民邮电出版社 20 余年的 IT 专业优质出版资源和编辑策划团队，打造传统出版与电子出版和自出版结合、纸质书与电子书结合、传统印刷与 POD（按需印刷）结合的出版平台，提供最新技术资讯，为作者和读者打造交流互动的平台。

社区里都有什么？

购买图书

我们出版的图书涵盖主流 IT 技术，在编程语言、Web 技术、数据科学等领域有众多经典畅销图书。社区现已上线图书 1000 余种，电子书 400 多种，部分新书实现纸书、电子书同步出版。我们还会定期发布新书书讯。

下载资源

社区内提供随书附赠的资源，如书中的案例或程序源代码。
另外，社区还提供了大量的免费电子书，只要注册成为社区用户就可以免费下载。

与作译者互动

很多图书的作译者已经入驻社区，您可以关注他们，咨询技术问题；可以阅读不断更新的技术文章，听作译者和编辑畅聊好书背后有趣的故事；还可以参与社区的作者访谈栏目，向您关注的作者提出采访题目。

灵活优惠的购书

您可以方便地下单购买纸质图书或电子图书，纸质图书直接从人民邮电出版社书库发货，电子书提供多种阅读格式。

对于重磅新书，社区提供预售和新书首发服务，用户可以第一时间买到心仪的新书。

用户账户中的积分可以用于购书优惠。100 积分 =1 元，购买图书时，在 里填入可使用的积分数值，即可扣减相应金额。

特 别 优 惠

购买本书的读者专享异步社区购书优惠券。

使用方法：注册成为社区用户，在下单购书时输入 S4XC5 使用优惠码 ，然后点击"使用优惠码"，即可在原折扣基础上享受全单9折优惠。（订单满39元即可使用，本优惠券只可使用一次）

纸电图书组合购买

社区独家提供纸质图书和电子书组合购买方式，价格优惠，一次购买，多种阅读选择。

社区里还可以做什么？

提交勘误

您可以在图书页面下方提交勘误，每条勘误被确认后可以获得100积分。热心勘误的读者还有机会参与书稿的审校和翻译工作。

写作

社区提供基于 Markdown 的写作环境，喜欢写作的您可以在此一试身手，在社区里分享您的技术心得和读书体会，更可以体验自出版的乐趣，轻松实现出版的梦想。

如果成为社区认证作译者，还可以享受异步社区提供的作者专享特色服务。

会议活动早知道

您可以掌握 IT 圈的技术会议资讯，更有机会免费获赠大会门票。

加入异步

扫描任意二维码都能找到我们：

异步社区

微信服务号

微信订阅号

官方微博

QQ 群：436746675

社区网址：www.epubit.com.cn

投稿 & 咨询：contact@epubit.com.cn